Cities, Nationalism, and Democratization

From Jerusalem to Johannesburg, Mumbai to Beirut, and from Sarajevo to Baghdad, numerous cities across the world have faced intense intercommunal conflict and violence. Many of these cities are in countries that are seeking to advance democratization or to reinforce past democratic gains. What can be done in cities to address deep-rooted nationalistic group conflict? How do social and political dynamics in a city affect a society's larger transition toward democracy? Bollens examines these questions in his extensive study of Barcelona and Basque Country (Spain) and Sarajevo and Mostar (Bosnia and Herzegovina).

Cities, Nationalism, and Democratization provides a theoretically informed, practice-oriented account of intercultural conflict and coexistence in cities. Bollens uses a wide-ranging set of over 100 interviews with local political and community leaders to investigate how urban policies can trigger "pushes from below" that help nation-states address social and political challenges. The book brings the city and the urban scale into contemporary debates about democratic transformations in ethnically diverse countries. It connects the city, on conceptual and pragmatic levels, to two leading issues of today – the existence of competing and potentially destructive nationalistic allegiances and the limitations of democracy in multinational societies.

Bollens finds that cities and urbanists are not necessarily hemmed in by ethnic conflict and political gridlock, but can be proactive agents that stimulate progress in societal normalization. The fuller potential of cities is in their ability to catalyze multinational democratization. Alternately, if cities are left unprotected and unmanaged, ethnic antagonists can fragment the city's collective interests in ways that slow down and confine the advancement of sustainable democracy. This book will be helpful to scholars, international organizations, and grassroots organizations in understanding why and how the peace-constitutive city emerges in some cases while it is misplaced and neglected in others.

Scott A. Bollens is a Professor in the Department of Planning, Policy, and Design, University of California, Irvine. He studies urbanism and inter-group conflict.

Questioning Cities

Edited by Gary Bridge, *University of Bristol*, UK and
Sophie Watson, *The Open University*, UK

The 'Questioning Cities' series brings together an unusual mix of urban scholars under the title. Rather than taking a broadly economic approach, planning approach or more socio-cultural approach, it aims to include titles from a multi-disciplinary field of those interested in critical urban analysis. The series thus includes authors who draw on contemporary social, urban and critical theory to explore different aspects of the city. It is not therefore a series made up of books which are largely case studies of different cities and predominantly descriptive. It seeks instead to extend current debates, through in most cases, excellent empirical work, and to develop sophisticated understandings of the city from a number of disciplines including geography, sociology, politics, planning, cultural studies, philosophy and literature. The series also aims to be thoroughly international where possible, to be innovative, to surprise, and to challenge received wisdom in urban studies. Overall it will encourage a multi-disciplinary and international dialogue always bearing in mind that simple description or empirical observation which is not located within a broader theoretical framework would not – for this series at least – be enough.

Global Metropolitan
John Rennie Short

Reason in the City of Difference
Gary Bridge

In the Nature of Cities: Urban Political Ecology and the Politics of Urban Metabolism
Erik Swyngedouw, Maria Kaika, Nik Heynen

Ordinary Cities: Between Modernity and Development
Jennifer Robinson

Urban Space and Cityscapes
Christoph Lindner

City Publics: The (dis)enchantments of Urban Encounters
Sophie Watson

Small Cities: Urban Experience beyond the metropolis
David Bell and Mark Jayne

Cities and Race: America's new Black Ghetto
David Wilson

Cities in Globalization: Practices, Policies and Theories
Peter J. Taylor, Ben Derudder, Piet Saey and Frank Witlox

Cities, Nationalism, and Democratization
Scott A. Bollens

Cities, Nationalism, and Democratization

Scott A. Bollens

Routledge
Taylor & Francis Group

LONDON AND NEW YORK

First published 2007
by Routledge
2 Park Square, Milton Park, Abingdon, Oxfordshire OX14 4RN

Simultaneously published in the USA and Canada
by Routledge
711 Third Avenue, New York, NY 10017

First issued in paperback 2014

Routledge is an imprint of the Taylor & Francis Group, an informa business

© 2007 Scott A. Bollens

Typeset in Times New Roman by Keyword Group

British Library Cataloguing in Publication Data
A catalogue record for this book is available from the British Library

Library of Congress Cataloging in Publication Data
A catalog record for this book has been requested

ISBN13: 978-1-138-86717-8 (pbk)
ISBN13: 978-0-415-41947-5 (hbk)

To my mother
In memory of a boundless spirit
Keeping the windows open and the birds singing

Contents

Illustrations

Preface

Since February 1987, when I was a student rapporteur for a week-long session on divided cities at the Salzburg Seminar in Austria, I have been fixated and challenged by the study of the city amidst nationalistic conflict, more specifically the proposition that local political and urban dynamics expose much about human nature and provide potential answers for working our way out of the intensifying group-based tensions in our world. This work is the result of about four years of research and analysis – beginning 2002 with the planning of sabbatical field research in Barcelona, Basque cities, Sarajevo and Mostar and ending with the completion of this book in 2006. The research is based on multiple sources, principally over 100 interviews that I conducted in these cities with politicians, bureaucrats, international officials, urbanists, community advocates, artists, academics, and others who live in these cities and love them.

I extend my gratitude to all the interviewees who graciously provided me with their opinions and their time. Specific appreciation is extended in Barcelona to city connoisseur Jordi Borja, Oriol Nel-lo, Manuel Solá-Morales, Paul Lutzker, Andreu Ulied, Ian Goldring, and Ignacio Pérez. I will always remember the captivating city of Barcelona for what I lost and for what I learned. My host institution from August 2003 to July 2004 was the University of Barcelona, Department of Constitutional Law and Political Science. Professor Pere Vilanova played a fundamental role in constructing this comparative Spanish–Bosnia project by providing background information and contacts for potential interviewees in both countries. Besides being one of the leading political scientists in Spain on issues of nationality and governance, Professor Vilanova had an inside connection to Bosnia, having served as Head of the Legal Office, European Union Administration for Mostar, April/July 1996. For logistical support in Barcelona, I thank Ana Palau. Funding for the research sabbatical came from my salary compensation, and a grant by the University of California, Irvine Academic Senate Council on Research, Computing, and Library Resources (CORCLR SIIG 2002-2003-2).

In Sarajevo, I wish to thank Javier Mier and Gerd Wochein for providing key points of access for my Bosnian research, Ferida Durakovic for her words and poetry, and "warrior of light" Morris Power. For Elvir Kulin, may you give despite knowing. In Mostar, Nigel Moore and Murray McCullough provided insight into the international community while Marica Raspudić and Zoran Bosnjak did the same for the Bosnian community. For Muhamed Hamica Nametak, may your puppets be reunited in the future. In the Basque region, I thank Pedro Arias, Victor Urrutia, and Francisco Llera for their key insights and Xabier Unzurrunzaga for his robust nationalistic soul.

Claudia, thank you for your love and daring; Damon and Denali, thank you for your sabbatical patience; Toby, thank you for sharing the joy of the path.

Although the thoughts and conclusions reported here are the synthesis of the thinking of many individuals, responsibility for errors lies solely with the author.

Acknowledgements

The author would like to thank the following for granting permission to reproduce images in this work:

Figure 2.2   Preston, Paul. 1993. *Franco: A Biography*. London: HarperCollins.

Figure 2.3   Burg, Steven L. and Paul S. Shoup. 1999. *The War in Bosnia-Herzegovina: Ethnic Conflict and International Intervention*. Armonk, NY: M.E. Sharpe: 365. Copyright 1999 by M.E. Sharpe, Inc. Reprinted by permission. All Rights Reserved. Not for Reproduction.

Figure 3.2   Institut d'Estudis Regionals i Metropolitans. 2002. *Enquesta de la Regio de Barcelona 2000 Informe General (Survey of the Region of Barcelona 2000 General Report)*. Barcelona: Institut. Page 22.

Figure 3.3   Institut d'Estudis Regionals i Metropolitans. 1990. *Enquesta de la Regio de Barcelona 1990*. Volume 1. Conditions of Life and Habits of the Population. Barcelona: Institut. Mapa 3, pagina XXVI.

Figure 3.4   Institut d'Estudis Regionals i Metropolitans. 1990. *Enquesta de la Regio de Barcelona 1990*. Volume 1. Conditions of Life and Habits of the Population. Barcelona: Institut. Mapa 7, pagina XXX.

Figure 4.2   United Nations, Department of Public Information, Cartographic Section, New York, NY. Map no. 3689, Rev. 10, February 2003.

Figure 4.3   King, Curtis S. "The Siege of Sarajevo, 1992-1995." Pp. 235–290 in Robertson, William G. and Lawrence A. Yates. 2003. *Block by Block: The Challenges of Urban Operations*. Fort Leavenworth, KS: Combat Studies Institute, US Army Command and General Staff College Press.

Figure 4.4   Burg, Steven L. and Paul S. Shoup. 1999. *The War in Bosnia-Herzegovina: Ethnic Conflict and International Intervention*. Armonk, NY: M.E. Sharpe: 376. Copyright 1999 by M.E. Sharpe, Inc. Reprinted by permission. All Rights Reserved. Not for Reproduction.

Figure 4.5   Map adapted from King, Curtis S. "The Siege of Sarajevo, 1992-1995." Pp. 235–290 in Robertson, William G. and

Lawrence A. Yates. 2003. *Block by Block: The Challenges of Urban Operations*. Fort Leavenworth, KS: Combat Studies Institute, US Army Command and General Staff College Press.

Figure 5.3   *The Basques: The Franco Years and Beyond.* Copyright 1979 by Robert P. Clark. All rights reserved. Reproduced with the permission of the University of Nevada Press.

Figure 5.4   Gobierno Vasco 2002. *Euskal Hiria*. 2002. Basque Department of Territorial Planning and the Environment. Vitoria: Central Publishing Services for the Basque Government. Page 77.

Figure 5.5   Pelli Clarke Pelli Architects. New Haven, Connecticut. Frank Castantino, illustrator.

Figure 5.6   Gobierno Vasco 2002. *Euskal Hiria*. 2002. Basque Department of Territorial Planning and the Environment. Vitoria: Central Publishing Services for the Basque Government. Page 47.

Figure 6.2   Yarwood, John. 1999. *Rebuilding Mostar: Urban Reconstruction in a War Zone*. Town Planning Review Special Studies No. 3. Liverpool: Liverpool University Press.

1 The promise of the city[1]

This is a study of cities in societies that have endured inter-group conflict, war, and major societal transformations. I test the proposition that cities in such societies are not necessarily inert receptacles dependent for change upon larger political and constitutional re-configurations. Rather, they may be critical spatial, economic, and psychological contributors to national ethnic stability and reconciliation. I examine in this book the capacity of urbanism to effectively address inter-group conflict in nationalistic settings and I probe the role of cities and urban policymakers in guiding societies and citizens during times of political change. In the first case, I study whether urban planning and policymaking can create built environments, provide economic opportunities, and deliver urban services in ways to create physical and psychological city spaces conducive to inter-group coexistence. In the second case, I study the utility of urbanism in leading, supporting, and/or reinforcing societal and political change, and I look for cases where urbanism constitutes a source of light that can help guide a society through the darkness of past memories and the uncertainty of the future.

I hypothesize that cities can be key elements in conflict – a target in attempts to destroy the fabric of a society, but also a necessary foundation on which to build a democratic, sustainable and peaceful society. I view cities as key bridges between broader ideologies (both malevolent and benign) and the psychological and material welfare of a society's citizens. The city is important in peace-building because it is in the streets and neighborhoods of urban agglomerations that there is the negotiation over, and clarification of, abstract concepts such as democracy, fairness, and tolerance. Debates over proposed projects and discussion of physical place provide opportunities to anchor and negotiate dissonant meanings in a post-conflict society; indeed, there are few opportunities outside debates over urban life where these antagonistic impulses take such concrete forms in need of pragmatic negotiation. Peace-building in cities seeks not the well-publicized handshakes of national political elites, but rather the more mundane, yet ultimately more meaningful, handshakes and smiles of ethnically diverse urban neighbors as they confront each other in their daily interactions. As microcosms of broader societal fault-lines and tensions affecting a nation, cities are laboratories within which progressive inter-group strategies may be attempted and evaluated. By discovering and addressing in progressive ways peoples' interactions in streets, neighborhoods, and cities, political leaders can develop policies that engage inter-group conflict at its living roots and inspire a more sustainable peace than one imposed through diplomatic formulas. I do not argue that the ultimate causes of inter-group conflict lie in cities; those causes in many politically contested cities lie in historic, religious, and territorial claims and counter-claims. What I do argue is that the most immediate and existential foundations of inter-group conflict frequently lie in daily life and across local ethnic divides and, importantly, that it is at this micro-level that antagonisms are most amenable to meaningful and practical strategies aimed at their amelioration.

Some cities in transitional and contested societies will play a peace-constitutive role. Others will not. Where local policies promote inter-group tolerance and accommodation, the city will help to anchor larger national peacemaking and building. Where local policies are impediments to advances in inter-group relations, the city will restrict and confine larger national peacemaking. I suspect that there may be a range of roles that urban regions play in societal rebuilding, and I seek to explain why some cities play a progressive role in shaping new societal paths while others do not. In those cases where urban-based peace-building is absent, I want to understand how and why the city was limited in contributing to peace, why and how such a role was misplaced or neglected, and how it may be resurrected. In these negative examples, political elites after active conflict has ended may carry on war through other means and exploit contested cities so that the general public interest fragments and collapses. Such a political attack on the capacity of cities to catalyze a future of mutual coexistence that is not wanted by ethnically entrenched political elites points not to the impotence of cities in the face of national conflict, but to the latent power of cities to influence societal change.

My hypothesis that cities can be semi-autonomous catalysts amidst larger societal conflict runs counter to prevailing state-centric notions that it is the high, diplomatic politics of nation-states that matter and that the roles of cities are limited and derivative. The academic literature on ethnic conflict management is strongly predisposed toward emphasis on the "high politics" of states and their promotion and protection of national interests. In this understanding, urban peace-building interventions must await advances in national peacemaking and in this way reflect and reinforce larger societal progress. I assuredly do not endeavor to dismiss the importance of national political factors in conflict management. Rather, I seek to develop a more sophisticated and nuanced understanding of the city vis-à-vis the state. I attempt this by observing whether there is variability in the roles of cities in nationalistic societies and by investigating the dynamics underlying either the potency or weakness of urban strategies.

MULTI-NATIONALISM AND DEMOCRATIZATION

This study provides an urban-based, grassroots perspective on two of the leading challenges of today: (1) the existence of competing nationalistic allegiances that can tear a society apart, and (2) the possibilities and limitations of democratization in multinational societies. Cities are fulcrums that can help move forward both multinational tolerance and democratization. To the extent that multiple cultures are effectively accommodated within the city, the prospects for a fuller democratic society are likely improved. Dahl (1998: 185) lays bare this connection, asserting that the "nature and quality of democracy will greatly depend on the arrangements that democratic counties develop for dealing with the cultural diversity of their people." Amidst the uncertainty inherent in a societal transition to democracy, the ability of local policies to address issues of group identity, fairness, freedom of expression, and opportunity can create the conditions upon which fuller, more genuine democratic accords can be brokered.

Democratization is no panacea (Sorenson 1998; Snyder 2000). Even if one accepts the legitimacy of democracy as a worthy societal goal, there remains the danger that the democratization process itself can be structured or manipulated by political leaders in ways that severely restrict and weaken the eventual democratic state. Further, as Snyder (2000) has shown, democratization can increase the risk of nationalistic conflict as well as avert it. The early years of building a democratic state tend to be the most tenuous because democratic governance is easier to start than to institutionalize (United Nations Development Program 2002). Democratization is subject to reversals and stagnation if the right conditions are not continually nurtured. Left open to influence by ruthless forces, democratization can lead to disaster as nationalist elites expertly utilize new opportunities to exacerbate nationalist fervor and inter-group conflict (Snyder 2000).

I investigate in this book an under-studied condition that may be conducive to healthy democratization and civic nationalism – the existence of local policies and principles that foster inter-group tolerance and mutual respect. My research connects urbanism and cities – on conceptual and pragmatic levels – to the possibilities and limitations of democratization discussed and analyzed widely today. I link urbanism to the dynamics and requirements of the different phases of democratization as they occur in a divided-society context. Debates in the literature about democratization confront issues of group identity (in particular, whether and how identity should be accommodated institutionally and culturally in a multinational society) and challenges of timing and phasing of democratic development (envisioning a fragile and reversible process of nondemocratic breakdown, democratic establishment, democratic consolidation). These issues of identity and phasing in the democratization debates and literature – until now not explicitly applied to the urban setting – provide the analytic lenses through which I examine my nationalistically robust and democratizing case study cities.

Without progressive and peace-constitutive city policies, national and international agreements that create democracy, while absolutely essential, in fact impose a set of abstract and often remote rules and institutions on the urban landscape. Such national-level negotiations often result in agreements at the political level, not at the level of daily interaction between ethnic groups and individuals. In contrast, urban strategies are capable of addressing the complex spatial, social-psychological, and organizational challenges of living together or alongside each other under a new political dispensation. Certainly, progressive urban actions that occur outside a framework or process of national peacemaking would likely fail. The argument here, rather, is that national political negotiations that lack an urban component are missing a key co-contributor to the formulation and operationalization of new political goals. Such a national peace, arranged by diplomats and societal elites, would be one detached from the practical and inflammatory challenges of inter-group and territorial relations. By literally bringing democracy to the streets, local policies can be central to the construction of new place-based political identities and possibilities for inter-group tolerance and acceptance.

Studying how city policymakers engage with inter-group issues presents a difficult challenge. Often, policies addressing ethnic, racial, and other urban groups are enacted by a city in incremental ways, are layered atop histories of multiple types of other city policies, and are thus hard to isolate and analyze. I looked for cities that needed to frontally face these ornery group-based issues. I found such cities in multinational societies that have experienced transitional periods of major societal uncertainty due to regime change or violent conflict. In these cities, the societal uncertainty associated with political transition forces policymakers to make an active and less haphazard decision about how they will address ethnicity, race, and nationality in the new post-transition society.

The unraveled nature of the cities and societies in this book makes them, in my opinion, clearer as objects of study. In some respects, cities that have gone through major societal disruptions and transformations may be said to be extreme cases. Far from being extraneous to the study of contemporary urbanity, however, such cities are central to debates about urbanism, democracy, and cultural diversity precisely because these challenges are fundamental to their future quality of existence.[2] Lessons from the case study cities in this book have wide relevance in today's urban world. Indeed, the ethnic fracturing of many cities in North America and Western Europe owing to changing demographics, cultural radicalization, and migration creates situations of "public interest" fragility and cleavage similar to my case studies. In studying creative practical approaches toward difficult issues of cultural management, this work seeks to provide guidance to the many urban leaders and professionals who increasingly are struggling to address multiple publics and contrasting cultural views of city life and function.

I investigate four cases, two in Spain (Basque Country and Barcelona) and two in Bosnia-Herzegovina (Sarajevo and Mostar). The cases examine the role of urbanism in a society with a 25-year record of regional autonomy (post-Franco Spain) and in a society immersed in reconstruction after war (post-1995 Bosnia). The Spanish case studies of urban planning and revitalization exist within a national framework that provides regional autonomy as a way to accommodate nationalistic aspirations. The Bosnian case studies are in a country of *de facto* division and I examine the spatial elements of reconstruction efforts. In both case studies, I look at how planning strategies have interacted with political reform. I use the urban arena as a lens through which to gauge the effectiveness of urban policy as part of subnational peace-building and the accommodation of inter-group differences. This effort is an extension of my earlier field research on urban planning in divided societies in Israel and West Bank (Jerusalem); Northern Ireland (Belfast), and South Africa (Johannesburg) (Bollens 1999, 2000).

Both Spain and Bosnia have experienced major societal transformations and present intriguing opportunities to understand urbanism amidst uncertainty and flux. The transitions came about through violent nationalistic conflict (Bosnia) and nonviolent political regime change (Spain). These two countries have experienced differing trajectories along the three phases of democratization, as outlined by Rustow (1970) and Sorensen (1998). In the first phase, there is a breakdown in the nondemocratic regime. In Spain and former Yugoslavia (Bosnia), this occurred when their authoritarian leaders died (Franco in 1975; Tito in 1980). In a second phase, there is the beginning of the establishment of a democratic order. In Spain, this took place between 1975 and 1979, ending with the popular approval of a new national constitution and regional autonomy statutes. In Yugoslavia, the period from 1980 to 1992 was a false start for this second phase as efforts to democratize and restructure the country unraveled into the 1992–1995 wars. Since 1995, under international community supervision, this phase of creating democracy has restarted. In the third phase of democratization, a new democracy is further developed and consolidated and democracy becomes ingrained in the political culture. Democratic consolidation and maturation has occurred in Spain since the 1980s, while in Bosnia it is arguable whether it has yet begun ten years after the end of war.

After significant societal conflict or trauma (as experienced in the Bosnian war and the decades of Franco authoritarian repression), it becomes necessary for a country to examine the inter-group divisions that led to, or were intensified by, such trauma. In the second, formative phase of democratization, there likely will be efforts to reformulate basic governance structures in order to more effectively advance tolerable coexistence. Indeed, Sorensen (1998) asserts without a democratic resolution of how to deal with ethnic and other cleavages in society, the chances of breakdown or reversal in the democratization process will increase. I seek to fill important gaps in the study of conflict by focusing on the *local* dynamics and outcomes of efforts to reconstitute substate societies and cities. A city focus enables a finer-tuned analysis of the practical, on-the-ground dimensions of building peace, including the intergovernmental (local-regional-national) issues involved in policy formulation and implementation. Emphasis on the local arena promises to achieve a level of specificity and groundedness not found in studies of national-level constitutional and political reform (such as, for example, Roeder and Rothchild 2005; G. Gagnon and Tilly 2001; Lapidoth 1996; Newman 1996; O'Leary and McGarry 1995; Lijphart 1968; Nordlinger 1972). Through analysis of the urban system, I seek to contribute on a theoretical level to a better understanding of the relation between planning, power, and societal transformation and to contribute principles at a practical level for urban interventions amidst societal transitions and inter-group tensions.

Studying cities and urban policy in circumstances where inter-group conflict, war, and societal transition have been facts of life allows us glimpses into how local public authority is used in contexts of uncertainty, turbulence, and disruption. Both during and subsequent to a society's reconstitution, how does urban policy and governance address nationalistic tensions that have been a central part of a society's traumatic story? I make the claim here, as in earlier research (Bollens 1999, 2000, 2002), that extreme circumstances reveal ordinary truths – that these unsettled urban contexts illuminate the basic relationships between urban policy and political power far better than in more mature, settled contexts.

MULTINATIONAL CITIES

This theoretically-informed, practice-oriented account of intercultural conflict and co-existence is significant due to the increasing vulnerability of cities throughout the world to ethnic and nationalistic challenges driven by group identity-based claims and immigration. It is also important due to the increasing importance of subnational governance as a means to address issues of ethnic coexistence and democracy within a world where many experts now view the nation-state as decreasingly the territorial answer to the problem of human political, economic, and social organization.

A troubling number of cities across the world are prone to intense intercommunal conflict and violence reflecting ethnic or nationalist fractures. In these cities and societies, ethnic identity[3] and nationalism[4] combine to create pressures for group rights, autonomy, or even territorial separation. Such politicized multiculturalism constitutes a "challenge to the ethical settlement of the city" (Keith 2005: 8). Political control of multinational cities can become contested as nationalists push to create a political system that expresses and protects their distinctive group characteristics. Whereas in most cities there is a belief maintained by all groups that the existing system of governance is properly configured and capable of producing fair outcomes, assuming adequate political participation and representation of minority interests, governance amidst severe and unresolved multicultural differences can be viewed by at least one identifiable group in the city as artificial, imposed, or illegitimate. Characterized by ethnic/nationalist saturation of what are typically mundane urban management issues, the unsettled nature of such cities "reveals the contested and limited nature of the national settlement in its schoolrooms and town halls" (Keith 2005: 3).

Cities such as Jerusalem, Belfast, Johannesburg, Nicosia, Montreal, Algiers, Grozny, Mumbai, Beirut, Brussels, and now Baghdad are urban arenas susceptible to inter-group conflict and violence associated with ethnic or political differences. In cases such as Jerusalem and Belfast, a city is a focal point or magnet for unresolved nationalistic ethnic conflict. In other cases (such as certain Indian or British cities), a city is not the primary cause of inter-group conflict, but becomes a platform for the expression of conflicting sovereignty claims involving areas outside the urban region or for tensions related to foreign immigration. In cases such as Johannesburg and Beirut, the management of cities holds the key to sustainable coexistence of antagonistic ethnic groups subsequent to cessation of overt hostilities. In cities such as Brussels and Montreal, there have been effective efforts to defuse nationalistic conflict through power-sharing governance and accommodation to group cultural and linguistic differences. Additionally, cities are often centers of democratic thought and action, and can be focal points of opposition to autocratic regimes. Largely peaceful urban revolutions and protests in Ukraine (2005), Lebanon (2005), Belarus (2006), and Nepal (2006) attest to the power of urban spaces in contemporary politics.

As we witness changes in the scale of world conflict from international to intrastate, urban centers of ethnic proximity and diversity assume salience to those studying and seeking to resolve contemporary conflict. Increasingly, cities are the arenas within which decision-makers face multiple and unprecedented social challenges connected to group identity-based claims and immigration. Subnational governance at urban, metropolitan, and regional levels appears increasingly to be the focal point in our attempts to address issues of ethnic coexistence, interaction, and democracy within a globalizing world. Urban areas that endure inter-group conflict and major transformations are doing so now in a world that is increasingly linked economically, socially, and politically (Sassen 2000, 1991; LeGales 2002; Loughlin 2001; United Nations 1999). This means that nationalist groups in a city have greater international avenues available to them through which they can seek to spread their political claims. Concurrently, processes of democratization now are more likely than before to be influenced and shaped by international and European political and economic considerations. And, it appears that what happens in cities – in terms of their political organization, immigration policies, and economic structures – affects the nature of the globalization process itself. As stated by McNeill (1999, 110), cities are "crucibles in which global processes can be grounded." Through economic, infrastructure, cultural, and human resource policies, cities can create productive niches relative to the transnational flow of goods, information, and finances. At the same time, urban social fragmentation and political conflict can obstruct the realization of urban benefits from economic globalization (Sassen 2000; Caldeira 2000; Appadurai 1996). Such urban economic hardships and break-downs in social structure are increasingly demanding reformulations of urban policy dealing with housing, economic opportunity, and cultural expression.

Nation-states appear increasingly ill suited to meet the challenges of contemporary sub-state ethnic divisions through centralized state action. Accordingly, there have arisen national political reform strategies such as local and regional autonomy, decentralization, power sharing, and federalism that devolve greater powers to sub-national units comprised of territorially-based ethnic groups (Lapidoth 1996). These reform strategies aim to provide minorities a measure of state power, offer minorities better prospects of preserving their culture, increase opportunities for new political coalitions across ethnic groups, and provide breathing spaces for a possibly fragmenting state to work out new constitutional arrangements. Such strategies are not without critics, who assert that such political restructuring may be a springboard to secession, entrenched ethnic identity or created new forms of identity, and compromise what are perceived to be the fundamental values of the state (Ghai 2000). Whereas such political restructuring seems essential to peacemaking in troubled societies, it usually is the result of elite agreements that are not usually sensitive to social and political dynamics at the grassroots level. Roeder and Rothchild (2005) conclude that elite-based political reform in ethnically divided societies often creates incentives for the escalation of conflict that threaten future democracy and peace. Instead, they recommend as more conducive to the long-term consolidation of democracy in such countries a more grassroots, bottom-up construction of multiple nodes of authority.

I hypothesize that urban and regional policy strategies are a critical part in the advancement of a larger peace. They are more capable than national accords of addressing the complex spatial and social-psychological attributes of inter-group relations. If effective, local and regional strategies can facilitate tolerance and operationalize what larger peace means, and thus can reinforce larger peace agreements. Within ethnically tense and fragmenting states, urban management of ethnic competition has profound consequences for the national, and ultimately, international level (Ashkenasi 1988). I deepen the argument that cities are critical agents within our globalizing world by observing that it is within cities that the great challenges of inter-group coexistence, tolerance, and multinational democracy will be addressed and negotiated over the next decades. I believe, as Murtagh (2002) describes, that the immediate causes of inter-group conflict lie not in the macro-level substantive matters of states and empires, but rather in the hidden micro-level roots and tangled social infrastructures of opposing groups in neighborhoods, communities, and cities.

Challenges regarding identity, citizenship, and belonging in a globalizing world will need to be addressed most immediately at the local level; our degree of progress at this grassroots level will either fortify or confine our ability to address these issues at broader geographies both within and between states. The promise of cities is that they constitute "privileged places for democratic innovation" (Borja and Castells 1997: 246). This potential role of cities as arenas of peace-building is consistent with cities' historical function in providing the innovative milieu that carries societies forward; as Hall (1998: 7) argues, cities "have throughout history been the places that ignited the sacred flame of the human intelligence and the human imagination."

Urban and regional policies comprise an important layer to study, in order to both better understand the trajectory and effectiveness of national peace-building efforts and to uncover local processes and opportunities where peace-promoting policy intervention is indicated. Urban and regional policies have direct and tangible influences on material and psychological conditions that have been linked by scholars such as Gurr (1993) and Burton (1990) to inter-group stability or volatility. In particular, urban policies affect conditions, such as territoriality, economic distribution, policymaking access, and group identity, that can exacerbate or moderate inter-group tension (Murphy 1989; Sack 1986; Stanovcic 1992). Territorially, cities can be important symbolic and military battlegrounds and flashpoints for violence between antagonistic ethnic groups seeking sovereignty, autonomy or independence. Economically, they are frequently focal points of urban and regional economies dependent on multi-ethnic contacts, social and cultural centers and platforms for political expression, and potential centers of grievance and mobilization. Politically, cities can include or exclude minority groups from formal and informal participation processes, and they are arenas where the size and concentration of a subordinate population can present the most direct political threat to the state. In terms of group identity, they provide the locus of everyday interaction where ethnicity and identity can be created and re-created (Eriksen 1993) and they are suppliers of important religious and cultural symbols. If identity conflicts are not managed effectively to promote mutual coexistence, cities become vulnerable organisms subject to economic stagnation, demographic disintegration, cultural suppression, and ideological and political excesses violent in nature.

URBANISM, GROUP CONFLICT, AND POLITICAL CHANGE

This project seeks on theoretical and practical levels to understand how urban policy and governance copes with inter-group differences and how it operates amidst the uncertainty of political transitions.

Group conflict

The debate over the effects of group identity politics within democratic societies – and whether group affiliation impedes or facilitates democratic expression – is ongoing (see, for example, Gutmann 2003; Young 1990). The challenge of how states in the world that are liberal or based on the value of individual rights can acknowledge the group presence is problematic (Weisbrod 2002). Many experts believe a balance is needed between group and individual rights. Young (2000: 7, 9) views group-based aspirations and expressions as a valuable political resource that appropriately pluralizes discourse. She sees value in a "differentiated solidarity," where universal and individual-based justice is combined with neighborhood and community-based participatory institutions differentiated by group identities. Similarly, Rex (1996: 2) suggests a "democratic or egalitarian multiculturalism" that couples recognition of cultural diversity with the promotion of equality of individuals. Unlike Young, however, such a cultural diversity is more confined to the private domain and less a feature of the public domain, which for Rex should promote a shared political culture of equal, not differential, rights. Borja and Castells (1997) assert that city residents' ability to maintain distinct cultural identities stimulates a sense of belonging that is needed amidst globalization; at the same, communication between cultures must be present, lest there be cultural fragmentation and local tribalism. Lynch (1981: 50) linked this debate directly to the spatial form of the city. While asserting that desired urban spatial qualities should have a degree of generalizability across cultures, he states that city form "should be able to deal with plural and conflicting interests."

In this book, I explore the challenge of group identity as it is rooted in the local, lived experience of the city and ask whether urban policy can effectively address inter-group differences and accommodate group desires while maintaining some city-wide collective allegiance or identity. For some urban scholars, the key is for practitioners to become more attuned to group identity as a criterion within planning processes and decisions (Neill 2004; Amin 2002; Umemoto 2001; Burayidi 2000; Sandercock 1998). For others, the critical objective is for planners to recognize but also help transcend such urban and societal divisions (Marcuse and van Kempen 2002; Baum 2000). This balance between recognizing and transcending group identity needs is surely a significant practical challenge, notwithstanding the theoretical and philosophical debates.

The physical, economic, social, and political structuring of a city shapes inter-group relations, and this is a phenomenon that policymakers can choose to constructively address or passively avoid. The built environment of a city can create and cement division of everyday practices through peace walls, politically motivated street alignments, double entry factories, and spatial buffer zones. These urban divisions reflect larger societal patterns and are the visible indicators of the unevenness and inequity of power relations that exist within a society at-large. How cities are structured can impede or facilitate the capacity of a society to move forward on questions of multinational group relations. Yet, I suggest that cities are not solely mirrors of larger societal dynamics and that there exist spaces of autonomy within which urban politicians and practitioners can act. Cities have their own spatial, political and social dynamics, certainly influenced by extra-urban forces, but never fully controlled by them. While extra-urban influences (such as peoples' attitudes and national economic conditions) are assuredly important in shaping urban conflict, urban interventions can independently shape the configuration of those social and spatial resources in the city that ethnic residents use as binding materials to bolster their collective identities. Amin (2003: 17) argues that beneath global and national influences, "additional local factors and the particularities of place explain spatial variation in the form and intensity of racial and ethnic inequalities." In a similar vein, Varshney (2002) argues that, while national and global level factors provide an important context, ethnic violence in the Indian cities he studied was only activated when these national level factors joined with local violence-facilitating processes. Because catalysts toward both ethnic conflict and tolerance reside within the urban landscape, cities are key points of intervention.

A city's capacity to address issues of group identity can run at different speeds than a society's at large. Due to the need to effectuate cross-ethnic political compromises in circumstances of inter-group proximity and economic inter-dependence, a city may be an urban catalyst that anticipates and stimulates broader societal progress. The urban built environment in this case may provide models that illustrate how everyday practices of ethnic division can be constructively overcome and transcended. On the other hand, the disjunction between urban and societal trajectories can also mean that a city may be a drag on national attempts at peacemaking. In this case, despite fragile progress nationally, a city may remain a flashpoint for violence due to the city's ethnic thickness and local history of antagonisms. The built environment in this case may reinforce antagonisms and obstruct local compromises for years after advances in national peacemaking.

I endeavor to find and describe local strategies that engage with issues of inter-group conflict and those that don't, to explain the motivations underlying these strategies, and to examine the effects of these strategies and interventions on cultural difference. These can be local government strategies that involve spatial development, housing, and economic opportunity or that influence forms of inter-ethnic civic engagement. Urban ecological and social ecological frameworks are useful in postulating possible linkages between the urban environment and the magnitude of inter-group tension and conflict.[5] Viewing a city as an ecosystem of interrelated parts and processes, Ball-Rokeach (1980) points out how malfunctioning of the system in terms of overcrowding or hampering of the territorial foundations of group identity is conducive to increased tension and violence. I will investigate indicators of urban ecosystem malfunction, including inter-ethnic political fragmentation and parallelism, hyper-segregation, and denigration of the collective public sphere. These are all indicators of a dual city that mutually excludes and separates, and thus perpetuates nationalistic antagonisms. I also search for indicators of healthy urban functioning conducive to peace-building. These include signs of centrality, accessibility and porosity of movement, vibrancy of a collective public sphere, support for urban ethnic identity, inclusiveness, and flexibility in urban form and activity patterns. I am not looking to simplistically categorize cities as either rigid or tolerant. Rather, critical to urban peace promotion is the ability of a city to balance the legitimate rights of an ethnic/nationalist group with the need to have an integrated collective city interest. In the face of ethnic pressures that can tear apart a city and society and amidst the uncertainty of political transitions, cities that integrate group-specific and city-wide interests appear best able to be catalysts toward both local and national peace.

Political change

Policymakers and community leaders in each of the cities that I studied have experienced societal turbulence and uncertainty in the form of significant political disruption and transitions. In the Spanish cases, the transition was a largely nonviolent yet difficult one, from an authoritarian dictatorship to a democratic regime. In the Bosnian cities, the transition is from a war-torn failed authoritarian and mixed-economy Yugoslavian state to a country more democratic and capitalistic. The larger transitions within which the four urban systems existed provide a unique opportunity to look at the role and effects of urban policy midst societal uncertainty. Since the Bosnian cases are still within transitional processes, this study also considers for Sarajevo and Mostar possible different urban futures and the role public policy might play in them.

I examine how policymakers during and after transitions seek to transform pre-transition geographic and physical legacies in ways that accommodate differing group aspirations and bring democracy to the streets. Cities do not change overnight because their physical structures have fixed and obdurate qualities (Hommels 2005). This concreteness of physical stock may make urban transformation seem slower at times than the pace of national political change. The goals of policymakers in new regimes to stabilize, normalize, and transform a city will encounter as obstructions the old spatial forms and processes created prior to the transitional period. Analyses of Eastern European socialist cities in transition to capitalism, for instance, suggest that socialist urban traits will slowly change in capitalist reconstruction (Szelenyi 1996; Harloe 1996). Cities in economic transition from socialism to capitalism have showed varied, or "path-dependent," shifts in their physical, social and economic restructuring (Stark 1990 and 1992; Putnam 1993). In political and urban transitions, legacies of past urban policy-making and development constrain and set parameters for contemporary urban reform. French occupation of the city of Algiers led to colonial urban planning which reified French urbanism and control, creating urban forms that exist to this day (Celik 1997). In Jerusalem, the 20 years of physical partition from 1948–1967 spatially separated Jewish west and Arab east Jerusalem populations, an obstacle to ethno-nationalist integration that remains 40 years later (Bollens 1996). When rebuilding of war-torn cities is attempted, policy officials engage not with a *tabula rasa*, but with urban forms and activity patterns erected both before and during urban warfare. In the rebuilding of German cities after World War II, radical new styles of German urbanization were not created; instead, the shape of reconstruction was influenced by long-term continuities in city-building practice dating to the early 1900s (Diefendorf 1993). Efforts during the 1990s to reconstruct debilitated Beirut needed to accommodate forms of urbanization – such as squatting, refugee camps, and self-sufficient neighborhood networks – which sustained individuals through crises (Yahya 1993). And, in the former Yugoslavia, political and physical post-war reconstruction faces the terrible legacy of urban "ethnic cleansing" that created new landscapes of demographic dominance and eradication.

Notwithstanding the constraining legacies of past urban development, political transitions afford opportunities for urbanism to seek significant transformation of a city's built and human landscapes. Examining the public planning function during and after such periods of societal uncertainty and transformation provides us with a relatively unfiltered view of how this function relates to axes of public and private power. During such times of flux and uncertainty, the power to guide society is less firmly entrenched and indeed may be up for grabs, compared to more stable state conditions. Political forces and interests compete and position themselves to assure that any movement toward democracy does not harm their interests (Przeworski 1991). Amidst societal uncertainty, these competing potential centers of power may structure, use, and exploit public planning and policymaking functions in ways to assist them in establishing authority or dominance. As such, urban interventions represent a process of organizational and societal adaptation to an uncertain and changing environment. In contrast to negotiated political agreements, they have more concrete and noticeable effects on peoples' lives. Amidst a process of democratization, urban policies can constitute leading implementation edges of new democratic goals and mechanisms and, importantly, can articulate early in societal transitions new city-building logics and expectations regarding how private power and public interest, and ethnic interest and city-wide needs, are to interact under a new regime. Alterman (2002) points out that during such uncertainty, the dilemmas in planning take on a sharpened edge and expose major unresolved issues. By studying transitional societies we may thus gain insight into how planning and urbanism (and their ability to present visions of the future city) fit into societal circumstances of unsettled power relationships, and how urban actions can contribute to societal betterment before a new "procrustean bed" of societal authority patterns is re-established.

Investigators have considered the roles of urban policy and decision-making amidst conditions of societal uncertainty or "turbulence." Alterman (2002) found that planning during societal crises could play a key role in focusing and searching for a critical path forward. Planning has an ability to translate an all-encompassing crisis into problems that could be conceptualized and managed by public policy. In this respect, crises provide a rare opportunity for positive and meaningful change by public planning. They may stimulate a reframing of fundamental societal problems and a "social learning" process wherein new directions and institutions are formulated based on a mutually educative experience involving multiple interests (Godschalk 1974; Bryson 1981; Morley and Shachar 1986). Friend and Hickling (1997: 1) conceptualize planning as a process of "choosing strategically through time" that allows for the creative management of multiple uncertainties. Schon and Nutt (1974) look at the practical requirements of such planning, asserting that the practice of planning should differ . . . in its skill base, its use of theoretical concepts, its effectiveness criteria, and in its professional self-definition . . . between conditions of stability and those of turbulence. New ways of knowing and an "ethic for the process of change itself" are needed to counter the natural tendency of organizations and sectors to actively resist change (Schon 1971: 11).

During societal change toward a functioning multinational democracy, I believe we can articulate specific spatial and physical characteristics of cities that urbanism should seek to foster or reinforce. For example, flexibility of urban form through spatial development that maximizes future options would appear more promoting of peace than walls, urban buffers, and other urban forms that unduly delineate physical segregation of groups and facilitate psychological separation. By preventing hardening and partitioning of the urban environment, there can be some mixing of ethnic populations (if and when members of the respective groups choose) and normalization of urban fabric. What should be avoided are urban interventions that unnecessarily partition the urban landscape culturally and thus lock in and overly politicize inter-group differences into the foreseeable future.[6] Governmentally and spatially, an "integrated public sphere" is an important element to preserve amidst political change and uncertainty (Snyder 2000). Governmentally, this means that ethnic groups should share control of the institutions of local governance. Spatially, this implies the protection or development of public and shared spaces perceived by all to be of neutral or cross-ethnic territoriality.

One of urbanism's greatest contributions midst societal uncertainty may occur even before the formal institution of a new governing framework. This role is to highlight urban issues as symptoms of root issues of political conflict – power imbalances and disempowerment – that will need to be resolved in negotiations concerning a new governing dispensation. Benvenisti (1986) and Bollens (2000) label this planning strategy a "resolver" role. An example of this planning role is described in Bollens (1999). In Johannesburg, South Africa, urban and metropolitan leaders during the 1991–1995 transition transcended a sole emphasis on urban symptoms of racial polarization and targeted the need to radically transform the basic parameters of apartheid-based urban governance. Nongovernmental and opposition organizations linked city-building issues in Johannesburg that involved day-to-day consumption problems and the black boycotting of service charges to root political empowerment issues. In this role, urban deficiencies, far from being mundane and unimportant politically, are positioned as basic reflections of core political disabilities. Urbanism is connected centrally to politics, bringing in diverse and usually atomized grassroots organizations and interests into the larger political struggle.

Notes

1. I borrow and adapt this useful expression from Tajbakhsh (2001) and Sennett (1970).
2. I am influenced by Nordstrom's (1997) observation that what we relegate to the margins of lived experience and theory often speaks most fundamentally to core aspects of human existence.
3. Ethnic groups are composed of people who share a distinctive and enduring collective identity based on shared experiences or cultural traits (Gurr and Harff 1994). Such group awareness can be crystallized through such factors as shared struggle, territorial identity, "ethnic chosenness," or religion (A. Smith 1993).
4. Nationalism is defined here as in Snyder (1993): a doctrine wherein nationality is the most important line of cleavage for establishing membership in societal groups, and overrides or subsumes alternative criteria such as social class, economic class, or patronage networks.
5. See Berry and Kasarda (1977) and Stokols (1996) for useful explanations of these perspectives.
6. I am not arguing for an ethnically integrated city forced upon city residents but rather the discouragement of a rigid hyper-segregation sanctioned by government policy that overly restricts individual choice of residence, workplace, and shopping destination in the future. In multi-national cities, a degree of self-selected ethnic group segregation is conducive to the healthy maintenance of cultural group identity.

2 Spain, Bosnia, and the urban conflict-stability continuum

At first glance, Spanish and Bosnian societies seem like odd bedfellows in a comparative analysis of cities and urban planning. The transition in Bosnia was due to a gruesome war between 1992 and 1996 that killed about 100,000 people and displaced approximately two million people. The transition in Spain in the 1970s was a difficult one, yet one largely managed politically in its shift from Franco authoritarianism to a democratic and regionally decentralized state. The constitutional reordering in Bosnia was internationally imposed in 1995 through the Dayton Agreement, while it was more organically developed in Spain. In Bosnia, inter-group differences resulted in military conflict and violence, while in Spain differences between regional and state-based nationalism, with one important exception, have been worked through more in the political system. Societal crisis and transformation occurred in Bosnia in the 1990s and in Spain in the late 1970s.

Despite these differences, there are two compelling similarities between the Bosnian and Spanish cases during the last quarter of the twentieth century. Both have been exposed to periods of significant societal uncertainty and engagement in far-reaching constitutional and institutional reform; and both needed to cope during this political reordering with how to effectively address significant differences between identifiable nationality groups – in Bosnia's case, between the three antagonistic nationality groups of Bosniak (Muslim), Croat, and Serb; in Spain's case, between those who argue for greater regional autonomy in places like Catalonia and Pais Vasco and those who favor a more centralized Spanish state anchored in Madrid.

STUDYING CITIES IN UNSETTLED CONTEXTS

A project studying such a complex and multi-faceted subject as cities, political transitions, and nationalism must use an interdisciplinary approach to analyze urban and subnational patterns and processes of inter-group conflict and its management. I use the insights of political science to examine political and legal mechanisms used to diffuse conflict and the dynamics of communal conflict and nationalistic group mobilization. I use knowledge of urban and regional planning to study policies affecting local and metropolitan settlement patterns, geography to explore the spatial and territorial aspects of conflict, and social psychology to deepen the analysis of group identity and urban attributes which may facilitate or obstruct aggression.

I use diverse sources of information, including over 100 field research interviews, published and unpublished materials, and quantitative urban data. The primary information source for this study is the knowledge and experience of key individuals involved in urban policy and administration. The main research tool is the face-to-face interview. I use interviews to obtain objective information about urbanism, and to construct a grounded, ethnographic account of urban policymaking midst societal reconstruction and political strife. I am interested in the complex objective realities and influences in these cities, as well as in how the interviewees make sense of their everyday activities, professional roles, and organizational environment. I seek to understand the organizational, cultural, and historical context within which governmental and nongovernmental professionals operate. I wish to know how an interviewee relates to the potent nationalistic politics that surround him or her. I observe closely the interplay between the professional norms and values of many policymaking roles and the more emotion-filled ideological imperatives that impinge daily upon the professional's life; in other words, how urbanism and nationalism intermingle. I will discuss distortions and omissions I encountered in the interviews, as well as how interviewees emphasized certain issues and not others, and how they defined urban issues, constituents, and the political or practical limits of urbanism amidst nationalism.

Between April 2003 and July 2004, I interviewed 109 political leaders, planners, architects, community representatives, and academics in the city of Barcelona, three cities in the Spanish region of Basque Country, and in the cities of Sarajevo and Mostar in Bosnia and Herzegovina. I developed core interview lists, based on my primary field contacts, prior to the in-field research portion of the project. I identified additional interviewees after arrival based upon word-of-mouth referrals from initial discussants and academics at my host institution in Barcelona, and through local media. Table 2.1 exhibits the distribution of interviews by study site.

Interviews lasted 75 minutes, on average. About 90 percent of them were audiotaped and subsequently transcribed. In about 10 percent of the cases, I used a translator to facilitate discussion. I also investigated published and unpublished government plans and policy documents, political party platforms and initiatives, implementing regulations, and laws and enabling statutes in terms of how they address issues associated with inter-group difference at local, regional, and national levels. I also employed quantitative data concerning growth and housing trends and budgetary spending to supplement interview-based findings.

I focus on two primary themes: the role of cities and urban policy amidst inter-group differences and during political transitions. Table 2.2 lists the primary questions that shaped interviews and secondary research.

I use the term "urbanism" to refer to a diverse and broad set of urban policy and governance attributes – including both interventions by public authorities into the built and social landscape of cities (what cities do) and the institutional forms and

Table 2.1 Interviews conducted*

City	Interviews	Dates of visits
Barcelona	55**	April 2003, August 2003–July 2004
Basque Country	15	February 2004
Sarajevo	17	October 1999, April 2003, November 2003
Mostar	22	April 2002, November 2003, May 2004
TOTAL	109	

Notes: *See appendix for full lists of interviews in each case study.
**The higher number of interviews in Barcelona is because this city was my "home" for about 75 percent of the 10½ month sabbatical August 2003 through July 2004. The three other case study visits each ranged from 10 to 15 workdays in duration.

Table 2.2 Primary research questions

CITIES AND INTER-GROUP RELATIONS

What are the obstacles, challenges, and opportunities in managing nationalistic group-based differences within urban systems?

What urban and regional planning strategies are used in efforts to increase mutual co-existence of groups and accommodation of differing nationalistic aspirations?

To what extent are cultural or nationalistic issues incorporated as salient criteria in the design of urban policies?

How does urbanism affect nationalistic political projects, and how does nationalism influence urban processes and outcomes?

How have urban strategies and projects affected inter-group relations and prospects for mutual co-existence?

CITIES AND POLITICAL TRANSITIONS

What are the perceived, and actual, functions of urban and spatial development policies in reforming or reconstituting cities and societies?

Have such efforts at the urban level helped guide societies and citizens during times of political flux and change?

How does "democratic" urban policy and planning differ in goals, process and outcomes from urban planning conducted under pre-democratic regimes?

Through what means, and to what degree of success, does urban policy and planning attempt to express "democracy"?

After war, how are urban and spatial development policies connected to societal rehabilitation and reconstruction? (Bosnia)

What does urban planning and development in cases of societal disruption tell us about the relation between urbanism, power, and governance in societies that have not experienced societal trauma?

organizational processes of city governance (how cities are organized).[1] I employ the term "urbanist" in a way that broadly encompasses all individuals (within and outside government) involved in the anticipation of a city's or urban community's future and preparation for it. The category includes, within government, town and regional planners, urban administrators and policymakers, and national and regional-level urban policy officials. Outside government, it includes community leaders, project directors and staff within nongovernmental, community or voluntary sector organizations, scholars in urban and ethnic studies, and business leaders.

To understand the role of urbanism in contested settings, I frequently emphasize urban planning as an important analytical lens. I do this because the planning function of government, through its direct and tangible effects on ethnic geography, can clearly reveal the intent and role of a governing regime.[2] Urban plans and decision-making can be expressions of new and emerging patterns of power and lines of thought during and after a societal transition. They seek to operationalize in the urban sphere commonly used abstract concepts such as democracy, empowerment, equity, reconciliation, tolerance and cultural identity. Urban policies regarding land and real estate development, economic development, reconstruction, housing construction and allocation, refugee relocation, capital facility planning and social service delivery often have immediate and substantial impacts on ethnic neighborhoods and households. In its capacity to structure material and social-psychological attributes of the urban system, urban planning operationalizes political power in ways that are concrete and visible.

Urbanism (encompassing both how city governments intervene in the city's built landscape and how the city government is organized) affect numerous material and psychological conditions that Gurr (1993) and Burton (1990) have linked to inter-group stability or volatility. In particular, urban policies and governance affect four particular types of conditions – territoriality, economic distribution, policymaking access, and group identity – that can exacerbate or moderate inter-group tension (Murphy 1989; Sack 1986; Stanovcic 1992; Toft 2003) [see Figure 2.1].

Urban planning policies affect the ethnic conditions of the urban environment most concretely through influence on *control of land and territoriality* (Murphy 1989; Yiftachel 1992; Gurr 1993; Williams 1994). I use Sack's (1986: 19) definition of territoriality as "the attempt by an individual or group to affect, influence, or control people, phenomena, and relationships, by delimiting and asserting control over a geographic area." In other words, an urban regime or an ethnic group living in the city will seek control over urban space as a way to secure political control. In contested cities, partisan urban territorial policies can be implemented by a regime to enforce control over select urban ethnic groups (Sack 1986). These policies create urban frameworks that foment ethnic mistrust and conflict (Yiftachel 1992). Two common techniques of territorial control amidst ethnic tension aim to alter the spatial distribution of ethnic groups and to manipulate jurisdictional boundaries to politically incorporate or exclude particular ethnic residents (Coakley 1993). An empowered group can significantly manipulate ethnic geographies, land control, and the planning machinery itself in an attempt to dominate a subordinate group. A local government's regulatory and developmental efforts can significantly affect the demographic ratios between the two sides, change the scale of focus of planning

URBAN ETHNIC CONDITIONS

CONTROL OVER LAND / TERRITORIAL JURISDICTION
Settlement of vacant lands; control of settlement patterns; dispossession from land; return and relocation of displaced and refugee populations; control of land ownership; demarcation of planning and jurisdictional boundaries vis-à-vis ethnic settlement patterns.

DISTRIBUTION OF ECONOMIC BENEFITS AND COSTS
Magnitude and geographic distribution of urban services and spending; allocation of negative and positive effects of urbanization.

ACCESS TO POLICY-MAKING
Inclusion or exclusion from political process; formal and informal participation processes; presence and influence of nongovernmental organizations.

MAINTENANCE OF GROUP IDENTITY AND VIABILITY
Maintenance or threat to collective ethnic rights and identity; education, language, religious expression, cultural institutions.

V
V
V

DEGREE OF URBAN STABILITY OR CONFLICT

FACILITATE OR IMPEDE MOVEMENT TOWARD CO-EXISTENCE
Decrease or increase in organized resistance and political violence; loosening or compartmentalization of ethnic territoriality; lessening or widening of inter-ethnic disparities; greater or lesser political inclusion of all groups and inter-group cooperation; growing or eroding of respect for collective ethnic rights.

Figure 2.1   Urban ethnic conditions, stability, and conflict

efforts, and reinforce or modify the ethnic identity of specific geographic subareas. Urban policy aimed at advancing peace will need to confront these issues of urban territory and land control. In reconstituting a city during or after political transitions, a local government most likely will face a difficult balancing act – respecting territoriality when it is associated with healthy community identity and cohesiveness and working to lessen territoriality when it is obstructing development of a more normal functioning urban system.

Urbanism also can significantly influence the *distribution of economic benefits and costs* and the allocation of service benefits (Yiftachel 1992; Stanovcic 1992). Urban land use and growth policies affect accessibility and proximity of residents and communities to employment, retail and recreation, the distribution of land values, and the economic spin-offs (both positive and negative) of development. The planning and locating of economic activities can significantly shape both the daily urban behavior patterns and residential distributions of ethnic groups. The development of economic centers of activity can either integrate or separate the ethnic landscape. For example, major employment or commercial centers could be placed along ethnic territorial interfaces as a way to turn formerly "no-man's land" into mutually beneficial spaces of inter-group economic and social interactions. In contrast, economic development that occurs amidst a single group's territory can solidify or reinforce inter-group separation. Further, urban service and capital investment decisions – related to housing, roads, schools, and other community facilities – directly allocate urban advantages (and disadvantages) across ethnic communities. The power of urban government to allocate resources and advantages can strengthen inter-group inequalities across a contested city's ethnic geography by distributing them disproportionately to the neighborhoods and commercial areas of the empowered ethnic group. Urban policymaking aimed at mutual co-existence, in contrast, would allocate activities and spending in ways that decrease inequalities.

Urban policy and planning processes can have substantial effects on the distribution of local political power and access to policy-making (Stanovcic 1992; Gurr 1993). In many contested cities, there may be a legacy of hegemonic control by one ethnic group where the opposing group has been fully excluded from the political decision-making process. In reforming urban societies, several options to hegemonic control have been used or proposed. "Third-party intervention" removes contentious local government functions such as housing, employment, and services from control by either of the antagonistic parties and empowers a third-party overseer to manage the urban region. Urban "cantonization" is a method of governance that devolves selected municipal powers to neighborhood-based community councils or boroughs, which then advise the city government on issues impacting their ethnic neighborhood. "Consoci-ationalism" (Lijphart 1968; Nordlinger 1972) is based on accommodation or agreement between political elites over a governance arrangement capable of managing ethnic differences.[3] Of particular relevance to the urban level is the use of local power-sharing arrangements as part of a transition from a local author-itarian or "ethnic" state to some form of democracy, and the use of ethnic proportionality standards in decreasing the bias commonly seen in the urban policing of contested cities.

A final aspect of urban ethnic conditions – *maintenance of group identity* – is critical to inter-group urban relations. In order to attenuate group conflict and advance the prospects for mutual coexistence, a city must be able to affect not only the material conditions, but also the psychological and identity-related conditions, of its different ethnic groups (Neill 2004). The different ethnic groups in an urban system look for breathing room and secure places that will protect and enhance their cultural expression and identity. An ethnic group may use urban territoriality within the city's neighborhoods to acquire or maintain such breathing room. Key collective ethnic rights relate to education, language, press, cultural institutions, and religious beliefs and customs. Because these group rights are connected to potent ideological content, the ability to exercise these rights can be viewed as a critical barometer by an ethnic group of an urban government's treatment of their rights. Psychological needs pertaining to group viability and cultural identity can be as important as objective needs pertaining to land, housing, and economic opportunities. Urban public policy can affect important forms of ethnic expression through its influence on public education (particularly dealing with language) or through its regulatory control over the urban side-effects (such as noise or traffic disruption) of religious observances. Urban service delivery decisions dealing with the location of proposed new religious, educational, and cultural institutions, or the closing down of ones deemed obsolete, can indicate to urban residents the government's projected ethnic trajectories of specific neighborhoods and can substantially bolster or threaten ethnic group identity. The identity and psychological needs of an urban ethnic group can also be consolidated or fragmented by where city government places its jurisdictional or planning boundaries. A city can gerrymander its borders to exclude a major share of a group's population; or, it may enact land use regulations that restrict and spatially fragment ethnic neighborhoods within the city. More productive to urban peace building, cities can redraw boundaries to bring antagonistic sides together under a single city government having a common fiscal base. Over time, this may promote a new transcendent cross-group identity as municipal residents.

I hypothesize that these urban ethnic conditions of land control, economic distribution, policymaking access, and group identity are important influences on the degree of urban stability or conflict. Further, because they are influenced by local government decisions and policies, I suggest that these conditions can be important gauges of the extent to which a city's policies are progressing toward peaceful inter-group coexistence. Movement toward tolerance in a city can be indicated by increased flexibility or transcendence of ethnic geography, lessening of actual and perceived inequalities across ethnic groups, greater inter-ethnic political inclusion and inter-group cooperation, and growing tolerance and respect for collective ethnic rights. In contrast, signs of urban peace impedance include ethnic territorial hardening, solidification of urban material inequalities, an ethnic group's nonparticipation in political structures and cooperative ventures, a public sector disrespect of a cultural group's identity, and most palpably a continuing sense of tension, intimidation, and potential conflict on city streets and in political chambers.

In a city progressing toward some form of mutual co-existence after a legacy of inter-group conflict, I do not expect to see street life change from inter-group antagonism and tension to active engagement with "the other." Rather, a realistic social-psychological objective in these cities is for there to be a type of urban "indifference" that may be necessary to achieve peaceful co-existence (Allen 1999; Simmel 1908).[4] Although this goal of interpersonal or inter-group "indifference" may seem modest, a city able to accommodate diverse populations and immigrant flows may over time be able to effectively balance respect for group rights with the development of a more inclusive, or civic, nationalism.

My goals during this project are descriptive and prescriptive. I wish to investigate the existing realities of cities, policymakers, and residents during and after societal transitions. And I seek to understand urban dynamics in societies where nationalistic group identity poses daily challenges to decision-makers in their design of city programs and to residents in their coping with the day-to-day life of competing nationalisms. Beyond understanding, I seek to discover policy paths through which cities and urban decision-makers can lessen deep-rooted group conflicts in meaningful and long-term ways. I look for circumstances in which urban-based strategies provide prototypes or lessons for de-escalating inter-group confrontations at extra-urban (i.e. regional and national) levels. Based on earlier research in the politically contested cities of Jerusalem, Belfast, and Johannesburg, I contend that cities do matter amidst broader inter-group conflicts and that actions in cities can escalate or ameliorate inter-group tension. City policies have made a difference in effecting change in inter-group relations in Northern Ireland, South Africa, and Israel. In Jerusalem, for example, development decisions by the Israeli government that sought to physically and symbolically extend its territorial claims in and around Jerusalem have created a psychologically divided urban landscape that exacerbates sovereignty-based ethnic-nationalist conflict (Bollens 2000). More positively, in Johannesburg during the transition from apartheid in the early 1990s, community and city decision-makers successfully connected daily urban problems to root empower-ment and political issues and provided fuel for later negotiations over broader transformation structures and processes (Bollens 1999).

I hypothesize that city policies and interventions have broader impacts that reach beyond city and metropolitan borders. Progressive urban policies which seek mutual accommodation of competing nationalisms can contribute not only toward the alleviation of conflict in the city, but can provide streams of policy innovation which, when joined with progress in national-level negotiation and diplomacy, are able to encourage constructive engagement with the underlying causes of inter-group conflicts. I wish to know why specific cities are active agents in larger societal transformation processes and why others are passive or even obstructive agents. In both types of cities, I wish to identify key processes and actors that lead to a city's activeness, passivity, or obstructiveness. Armed with this knowledge, negotiators and third-party intermediaries in transitional societies may be better able to incorporate urban peace-building components into larger national efforts at peacemaking and political reform.

OVERVIEW OF THE CASES

Spain: regional self-governance in a multinational state

I examine in the Spanish cases tensions that exist between the regional nationalisms of Catalonia and the Basque Country, on the one hand, and centralist Spanish nationalism, on the other. I focus on the political transition, begun in 1975, from Franco authoritarianism to Spanish democracy. After Franco's death in 1975, a broad political consensus during the transition emerged that resulted in the approval of the 1978 Spanish Constitution (Moreno 1997). This constitution created a quasi-federalism whereby powers are shared between the central government and the governments of 17 autonomous communities (*comunidades autonomas*), two of which govern the Basque Country (*Pais Vasco*) and the region of Catalonia (see Figure 2.2). The constitution created a state balanced between two views: (1) the idea of an indivisible Spanish nation-state, and (2) Spain as an ensemble of diverse people, historic nations, and regions (Moreno 1997).[5] This granting of significant regional autonomy has helped defuse regional separatism, although nationalist violence remained a fact of life concerning the Basque issue. Regional governments in Spain have considerable responsibility for healthcare, education, urban planning, social services, and cultural activities. Each has its own legislature, political institutions, bureaucracies, public services, and in the case of the Basque Country, its own police force. The regions have some financial autonomy, with revenue coming from the collection of national taxes delegated wholly or partially to the regions, local taxes, and other state income.

Within Catalonia is the city of Barcelona, the second largest city in Spain and an economic powerhouse over the past 20 years (dropping unemployment rate from 18 to 7 percent from 1986 to 1999). There is a distinct historically-rooted regional identity, symbolized by the presence of the regional legislature and executive (the "Generalitat"), and reinforced through the area's use of a distinct language, Catalan. The city contains approximately 1.5 million people (about one-third the population of the metropolitan region at-large) and is known to many as a cosmopolitan place of innovative public sector activity and cultural vibrancy. The ambitious city projects and strategies of Barcelona over the past two decades serve as a reference point for other European cities seeking greater opportunities amidst the restructuring of European governance (Le Gales 2002). Barcelona is a valuable model of an active municipality connecting to both the cultural heritage of its region and to the growing web of international organizational linkages related to economic globalization and European political integration.

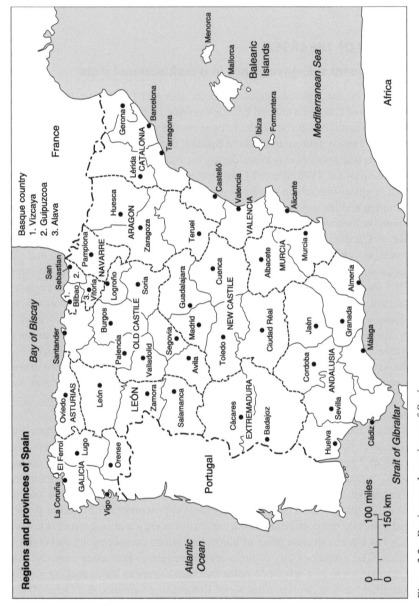

Figure 2.2   Regions and provinces of Spain

Despite a strong sense of regional identity, regional and local politics in Catalonia since Spanish democracy have displayed internal fault-lines and complexities. For the first 25 years of regional autonomy, a center-right coalition nationalist party, *Convergencia I Unio (CiU)*, controlled regional government.[6] Meanwhile, the City, or *Adjuntament*, of Barcelona has been in the hands of the Socialist Party of Catalonia (PSC). The CiU party consists of strong proponents of Catalan autonomy and see themselves as representing a separate Catalan "nation," although they do not seek to withdraw from the Kingdom of Spain. The PSC party, in contrast, advocates greater regional autonomy in the context of a more decentralized system of federalism for all of Spain. Having less of an electoral impact in Catalonia has been the Popular Party, a conservative party of Spanish centralism that commonly competes with the Socialist Party for control of the Spanish national government. Within Catalonia, the Catalan nationalist party of CiU and the Socialist Party have been led by charismatic personalities – Jordi Pujol for CiU, Pasqual Maragall for PSC – and this has led to frequent high political drama and policy and financial conflicts between the Generalitat and the Adjuntament. In addition, and a phenomenon that is significant to my study, the traditional constituencies of these two parties – rural and smaller town support for the CiU, urban and metropolitan support for the PSC – have led to differing views of Catalan nationalism and the appropriate role of Barcelona city vis-à-vis the region at-large.

Basque Country (*Pais Vasco*) is the region in Spain with the greatest amount of financial autonomy, a result of an economic compact between Spain and the region in 1978. The *Partido Nacionalisto Vasco* (PNV) political party supports enhanced regional autonomy but rejects violence and is considered a moderate nationalist party. Since the transition, the PNV has had a significantly larger political constituency in the Basque Country than the *Herri Batasuna* (People's Unity) political party. Batasuna has been aligned with the *Euskadi ta Askatasuna* (ETA) [Basque Fatherland and Liberty] paramilitary group, a leftist separatist group that uses violence as a political tactic. Batasuna usually has not achieved more than 10 percent of the regional vote. However, because these votes came disproportionately from the many smaller and rural towns, nearly 900 town councilors in the Basque Country and neighboring Navarra region were at one time party members. In public polling of Basque residents, about 40 percent of respondents label themselves as Basque "nationalists;" about 50 percent "non-nationalists" (Euskobarometro 2005). While many Basques (about 60 percent) favor regional autonomy or even independence from Spain, a strong majority rejects terrorism and violence as appropriate means toward those ends.

My research visit in February 2004 to Basque Country occurred during a time of substantial political uncertainty. About 18 months before my visit, in Fall 2002, the political party Batasuna was suspended for three years by a Spanish judge due to its links to the ETA and then was outlawed altogether by the Spanish Parliament. In addition, Basque regional politicians were debating in 2003 and 2004 a controversial plan to make the region a "freely associated state," having a separate court system and separate representation in the

European Union, that would be able to negotiate a new relationship with the Spanish state. Subsequent to my visit, in 2006, ETA put forth what it called a permanent ceasefire, declaring an end to its decades of violence. In my study, I focus on the three major cities in the Basque Country – Bilbao, Vitoria-Gasteiz, and San Sebastian – as lenses through which to gauge the effectiveness of regional peace-building in accommodating Basque aspirations in Spain. Bilbao is the largest city in the region (about 500,000 city population; total regional population is about 2.1 million), Spain's busiest port, historically industrialized and now a focal point for physical and cultural revitalization. Vitoria-Gasteiz, with population over 200,000, is the capital of the Basque Country. San Sebastian (about 180,000 residents) was considered the stronghold of Batasuna and ETA supporters.

Bosnia and Herzegovina: a vulnerable political geography

I examine in the Bosnian cases the antagonisms between three nationality groups – Bosniaks (Muslims), Bosnian Serbs, and Bosnian Croats – and I emphasize the 10-year transition period since the end of the Bosnian war in 1995. Bosnia and Herzegovina (BiH), an independent state formed out of the hell and trauma of the 1992–1995 Bosnian War, is attempting, with strong United Nations oversight, to re-create itself as a loose confederation of two entities – one Muslim and Croat, the other Serb – whose boundaries were created largely by war and ethnic cleansing (see Figure 2.3). The Bosnian War killed 100,000 people and forced half the country's four million people to flee their homes to friendlier locales within the state (one million people "internally displaced") or to other countries entirely (one million "refugees"). Ten years after the war, BiH is dealing more with reconstruction and peace implementation than engaging in a fuller transition agenda dealing with human and economic development (Commission of the European Communities 2003a). Dealing with the aftermath of war remains in many ways a primary occupation of the international community and domestic governments. Gross Domestic Product (GDP) for BiH remained in 2003 below half its pre-war level, GDP per capita in 2002 stood at a paltry $1,800, and the country has a chronic dependence on international assistance. The European Commission and EU member states contributed over 3.6 billion euros in assistance in the first eight post-war years. The political organization of BiH is complex in structure and has faced many difficulties. The Dayton Accords (General Framework Agreement for Peace in Bosnia and Herzegovina) signed in 1995 provided for the continuity of Bosnia and Herzegovina as a state but created two constituent entities of ethnically separated populations – the Federation of Bosnia and Herzegovina (with a post-war Bosniak [Muslim] – and Croat majority) on 51 percent of the land, and Republika Srpska (mostly populated by Bosnian Serbs) on 49 percent of the land (Burg and Shoup 1999). Each autonomous entity has its own legislative and administrative structures. The Bosniak-Croat Federation has a mixed system with a president and a parliament. Republika Srpska has a parliament-president system.

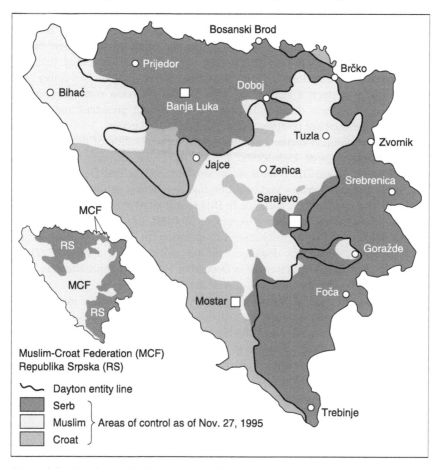

Figure 2.3 Bosnia and the Dayton Accord Boundaries

At the state level, central governing institutions are weak. Legislative and executive decision-making for the state requires broad agreement and consensus to function. Given the existing animosity and absence of trust, such consensus does not exist. The state Parliamentary Assembly has two chambers – the House of Peoples and the House of Representatives – that both use electoral balance and power-sharing mechanisms. Representatives of each of the "constituent peoples" – Muslim Bosniaks, Bosnian Croats, and Bosnian Serbs – have the right to declare any prospective decision in the Parliament Assembly "destructive of a vital interest" in which case concurrent majorities within each of the ethnic group representatives is needed. There is also a three-member Presidency composed of a directly elected leader from each ethnic group. A majority decision is possible; however, decisions "destructive of a vital interest" can be referred to ethnic group representatives, where a vote of two-thirds of the relevant group renders the decision null and void. With a vacuum of power at the state level, there is a concentration of power within the two ethnically demarcated autonomous entities, and this appears to be a major impediment to Bosnia's transition to a multi-ethnic democracy. Bosnia needs to strengthen central institutions as well as to provide effective governance at sub-entity (cantonal and municipal) levels. Bosniak, Serb, and Croat leaders signed in November 2005 an agreement to seek a strong and more unified Bosnian state government, envisioning a process that would create a single-state president and perhaps a strong prime minister position and strengthened parliament.

Nationalist political parties have dominated representation of their respective ethnic groups – Croat Democratic Union (HDZ) for Bosnian Croats, Party of Democratic Action (SDA) for Bosnian Muslims, and the Serb Democratic Party (SDS) for Bosnian Serbs – and coalition-building across nationalities has been difficult. The international presence in BiH is still large and includes the United Nations (the de facto civilian government of BiH), the European Union, and the Organization for Security and Co-operation in Europe. Militarily, the international presence has declined over the decade, from a high of 60,000 NATO forces sent in to end the war in December 1995 to the approximately 7,000 troops in 2004 under the direction now of the European Union.

I examine the case of Sarajevo, a post-war Bosnian Muslim majority city located at the frontier between the two new autonomous entities, and Mostar, a case of obstructive local power sharing between Bosnian Muslims and Bosnian Croats. These two cases promise to illuminate the local strategies used after breakdown of a previous political order. Sarajevo, with a mixed ethnic population of 540,000 Bosnian Muslims (40%), Bosnian Serbs (30%), and Bosnian Croats (20%) in 1991, is a now approximately 80% Muslim city of about 340,000 population. The siege of the city by Bosnian Serb and Serbian Militias lasted 1,395 days, killed 11,000 residents of the city, 1,600 children, and damaged or destroyed about 60 percent of the city's buildings. Although many Bosnian Serbs stayed in the city during the war in defense of the bombarded concept of multi-ethnicity, substantial numbers fled after the Dayton accord fearing retaliation. During and after the war, Muslim refugees from ethnically cleansed eastern Bosnia (now Republika Srpska) migrated in large numbers to inhabit shelled and burned-out flats in war-torn neighborhoods. Sarajevo today, like Jerusalem since 1948, is a frontier city – an urban interstice – between opposing political territories. The boundaries between the Dayton-created Muslim-Croat Federation and Republika Srpska entities (in international language, Inter-Entity Boundary Lines or IEBLs) are drawn just outside the city's southeastern parts. Programs of physical reconstruction and refugee relocation confront the international community and policymakers with challenges about whether the ethnic geography created out of the ashes of war should be transformed or reinforced, calling into question whether policymakers should emphasize re-emergence of the city's former multiculturalism.

Mostar is a case of attempted local power sharing between Bosnian Muslims and Bosnian Croats, enemies themselves in a "war within a war" in this part of the state and now spatially segregated in the city. The Rome Agreement and an

Interim Statute of 1995 created a central zone, consisting of a common strip of land around the former confrontation line within the city, which was to be administered by a city council and administration, where Muslims, Croats, and "other" groups were to have equal representation. Outside this central zone, the post-war city was carved into six municipalities – three Bosnian Croat and three Bosnian Muslim – which have held onto many decision-making powers and created obstacles to effective governance within the city's former boundaries. Urban planning, both within the central zone and in the six municipalities, has been a hotly contested competency exploited for partisan advantage. Beginning March 2004, in response to a dictate by the UN High Representative in BiH, the city of Mostar was unified by merging the six former Mostar municipalities and creating a city-wide administration and city council. Mostar's governability remains today a significant test of the future stability of the Bosniak-Croat part of Bosnia and Herzegovina.

AN URBAN CONFLICT-STABILITY CONTINUUM

Cities, and the societies within which they are part, are complex entities and a social scientist is certainly at risk when making comparisons. Yet, it is only through comparative research that we can transcend the assumptions and values of single-country studies of urban and regional phenomena. Alterman (1992, 39) identifies the need to develop a "systematic body of knowledge about the contexts in which planning practice and planning education occur cross-nationally." Among the myriad and multiple dimensions along which cities differ, I focus on the placement of urban regions along an "urban conflict-stability" continuum as a way to provide a necessary comparative context (see Figure 2.4). I place cities along this scale depending upon whether the city is experiencing active conflict, a suspended condition of static non-violence, movement toward peace, or urban stability/normalcy. Examining cities along such a continuum may provide insight into the range of possible interventions by city governing regimes amidst inter-group differences.

The urban conflict-stability continuum, as proposed, is not intended to be a comprehensive measuring tool but rather a useful heuristic model. It enables us to

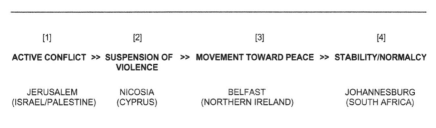

[1]	[2]	[3]	[4]
ACTIVE CONFLICT >>	SUSPENSION OF >> VIOLENCE	MOVEMENT TOWARD PEACE >>	STABILITY/NORMALCY
JERUSALEM (ISRAEL/PALESTINE)	NICOSIA (CYPRUS)	BELFAST (NORTHERN IRELAND)	JOHANNESBURG (SOUTH AFRICA)

Figure 2.4   Urban conflict–urban stability continuum
Note: Cities are placed on this continuum based on one primary criterion—the extent to which intergroup conflict over *root political/nationalistic issues* has been effectively addressed. City examples are from author's prior research.

think about the differences across types of contested cites and what these differences mean for urban intervention and national peacemaking. In placing cities along the continuum, I focus on a sole overriding criterion among multiple urban dimensions – the degree that active inter-group conflict over *root political issues* has been effectively addressed. Similar to the way that the United Nations tracks countries in terms of their human development, I posit that cities may be classifiable in terms of their vulnerability to conflict (UNDP 2002a).[7] To help introduce this comparative framework to the reader, I illustrate it using findings from my earlier research involving four politically contested cities.[8]

In category [1] cities, hostility, antagonism, tension, and at times overt violence, exist between urban groups. This is so because the root political issues of the broader nationalistic conflict remain unresolved. In such a circumstance, the city becomes a flashpoint, platform, and/or independent focus of broader conflict. When there is such active urban conflict and a vulnerability of the urban arena to deeper nationalistic currents, urban policy and planning approaches are likely to become rigid, defensive and partisan in efforts to protect the governing group in this unstable environment. This is what I have found in Jerusalem, where strife and conflict remain a fact of life in both the city proper and in the interface areas bordering Israel proper and the West Bank. The Israeli government, pursuing goals of unification and security, has utilized land use planning as a territorial tool to extend the reach of its disputed public authority. Urban growth and development policies, equating land with political control, have spurred territorial extensions that have penetrated and diminished minority land control. These actions have facilitated the pace and increased the magnitude of Jewish development to maintain a Jewish majority in the city. They have also influenced the location of new Jewish development in annexed areas to create an obstacle to the "re-division" of the city, and have restricted Arab growth and development to weaken their claims to reunified Jerusalem. The building of the Israeli separation barrier since June 2002, including what will be a 40-mile wall extending alongside and beyond the city's politically contested municipal boundaries, is the most recent and imposing manifestation of Israel's use of planning as political tactic.

In category [2] cities, there is tenuous cessation or suspension of urban strife but not much more. A city is marked more by the absence of war than the presence of peace. After the ending of overt conflict, there is likely to remain deep segregation or partitioning of ethnic groups in the city, local politics may persist in parallel worlds, and there may still be tension on the streets. This is because the legacies of overt conflict live on far past the duration of open hostilities themselves. In category [2] cities, however, this potential for inter-group differences to inflame violent actions is lessened somewhat due to a negotiated agreement between nationalist elites and/or intervention by a third-party mediator. Although this is a significant advance, suspension of overt conflict is only a starting point in urban peace-building and requires important steps in the future that bring positive changes to a city in the forms of tolerance, openness, accommodation, and democratic and open participation. Without these movements toward peace on the ground, a city will stagnate and be vulnerable to regressive violent and political acts.

The city of Nicosia on the island of Cyprus represents well this moribund and unsure space between active conflict and movement toward peace-building. The city and island have been for over 30 years cleanly separated through a physical partitioning into opposing sides. The wall has suspended overt conflict, yet genuine political and social reconciliation dealing with the root causes of the Cyprus problem remains elusive. Southern Nicosia is the capital city and seat of government of the Republic of Cyprus, the part of the island inhabited by the Greek Cypriot population. Northern Nicosia is the capital of the Turkish Cypriot "Turkish Republic of Northern Cyprus," officially declared in 1983 and unrecognized internationally. Intercommunal violence in 1958 and 1964 had necessitated foreign intervention and de facto ethnic enclaves as ways to stabilize the strife-torn city. Then, as mainland Greek and Turkish political agendas pierced the island, the Turkish army entered Cyprus and an estimated 175,000 Greek Cypriots were displaced from the north and about 40,000 Turkish Cypriots from the south. Since 1974, the Green Line has separated the city – a United Nations maintained buffer zone built upon ethnic demarcation lines first drawn in the early 1960s. Today, lacking special permission, none of the 650,000 Greek Cypriots to the south may enter the north and none of the 190,000 Turkish Cypriots to the north may enter the south.

In category [3] cities, there are efforts to transform urban conflict geographies to peace-promoting ones and to use urban and economic development policies to transcend ethnic and nationalist differences. Decisions regarding the built environment, provision of economic opportunities, and delivery of public services are done in ways that create and promote urban spaces (both physical and psychological) of inter-group coexistence. Examples may include providing flexibility in the urban landscape to facilitate mixing of different groups if and when they desire it, creation of cross-ethnic joint planning processes, establish-ment of economic enterprises in areas that link different ethnic communities, provision of public spaces that bridge ethnic territories, sensitive oversight of the location of proposed development projects explicitly linked with one ethnic or religious group (churches, mosques, community centers), post-war reconstruction and relocation decisions that do not solidify war-time geographies, and provision of community and youth services in ways that bring children together so they can learn from each other. Although category [3] cities show movement toward normalcy, local peace-building efforts remain experimental in the sense that full urban stability has not yet been reached. Remembrances of trauma and conflict remain below the surface, and they can be stimulated by local public policies that are not sufficiently sensitive to these scars. Up until the time when democracy is seen as the only game in town (category [4] cities), the movement of cities toward peace can be held hostage by the threat of political violence by paramilitary groups. Even when the threat is not actualized, the potential for such violence becomes part of the political debate in that city and society. The difficult challenges of urban peace building as part of a fragile movement toward national peacemaking is illuminated by the experience of Belfast and Northern Ireland. Amidst the on-again, off-again national progress toward Northern Ireland peace, British policymakers have made experimental efforts to modify the city's strong ethnic territoriality, yet progress is piecemeal and slow.

Category [4] cities represent a fundamental turning point where there is the consolidation of peace building, a beginning in the transcendence of inter-group differences, and the undertaking of fundamentally new directions in urban governance and policymaking. An important threshold is passed when nationalistic and inter-group differences take place solely within political and legislative channels with no or little threat of a resort to political violence. Regarding urban development specifically, category [4] cities are more able than category [3] cities to enact policies that fundamentally redistribute the costs and benefits of city growth, reverse growth ideologies that guided the former governing regime, and imprint on the urban landscape values such as public access, equality, and democratic participation. In the Johannesburg, South Africa case, the momentous transformation from "white rule" to majority-rule democracy in the mid-1990s meant that the root causes of political conflict were effectively addressed. This places Johannesburg further to the right on the continuum than Belfast, Nicosia, and Jerusalem, where political root issues remain partially or fully unresolved. Difficulties surely remain, including rampant crime and gross disparities in urban opportunities across race and income, which make it arguable that Johannesburg is properly placed at the right end of this scale. However, I am not alone in this positive comparative assessment of South Africa; Sisk and Stefes (2005) point to its political transformation as holding important lessons for places like Northern Ireland and Bosnia. Because they are further along the path toward political peace, Johannesburg policymakers have the space to consider and seek to remedy the gross and inhuman inequalities associated with state-sanctioned racial discrimination and state terrorism, and to confront the severe psychological pains and scars that permeate black African society. Indeed, the South African "peace" exposes a set of damaging and dehumanizing urban effects of inter-group conflict, problems that are not addressed, or are actively suppressed, in Jerusalem, Belfast, and Nicosia because root political issues are either exacerbated through planning (Jerusalem) or bypassed (Belfast and Nicosia). The apparent irony on the surface – that non-political criminality is much higher in Johannesburg than in Jerusalem, Belfast, or Nicosia – illustrates the severe after-effects of decades of immoral state policies and that inevitable societal disequilibrium will linger far after negotiated agreements start a country and city down the road of "peace" and "normalcy."

I will locate Mostar, Sarajevo, Basque Country, and Barcelona along this conflict-instability continuum. I expect that due to the differing lengths of time since each country's political transition, the Spanish urban areas will be placed to the right of the Bosnian cities. Thus, I will look for elements in the 25-year history of Spain's democratic evolution that may have lessons for Bosnia and its cities. Amidst numerous examples in the world pointing to the hardships associated with the transition to democracy, Spain's transition from authoritarianism to a healthy and functioning democracy has been phenomenal. Indeed, these 25 years constitute the first time that Spanish democracy has been self-sustaining (Edles 1998). Such a successful transition, albeit not without problems, thus merits attention for the lessons it may provide to other societies.[9] I also expect to find, beyond a simple contrast between Spanish and Bosnian cases, intrastate variation (that is, within each country, one city will be more advanced along the continuum than the other). To the extent that this is true, it will provide an opportunity to isolate those attributes, independent of national context, which have caused certain cities to be further ahead as peace builders while other cities lag in this capacity.

Notes

1. I am not using the term in its more sociological meaning (Wirth 1938) connoting a way of life or set of social practices engaged in by urban residents.
2. Although I define planning broadly in this study to encompass many types of public sector actions beyond the traditional definition of the city (town) and regional planning profession, there are certain urban activities I do not emphasize in order to maintain a coherent analytical focus. These include urban education and policing. Urban policing is a core issue in multinational societies that is outside the scope of this study. I focus on the planning-related and often land-based policies that structure opportunities and costs in contested cities, rather than the maintenance of societal order through police and military force.
3. Consociationalism has been conceptualized primarily as occurring at the national level. Nonetheless, elements of urban consociational democracy can be found in Brussels (Belgium) and Montreal (Canada).
4. The challenge of Sennett (1999, 130) to go beyond indifference on the city street toward a confrontation with the other that would "acknowledge other groups and take a risk with the boundaries of one's own identity" appears too difficult in cities coming out of conflict, at least in the short term. I suggest that "indifference" over a period of, say, one generation, might build the foundation for Sennett's fuller engagement. In the shorter term, Sennett's confrontations may need to be confined to structured and mediated interpersonal settings.
5. Article 2 of the Spanish Constitution recognizes both the unity of the Spanish nation and the right to autonomy of nationalities and regions (Agranoff and Gallarin 1997). Such ambiguity has left ample space for negotiation and debate between regions and the state in more than 25 years since adoption of the Constitution.
6. This monopoly on regional government political control ended for CiU October 2003 when the socialist party, in coalition with two other parties, gained power. This change in government during my research stay provided an opportunity to examine the contrasting regional visions, and proposed growth strategies, of the two parties.
7. In this UN report there is the ranking of countries in terms of their capacity to protect personal security (measured by the numbers of refugees and armaments, and victims of crime) and human and labor rights. Such measurement of urban refugees, crime, and inter-group relations would likely comprise an important part of a more comprehensive urban index of stability/instability. The increasing ability of terrorist groups to target cities worldwide brings another aspect into the measurement of urban stability. An attempt to measure cities in terms of their potential for this violence is Savitch and Ardashev (2001). This study uses three criteria – social breakdown, resource mobilization, and target-proneness – to assess the probability for 40 major cities of experiencing terror.

8. I studied Jerusalem, Nicosia, Belfast and Johannesburg using field research methods and interviews similar to those I used in the Spanish and Bosnian research. In each of these cities except Nicosia, I carried out extensive interviews over a three-month period of time. For Nicosia, my observations are based on more limited field research. This earlier research is reported in Bollens 1999, 2000, 2001.

9. The inclusion of Spanish city cases in this study came about because Spain's reputed transitional success attracted my attention as a likely "positive" example that would contrast with, and inform, the cases of more difficult transitions investigated in my earlier work. In addition, the Basque Country "exception" (i.e. its long struggle with political violence) within an otherwise successful national case builds comparative elements into the research design.

3 Barcelona: constructing democracy's urban terrain

Figure 3.1 State repression of Catalan nationalism

Since the end of the Spanish Civil War in 1939, war and overt conflict have not reached Barcelona. The city and its residents suffered under the authoritarian rule of Francisco Franco Bahamonde for over 35 years, experienced a political transition of great uncertainty for five years after Franco's death in 1975, and since 1980 have experienced democracy in the city and a high degree of regional autonomy for its home region of Catalonia (see Figure 3.2). For the past 25 years, there has been limited overt conflict owing to nationalism, no Catalan terrorism or paramilitaries, and mainly indirect references to independence on the part of political leaders. Yet, when one examines the

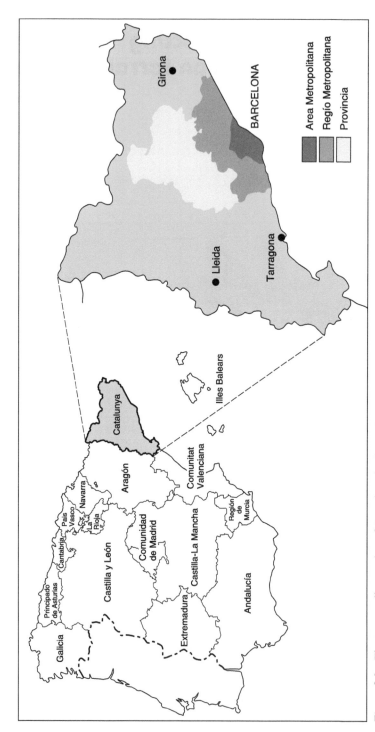

Figure 3.2 The metropolitan region of Barcelona in Catalonia and in Spain

history, talks to the people, and goes beneath the mesmerizing authenticity of Barcelona, one finds a deeply rooted Catalan nationalism based on the region's distinctive culture, language, and history which differs substantially from the centralist nationalism that has permeated the Spanish state for centuries. Politics in Catalonia are dominated by the "national question" – specifically the appropriate political relationship between Catalonia and the Spanish state. Since the 1978 Spanish Constitution and the 1979 Autonomy Statute created Catalonia as an autonomous region in a democratic Spain, the push and pull of nationalist politics has been an ever-present characteristic of Catalonia's social and political life.

The degree of Catalan regional autonomy since Franco has been far above many other models of state governance (Interviews: Ferran Requejo, professor of political science, Universitat Pompeu Fabra; Enric Fossas, Institut d'Estudis Autonomics). Nonetheless, many Catalans support the need for renegotiation of the autonomy statute to increase Catalan powers. In June 2006, this push for greater autonomy advanced, with 74 percent of Catalan voters approving a new agreement giving the region enhanced powers in taxation and judicial matters and providing it with "nation" status within the Spanish state.[1]

The fact that Barcelona may seem an odd companion in a book that includes the Basque Country and the former Yugoslavia means not that nationalist, group-based identity conflicts do not exist in Barcelona and Catalonia like they do in these other places, but rather that something has gone right in the northeastern part of Spain to channel these significant nationalist and group-based differences effectively into more constructive channels. I give attention in this story to why and how these things have gone right, while also giving due attention to the fact that the process of nationalism is an ever evolving one that can turn destructive as well as constructive.

CATALAN NATIONALISM, BARCELONA, AND THE FRANCO YEARS

Catalan nationalism

The contemporary manifestations and forms of Catalan identity and nationalism are evident even to an outsider. In the city of Barcelona and its urban region, group differences display themselves in terms of identity, urban spatial distribution, language, and economic status. The urban area contains in many respects dual identity groups. In a large survey of households in the province of Barcelona,[2] 44.7 percent of respondents felt "more Catalan than Spanish" or felt "purely Catalan," while only 18.3 percent felt more Spanish than Catalan or felt purely Spanish in identity (Institut d'Estudis Regionals i Metropolitans, 2002a, 413). There is also spatial segregation between the ethnic Catalan population and the large number of Spanish immigrants who have come to the urban region since the 1960s (and their offspring), the latter group residing disproportionately outside the city of Barcelona, particularly in the first suburban crown (or *corona*) [See Figure 3.3].

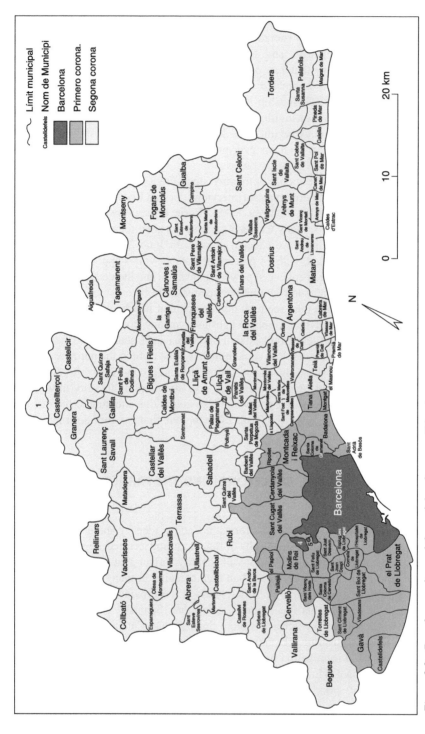

Figure 3.3 Barcelona region and the peripheral crowns (Rings)
(*Note*: Primera corona = first crown; Segona corona = second crown; Limit municipal = municipal boundary; Nom de Municipi = name of municipality)

Linguistic differences are evident and vary across urban space. For the urban region overall, 30 percent of the population considers Catalan their native language, 56 percent Castilian (Spanish), and 14 percent consider both Spanish and Catalan as their native languages (Institut d'Estudis Regionals i Metropolitans, 2002a, 402).[3] In the first crown "red ring" suburbs[4] this linguistic distribution becomes 68 percent Spanish-speaking and 24 percent Catalan-speaking. In Barcelona city proper, meanwhile, there exists a basic split of the linguistic groups in terms of percentage. Within Barcelona city, furthermore, the native language of residents varies significantly district to district. Catalan is the native language of 67 percent of respondents in the Gracia district and 62 percent of those living in Eixample, while it is the native language for only 22 percent in Nou Barris and 36 percent in the Horta-Guinado district (Institut d'Estudis Regionals i Metropolitans, 2002b, 50) [see Figure 3.4]. Economically, a middle class has arisen that consists of ethnic Catalans who are highly educated, speak Catalan as their primary language, and are well connected to societal networks (Institut d'Estudis Regionals i Metropolitans, 2002a, 416).

Despite the linguistic, economic, and spatial differences that contrast ethnic Catalan and immigrant populations, there is a degree of hybridization between the Catalan and Spanish immigrant populations. At the level of daytime "on the

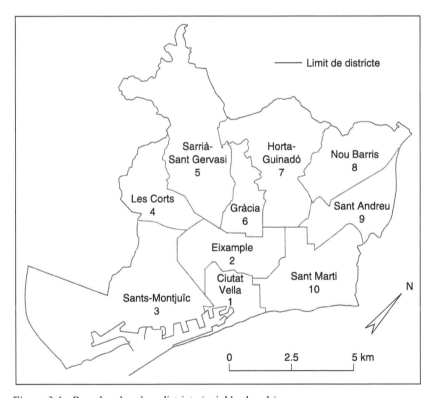

Figure 3.4   Barcelona's urban districts (neighborhoods)

street" functionality, the Barcelona area in many respects approximates a bilingual society. One-third of the surveyed population perceive a dual identity, feeling "as Catalan as they are Spanish." As a way to increase their chances economically and socially, a significant number of native Spanish-speaking persons have adopted a dual Catalan-Spanish identity (Institut d'Estudis Regionals i Metropolitans, 2002a). Since the Universal Exposition in 1888, Catalonia has had three major waves of Spanish immigrants – at the end of the nineteenth century, in the 1920s, and during the 1950s, 1960s, and 1970s. In the 1950–1975 period alone, about 1.4 million immigrants moved into Catalonia (Cabre and Pujades 1988). This has resulted today in a situation where about 60 percent of Catalonian residents are first-, second-, or third-generation immigrants from elsewhere in Spain (P. Vilanova, professor of political science, University of Barcelona, interview).

Regional nationalism has deep historic roots in Catalonia. It has long experienced semi-autonomous status, dating from the establishment of the Council of One Hundred in the thirteenth century. A regional governance structure, established in 1359, was composed of the "Generalitat," which includes the regional president, regional executive council, and the Parliament. The city of Barcelona, as port of the Kingdom of Aragon and the Principality of Catalonia, reached its golden age between the thirteenth and fifteenth centuries when Catalan shops and commerce dominated the Mediterranean. The merger of the confederate Crown of Aragon with the Kingdom of Castile in the fifteenth century is viewed as the beginning of centuries of decline of the Catalan nation that would not be reversed until the nineteenth century. Catalonia became effectively integrated into the Spanish state following its defeat in 1714 in the War of the Spanish Succession.[5] In 1716, the Decree of the *Nueva Planta* abolished Catalan political institutions, imposed Castilian law, and laid the grounds for attacks on the public use of the Catalan language in favor of Castilian. It was not until the short-lived democratic Republic era of 1931–1936 that self-government was to emerge again in Catalonia, in the form of an association of provincial councils or *Mancommunitat*.

Although perennial accounts trace Catalan nationalism back centuries, modernist accounts focus on the emergence of a Catalan political nationalism in the late nineteenth century (Etherington 2003). This nationalism drew on four main lines of thought (Hughes 1992) – republicanism or federalism, populism, regional conservatism, and a literary renaissance. Whereas the republican line of thought that emphasized creating Catalonia as a state within a state had links with socialist ideas and the goal of uniting industrial workers, the regional conservative view was associated with the moderate industrial bosses more concerned with protectionism for their products than breaking free of the Spanish state. Advocates of this regional conservative perspective helped to form the first Catalanist political groups, and then the first Catalan political party, the Lliga Regionalista, in 1901. There also existed within Catalan nationalist thought an internal division between urban modernists and rural traditionalists. These groups had contrasting views as to the contribution of urbanity to the regional nationalist project, with one group extolling the virtues of Barcelona's growth, the second warning of its spiritual dissipation. The cultural Renaissance (*Renaixenca*) beginning in the mid-1800s that sought to rediscover Catalan as a language had also provided a conservative traditionalist interpretation of the national homeland, wherein the rural hinterland emerges as the spiritual reserve of the country, the "real" Catalonia, in opposition to urban Catalonia and Barcelona (Etherington 2003). To conservative Catalan nationalists, the continual growth of Barcelona meant that the city had become by the late 19th century "excessive," "out of proportion," and "macrocephalic" in comparison with the rest of Catalonia, resulting in a Catalonia of "territorial imbalance." Such disequilibrium threatened rural- and small town-based Catalan nationalism by allowing outside influences to penetrate and erode the moral and social order of the fatherland ("la patria").

Barcelona under Franco (1939–1975)

Oh Catalonia, shady country of dark bushes and thick woods, land of our fathers! What hand of man would suffice to break the filial bond that unites us to you?

Pau Piferrer
1818–1848

The Franco years were long ones of political and cultural repression in Catalonia. Franco's ideology created a centralized Spanish state able, in his view, to further a unified and secure Spain impervious to the fragmenting forces that had led in the 1930s to unstable democratic rule and then the Civil War. Autocratic rule, or "autarky," which suppressed other views and interests, was viewed as the only way to hold together the central state. Accordingly, the Generalitat form of regional self-rule and other political institutions were abolished after Franco's victory in the Civil War (1936–1939) and Catalan autonomy came to an end in 1939. In its place, a centralized and hierarchical political and administrative apparatus based in Madrid took over to implement Franco *dictates*. A Civil Governor appointed by Franco now managed Catalonia. The region was now administratively divided into four small provinces created by, and reporting to, Franco's central administration. All political parties except one were outlawed. Individuals who had views not perceived as supportive of the Francoist *Movimiento Nacional* were vetted and dismissed from positions in public administration and industry. A huge police and "special investigations" apparatus, linked to the regime's single party, the *Falange*, forcefully imposed "social harmony" with the regime. The authoritarian regime practiced a "scorched culture" strategy, forbidding the teaching of Catalan in local schools and prohibiting the use of Catalan in public places (Montalban 1992). Catalan newspapers were closed, and radio and television programs in Catalan were banned. Labor was disciplined and controlled through their mandatory membership in state-sponsored unions or "vertical syndicates," which subordinated their interests to the needs of business owners and proprietors, restricted their ability to make demands and launch protests, and outright repressed them when need be (Molinero and Ysas 2002).

Being a focal point for a non-Spanish nationalism, Barcelona and Catalonia suffered disproportionately under Franco rule. The "anos grises" (gray years) in Barcelona under Franco were years of economic need, political repression, substantial in-migration, and a predatory speculative chaos instrumental in the city's "destructive reconstruction" (Busquets 2004). This Franco city also created objective conditions of substantial material disadvantage and inequality that eventually spawned an urban-based critique of the regime.

The early post-war years of Franco control in Barcelona were years of economic want, in-migration and housing shortages. Spain was economically and politically isolated in the 1940s and early 1950s. In these early years, the average weekly wage of an industrial worker in Barcelona was 100 pesetas (compared to a price of 30 pesetas for a dozen eggs) [Montalban 1992]. Throughout the 1940s in Barcelona, "a hundred thousand immigrants arrived with nowhere to sleep" (Montalban 1992). Such in-migration into the city proper, from southern Spain principally, became significant in the 1945–1955 period as migrants from the depressed rural south looked to the city for opportunities. Between 1940 and 1950, 165,000 immigrants came to the city and another 80,000 established themselves outside the city but within the county[6] (Salvado and Miro 1996). Another 195,000 immigrants moved to the city between 1950 and 1958. A significant housing shortage led to the creation of shantytowns in the city, and thousands of immigrants ended up in ghettos near the city's periphery. In 1957, there were as many as 12,500 shacks within the city. In these shantytowns, as many as 50,000 people lived spread out on Montjuic mountain and another 9,500 lived informally in Somorrostro along the beach in the Sant Marti district (Salvado and Miro 1996). A 1969 report estimated 20,000 shacks existing in the city (Vivienda 1969).

Spain's isolation began to lessen in 1953, when there was the signing of economic and military agreements with the United States; in 1955, when Spain was admitted to the United Nations; and in 1959, when the national Stabilization Plan allowed the peseta currency into the free market. For Barcelona and its urban sphere, the Stabilization Plan stimulated a takeoff of industrial and urban development. The Franco regime used a development pole model of economic growth, concentrating industry and the needed labor force in three urban areas – Barcelona, Bilbao, and Madrid (F. Muñoz, geography professor, Autonomous University of Barcelona, interview). Urbanization in Barcelona became connected to Franco's industrialization project. This resulted in the development of an industrial free trade zone near the port (Zona Franca), the building of massive new housing projects on vacant land at the city's periphery, and the stimulation of Spanish immigration through housing offers, whereby a household could own a flat after paying monthly installments for 15 years. Due to the industrial magnetism of Barcelona and the region, immigration skyrocketed to 800,000 immigrants into Catalonia during the 1961–1965 period. Of these, about 570,000 individuals located in Barcelona province (about 145,000 in the city, another 145,000 in the county). In the next five years (1966–1970), another

210,000 moved into the county and an additional 70,000 immigrants into the city proper.

The push to industrialize and accelerate urbanization had detrimental effects on the urban area through the creation of labor warehouses at the periphery and speculative and chaotic densification in the internal city. In the urban periphery, authorities saw the solution to the housing problem as the "housing estate," a unitary large-scale operation involving thousands of flats in high-rise tower blocks (or polygons) and usually built near the periphery where land was available. The Social Emergency Plan of 1957 listed the housing estates to be carried out and most of these were built. Examples include the Eucharistic Congress Housing with more than 2,700 homes (1952–1962), the Besos Southwest estate (late 1950s), Sant Ildefons (in the city of Cornella) with over 11,000 flats (1960), Bellvitge (1968) with 13,000 flats and some 30,000–40,000 residents (in the city of L'Hospitalet), Ciutat Meridiana at 4,000 flats, and La Mina (1971), built just outside the city border in Saint Adria. In the years before 1960, these "polygons" were built in or near the city; starting in the 1960s, however, they were developed primarily outside Barcelona city to the west in the delta area of the Llobregat River, in such suburban cities as Cornella de Llobregat, l'Hospitalet de Llobregat, and El Prat de Llobregat.

Residential development increased significantly in the twenty years after the Stabilization Plan. Whereas 11,000 housing units were built in the province in 1959, 53,000 units were built in 1966. In terms of polygon development, over 26,000 units were built between 1955 and 1965 in the city proper and another 18,000 outside the city in Barcelona county. Polygon development picked up substantially between 1965 and 1972 and also shifted in its location; only 9,800 polygon flats were built in the city but over 91,000 were created in the county. Development of the residential periphery of Barcelona (both within the city proper and in first ring suburban municipalities) created three radically different types of environments – neighborhoods of suburban expansion, neighborhoods of marginal urbanization and of self-built housing, and polygons of massive housing. What was common across types was an incompleteness of the built environment – a lack of public spaces and community facilities, of centrality and focus, and of coherency of image (Busquets 2004).

Within the city of Barcelona itself, a process of significant metamorphosis began in the late 1950s – one driven by rampant speculation and limited public sector control. Francoist mayor Josep Maria Porcioles, elected in 1957, gave free rein to private initiative and speculation in his efforts to absorb surplus population and to expand the city to take in overpopulated sections on its outskirts. By 1970, *Porciolismo* was viewed as an unqualified disaster, resulting in the revitalization of the city but at the expense of characterless densification. There had occurred for about fifteen years a ferocious level of speculation in existing city neighborhoods, facilitated by lenient limits on allowable floor area and height. These overly high net densities resulted in strained capacity for infrastructure, the disappearance of spaces set aside for parks and facilities, and, in the case of the largest Eixample district, abusive transformation of the historic 1859 Cerda plan and its closed block design.[7] Public services and spaces normally part of good city planning were restricted as private interests overshadowed public values. Due to limited building of new public schools and the poor conditions of existing ones, fully 46 percent of children in the city were schooled at home by the end of the 1960s (J. Solans, former Generalitat director of planning, interview). Speculative polygon development at the periphery also occurred in the second half of the 1960s. Land zoned for rural restriction would be sold to private developers, the land would be rezoned for urban and high-rise development, and the developer would appropriate the substantial "plusvalor" (the windfall, or unearned value, caused by the rezoning) for himself. This appropriation of "plusvalor" by speculators, and negligible public investment by municipal authorities, meant under provision of common neighborhood services and facilities such as sanitary services, roads, cultural facilities, schools, and parks.

By 1970, the city of Barcelona was home to 1.74 million people, and the metropolitan area (containing the first crown suburbs) contained 2.72 million, fully 53 percent of the total population in Catalonia. The Franco regime over a thirty-year period had imposed its identity upon the city in the forms of densification, speculation, characterless development, lack of public services and spaces, dehumanizing block tower construction at the periphery, and the silenced voice of residents. These outcomes constituted, in the words of Montalban (1992, 166), the "destructive reconstruction" of the city.

The accelerated urbanization and speculative chaos in Barcelona during the Franco years brought back into debate territorial imbalance criticisms that Barcelona was growing too "macrocephalic" and led in the 1950s and 1960s to two plans that sought some dispersion and decentralization of population.[8] The Plan Comarcal (County Plan) in 1953 was the first general urban plan for Barcelona since the Cerda plan of 1859 (Ajuntament of Barcelona 1953). It was the initial effort at planning for the metropolitan scale, applying not only to Barcelona city but also 25 adjacent cities.[9] It sought to institutionalize a supra-municipal perspective and identified growth extensions, suburbs, and garden cities in this larger area. Despite reasonable goals pertaining to the spatial containment of growth and the setting aside of protected areas, the Plan became more a visionary document than an effective planning instrument. Partial plans that were more specific and aimed at implementing this general plan were the important decision points about how the city was to be physically ordered and these became "the root of all speculation" (Salvado and Miro 1996: 143). Such plans were consistently revised to allow the building of massive polygon developments. Then, in the 1968 Master Plan for the Barcelona Metropolitan Area ("Provincial Plan"), there was the conceptualization of Barcelona as a "territorial city" midst a polycentric network of urban regional development (Provincial Commission for Urban Development 1968). In this plan, the study area expanded to encompass 164 towns. In contrast to the 1953 County Plan, which was a general plan of urban development, the 1968 Provincial Plan was a spatial plan without force of law. Nonetheless, it did influence subsequent plans in terms of protection of natural spaces, prevention of urban sprawl, and transportation planning.

The plans and ideas about city growth in the 1950–1970 period attempted to bring some modern approaches – open neighborhood units, natural land-scape units, supra-municipal planning – into a situation where city planning was in chaos, its collective aims eroded by private sector dominance and speculation. Most of these efforts were stymied and had to wait for up to 25 years (and a change of regime) to materialize (Katz 1996). Yet, the wait bore fruition, initially in the form of a new plan in 1976 that changed the prevail-ing logic of Barcelona's development, and then in the form of innovative city-building interventions that helped express democracy on the streets in the 1980s and 1990s.

URBANISM AND POLITICAL OPPOSITION

The capacity of critical urbanist thought in the 1960s and early 1970s to connect to neighborhood, democratic, and labor movements created an important springboard for the significant set of progressive actions that occurred during the transition from dictatorship to democracy from 1975 to 1980.[10] From the early 1960s to the death of Franco in 1975, urbanist criticism of the chaos and specu-lation in Barcelona constituted an indirect attack on Francoism and an indirect argument for a more democratic city. The technical shield of the city-building professions – planning, architecture, engineering – constituted an advantage because it provided protection for those on the political left to develop their urban arguments for a new way forward that would be more socially equitable, humane, and democratic. This technical shield was illustrated when urbanist activist Jordi Borja wrote *La Gran Barcelona* in 1971, a book critical of urbanism in Barcelona amidst the conflictive tensions of that period. The regime's response was that "your arguments are very good but you have a confusion between political and social conflict, on the one hand, and urbanism on the other. Urbanism is not a conflictive question" (J. Borja, planning consultant and former city deputy mayor, interview).

Critiques by urban planners and architects of the Franco city were important in and of themselves; yet, the power to transform was heightened when such technical knowledge of the city was connected to (1) local neighborhood-based actions that rallied for better conditions; (2) democratic movements within universities and unrecognized political parties; and (3) clandestine worker groups (J. Borja, interview; C. Navales, 1970s labor activist, interview). To understand the transition and democratic periods, one must understand the key role of urban social movements and professionals from 1960 to 1975 in prefiguring the democ-racy to come. Within the civic movements, political parties, and local adminis-trations that blossomed with democracy in the late 1970s, there were many individuals who had begun their public roles as architects and planners resisting Franco urban policies (M. Solá-Morales, interview). Furthermore, as described by Rowe (1997: 37, 61), the basic planning approaches and design concepts used in post-Franco Barcelona "were well developed long before they emerged publicly with the end of Franco's Spain."

Urbanist critique

In the early Franco years through the mid-1950s, the dictator imposed a will on architecture and city-building professions to build an "architecture of the new state" (M. Solá-Morales, architect, interview). Architects and city builders played important roles in helping to plan "rural cities" that the regime built for individuals who had collaborated with the military during the Civil War (J. Montaner, professor of architecture, interview). Architecture, particularly modernism, played a key utilitarian role in Franco's view of the state, which asserted that public authority should be used to transform the landscape through mass interventions. Organizationally, the Collegio de Architecturas (College of Architects) was created during this time. Financed by development proceeds, this professional organization had more power than similar organizations in Francoist Spain. Such a protective structure for architects in Spain does not exist elsewhere in Europe. In addition, Franquist legislation required professional architect involvement in all development projects (Montaner, interview). Even after this Fascist pretension of building a new state through architecture was abandoned, Franco still saw architecture as part of the economic growth and state-building process, in particular its utility for designing housing estates for the industrial labor force, and he thus supported the profession (M. Solá-Morales, interview).

Over time, however, urban professionals became more and more aware of how ill prepared the regime was to handle urban issues such as housing for immigrants and the provision of basic neighborhood services. Speculation-based urban chaos became more the accepted method of city building for the regime as it increasingly relied on the market to produce urban outcomes. Architects, town planners, and sociologists who studied such urban chaos in the late 1950s and early 1960s were among the first critics of the Franquist regime (Montalban 1992). Because city building was not considered as "political" as other professions, the regime tolerated dissent in urbanist commentary much more than in other fields. Urban professionals and architectural and engineering academics took to task speculative and predatory actions of developers. At the same time, town planners influenced by Parisian sociology – many of whom would participate in local government after the transition to democracy – pointed out the absence of resident voices in the making of neighborhoods and communities. It also helped the progressive cause that the regime thought of local governments as secondary institutions in Spain, thus making it easier to be critical of actions at this level (J. Borja, interview).

In the 1950s and 1960s, an opposition culture developed within the Catalan architecture and urbanist communities, one that contained both "idealists" who believed in the transforming ability of architecture and "realists" who sought reform amid the recognized obstructions imposed by the authoritarian regime

(Pizza and Rovira 2002).[11] Group R emerged in the 1950s; this was a group of architectural professionals who discussed alternatives under the Franco regime for improving the quality of life and the built environment. In the 1960s, several "Pequeno Congressos" (Little Congresses) were held; they allowed networking between progressive urban professionals from Catalonia, Madrid, Paris Vasco, Andalucia, Portugal, and elsewhere in Europe. These were important forums for the expression and diffusion of city-building ideas, especially among those urbanists within the architecture field. In the 1960s and 1970s, the regime became open to modern ideas as a way to catch up with the rest of the world after decades of stagnation. Architecture was viewed as part of this change in larger society and Francoism opened to some progressivism in built form in its later years, although fuller experimentation would need to wait until democracy was in place.

Systematic considerations of Barcelona's urban landscape highlighted problems and indicated the need for alternative paths. A path-breaking piece of 1960s criticism was by Oriol Bohigas, who criticized the evolution of Barcelona's development and the painful residential condition of immigrants in his book, *Barcelona: Between Cerda and Informal Shelter* (1963). The professional journal, *Cuadernos de Arquitectura y Urbanismo (Architecture and Planning Notebooks)*, dedicated an issue to *La arquitectura en peligro* (*"Architecture in Danger"*) in which it opposed predatory and speculative urban development practices that endangered culturally significant buildings. And in 1973, neighborhood associations and urban professionals came together to produce a "counter plan" to the official 1965 plan to redevelop the La Ribera waterfront zone.

The ability of architecture and urbanism to engage in dissent during the Franco years is traceable to unique characteristics of Spain and Catalonia. In Spain, architects have a noticeably more elevated status than most other European and North American countries. M. Solá-Morales (interview) asserts that this is because merchant cities such as Barcelona, where people do their business "with the city," tend to pay greater attention to public space and appearance than cities where economic power is more concentrated and insular. Within this context, architects developed a "root cultural relationship" with the city and its residents. Thus, it was incumbent upon architects and urbanists to challenge the various policies of the regime that were damaging the heritage and wellbeing of Barcelona. The College of Architecture, although a creation of the regime, had developed a progressive attitude (M. Solá-Morales, interview). Through conferences that enabled progressive debate, published material in its professional magazine, declarations against police actions, and invitations to critical observers such as painters Joan Miro and Pablo Picasso, the Collegio became a known source of opinion advocating change. In the last decade of the Franco regime, urban professionals could say more that was critical as the regime became less resistant to progressive deliberations and organizational activities. Progressive critique of built form and city life was allowed, strikes and more open manifestations of political views were not. In such a strained atmosphere, says J. Borja (interview), it was not difficult for progressives to decide what was important to do, but rather how to organize practical actions in the face of such rigidity.

Neighborhood organizations

> *In addition to fighting in workplaces, we recognized that we could fight as consumers, as neighbors, as residents because the new neighborhoods created in the 1960s and 1970s did not have markets, schools, and streets. This helped people to fight and to ask for what they needed. That was the beginning of this movement.*
>
> Marina Subirats
> City Councilor, Barcelona
> Former neighborhood activist (interview)

Criticism from those educated in the urban fields did not occur in a vacuum but alongside, and at times in active collaboration with, other forms of public engagement seeking greater equality and openness in Spanish society. One important source of public involvement was residents' growing voice and activity within neighborhood-based organizations. Individual organizations in neighborhoods first sprung up in the 1960s, but were commonly isolated efforts focused only on single areas. It was in the 1969 to 1975 period that many neighborhood associations (*associacions de veins*) were created, both in Barcelona city and in many of the new neighborhoods in the suburban periphery, and from the 1970s onward there was greater coordination of efforts across neighborhoods and increasing connections to the larger struggle for democratic local government.[12] By 1976, an estimated 120 neighborhood associations existed in Barcelona city proper alone (J. Borja, interview).

These neighborhood efforts did not arise to protest authoritarianism but rather to counter the urban symptoms created by a regime that cared little about creating humane and balanced communities. The Franco regime's sole interest in developing large-scale peripheral neighborhoods as labor storage sheds meant that services, public spaces, and the general quality of life were frequently inadequate. This led to grassroots efforts and urban protests aimed at improving daily conditions of neighborhood life. Neighborhood demands concerned adequate housing, health, and schooling, provision of parks, speculative construction of tower blocks, lack of cultural amenities, contamination and pollution, and the location of dangerous industries. In the city, speculative pressure in the city center (Ciutat Vella), the creation of a high school in Poble Nou, or a ring road in Nou Barris, or the preservation of an historic building in Sant Antoni would bring people into the fold who otherwise would stay away from clandestine political party or union activity. Between 1969 and 1975, the Technical Architect's Society documented eighty-three urban protests in the city, almost all led by residents' associations (Montalban 1992). Over the last ten years of the regime, its unbalanced and uncaring urbanism spawned a grassroots opposition culture in many residential districts.

When the efforts of these neighborhood organizations became aligned in the 1970s with progressive urbanists, democratic organizers, and labor activists, a fundamental source of energy for the democratic transition and beyond had been established. In the early 1960s, neighborhood organizations were primarily involved in daily issues and needs. By 1968 through 1974, these groups had evolved toward a wider criticism and articulation of the need for broader changes in society. The associations soon focused on the need for basic political freedoms. As expressed by Marina Subirats (interview), then a neighborhood organizer and now a city councilor, "after two or three years, people realized it was not just a question of the street, but a question of political rights and democracy. It was not possible to address the daily issues while the Franco system was in place because all things were organized in a restrictive way." Thus, specific demands for urban services and facilities were connected to broader calls for democratization, amnesty for political prisoners, the granting of autonomy for the Catalonia region, and recognition of Catalan language and culture.[13]

Instrumental in this evolution of neighborhood associations were architects, urban planners, lawyers, economists, and urban journalists who provided technical assistance and professional judgment to these groups. For example, Jordi Borja created a Center for Urban Studies in the early 1970s that educated and prepared young architecture professionals to assist the neighborhoods. Experts such as Borja and many others provided specific information not only about neighborhood problems, but also enabled these groups to move forward with alternative plans and visions that were contrary to the regime. As Camos and Parramon (2002, 214) describe, the professionals "made it possible for associations to advance from defensive positions to take the initiative in planning." Along with highlighting neighborhood problems and possible solutions, professionals "gave at these moments, if possible, a political orientation that spoke of democracy and freedom" (J. Borja, interview). Within academia, there was increasing attention within departments of architecture, geography, economics, and sociology to issues of the urban environment and political resistance. Most of the shared intellectual ideas about urban innovation arose from the Escola Technica Superior d'Arquitectura de Barcelona (ETSAB) at Universitat Polytechnic de Catalonia (UPC) in the 1970s, in particular from the writings and works by Manuel Solá-Morales and department head Oriol Bohigas.

Manuel Solá-Morales was one of the leading urbanist voices during the latter years of Franco. Educated in Rome and at Harvard University, he came back to Barcelona and in 1968 established a research unit within the UPC, the Urban Laboratory, which sought "to prepare us for the next situation when we could act" (Solá-Morales, interview). Seeing the futility of working with Franco realities, he stopped his professional practice at this time, stating, "Maybe you could do a nice building with an ugly regime but it is very difficult to do good urban planning with bad politics." Instead, he focused on working with university students to analyze and explain the urban problems of marginal and self-built housing, mass housing estates, and under-provided infrastructure. Significantly, he and his group also proposed alternatives to the Franco city. Such a consideration was useful for two reasons. First, several of these options became models for post-Franco development. Second, there was a pedagogic function to thinking about alternatives to Franco city form; to inculcate in students "a sympathy or empathy with urban problems and a consciousness of the city as being of the people," to teach them that there is "a political implication of any act of architecture," and that architecture "should be an act of service to the public" (interview).

The participation by activist professionals, together with emerging neighborhood support for labor issues and growing connections with clandestine democratic groups, meant that the neighborhood-based movement would become increasingly politicized through the years. As early as the 1960s, the banned political party PSUC (United Socialist Party of Catalonia) developed a strategy of "conquering spaces of liberty," which included advocacy for the establishment and legalization of neighborhood associations. It was not until the 1970s, however, that urban interests intermingled with broader political opposition. This occurred when urban movements connected themselves in 1971 to the Assembly of Catalonia, an underground organization of political parties, trade unions, church, and university representatives. The major players in this Assembly were the Comisiones Obreras (CC OO) [workers' commissions], covert worker groups who were effective recruiting agents for anti-Franco interests. The platform of the Assembly included the explicit political objective of gaining "the effective access of the people to economic and political power" (Molinero and Ysas 2002: 198).

These grassroots efforts by multiple sectors constituted a significant base of opposition to the Franco regime. Indeed, as provocatively stated by J. Borja (interview), "at the time of Franco's death in 1975 when the central government of Spain said that 'the street is of the state', the clear answer of the urban-based movement was 'no, the street is ours, it is of the people.' "

Labor activism, workplace, and community

The workplace was a key focal point of anti-Franco activism. It connected to and reinforced grassroots movements generated by neighborhood actions, urbanist critique, and political party organizing. A special quality of the factory was that it was the workplace to indigenous and immigrant employees alike and thus constituted an integrative force not found in many other parts of the urban area. In addition, factories were embedded in neighborhoods and communities and workplace grievances would seep out into the streets, there buttressed by a web of community support. The recognition of how citizen voice was silenced in both realms – workplace and community – catalyzed a united front across these different sectors of everyday experience.

The 1960s was a period of dramatic economic development in Spain and Catalonia, especially in contrast to the doldrums of the 1940s and 1950s. Particularly in the suburban areas outside Barcelona city, such as the Baix Llobregat area west of the city and the industrial towns of Sabadell and Terrassa, industrialization was intense and attracted thousands of migrants. Fifty-three percent of the active employment in Barcelona metropolitan area worked in industry in 1964, compared to a little above 30 percent for Spain overall (Pizza and Rovira 2002). Despite this burst of industrial and economic development, growth was uneven and certain sectors faced debilitating hardships. Employment in textile industries barely grew between 1962 and 1971, and such stagnancy would stimulate the emergence of the Catalan labor movement (Molinero and Ysas 2002). Employment in steel and metal industrial production increased, but technological innovation meant that fewer new employees were being added per unit of productivity. A general de-skilling of labor needs was evident due to production transformation and economic restructuring. In contrast to 1964, when 53 percent of employment was industrial in the metropolitan area and 30 percent was in tertiary or service employment, 41 percent was industrial by 1970 and fully 51 percent were less-skilled tertiary jobs. Amidst this restructuring, labor unrest grew in the latter years of authoritarianism.

The tight-fisted control of labor in Franco's Spain by the Organizacion Sindical Espanola (OSE, Spanish Syndicate Organization) deprived workers of all independent association, banned class-based unions, and through the "vertical syndicates" (basically, the state regime's union), workers found themselves under direct control of the Franco (Falangist) bureaucracy (Benjoechea 2002). As early as 1947, workers committees (called *jurados*) had begun to emerge from the Francoist syndicate organization, but it was not until the early 1960s that these legal workers' committees evolved into clandestine "workers' commissions" (*Comisiones Obreras*) that engaged increasingly in decentralized collective bargaining (Edles 1998). Preston (2004) marks 1962 as a turning point in the power of Falangist vertical syndicates when there was emergence of a new clandestine working class movement. Still, operating amidst the repressive machinery of the Franco state, efforts to organize for societal change took time. In 1966, only 16 strikes occurred in Barcelona province (Molinero and Ysas 2002).

By the end of the 1960s, labor demands related to wages and working conditions led to a spectacular jump in strikes in Barcelona province, from 65 in 1969 involving 16,000 strikers and 274,000 hours lost to 251 strikes one year later involving 72,000 strikers and 2.3 million hours lost (Molinero and Ysas 2002). From 1970 to 1974, over 200 strikes took place per year, in 1974 involving almost 200,000 strikers. A deadly cycle was established by 1973, with economic austerity measures by the state stimulating industrial unrest in Catalonia, which in turn was countered by greater state repression (Preston 2004). Not only the frequency but also the nature of the strikes was changing, moving from protests over solely economic demands to broader demands for social and political change.

"Industry more than any other institution contributed to the democratization process of the citizens," states Carles Navales, labor activist in the mid-1970s and subsequently city councilor for the city of Cornella de Llobregat (interview).[14] He points to two key roles of industry. First, it acted as a conduit for assimilating and integrating both native Catalans and Spanish immigrants at the workplace and thus crystallized a cross-ethnic working class solidarity which may not have developed in Barcelona's more segregated neighborhoods and cities. Second, the workplace was a location by the 1970s where opposition could more easily be organized, in comparison to neighborhood groups or political parties that were still barred. As such, workplace activism acted as a leading edge of protest that would spread outward into neighborhoods and community structures.

A 1974 strike in a crystal manufacturing plant in Cornella, which began when Mr Navales was laid off from his job, illustrates the opposition-based network of workplace, neighborhoods, church, and press that existed in the last years of Franco's life and regime. What was born in Cornella with this strike spread eventually across the urban region. Utilizing cooperation across these different sectors, this model was based on a common agreement or consciousness within the city about how social and economic conditions and needs go together. Such a unified front linking workplace and community was a substantial threat to the regime, leading a Spanish Francoist Interior Minister to once remark, "there are two problems in Spain – the Basque terrorist group ETA and Cornella" (C. Navales, interview). Clearly, change by 1974 was afoot and alive.

When Navales was laid off, approximately 900 people left their work and walked around town dressed in work clothes. Through these means, city residents could see them, ask them questions, and become informed. Through the 45 days of the strike, money was collected for the strikers through donations put in bowls near cashier's checkout stations in supermarkets and drug stores, receptacles that were conveniently hidden when police entered the premises. There was a developing solidarity in workplace and community between ethnic Catalans and immigrant Spanish. Indeed, most of the workers who went out on strike in support of Catalonian-born Navales were immigrants from southern Spain; "it was the fact that I was a fellow worker that created the solidarity; other things didn't matter" (C. Navales, interview). By this time in the 1970s, second-generation Spanish immigrants were on the scene and they, unlike their parents whose "life was their work," participated in both the firm and the city, and thus grew up more together with the local Catalans. These second-generation immigrants had places for meeting outside the firm, in community centers (*barrio casuals*) where cultural activities took place that increased cohesion and mixing of people. Another integrating factor was the church, which was increasingly parting ways with the regime. Not only left-wing priests, but also more and more the mainstream church was interested in democracy. Because the regime had a hard time confronting the church, neighborhood churches became particularly strategic locations for posting activist announcements.

In addition to workplace, neighborhood, and church, the press was a site through which activism asserted its demands, although in intricate ways to bypass the regime's censorship machine. Some reporters with empathy for the labor and democratic cause worked in local offices of newspapers such as Barcelona-based *La Vanguardia*. All material to be published needed to go through review bodies before being printed; if not, penalties and sanctions would result. However, these reporters learned that if they presented their material about social movements and strikes to censorship reviewers near the end of the business day when there were fewer people working, they would encounter less review. Such material would usually be written in language that was ambiguous to censorship officials, yet had potent meaning for pro-democracy readers. These news items would also at times be placed in red, and on the front page, in ways that made the readership understand the underlying story. If a story explicitly about local flooding and cholera risk in the neighborhoods was published, the newspaper would likely be sanctioned. However, if the reporter was able to query a regime official who would deny the problem, then the newspaper would publish a story that "the official said there is no risk of cholera." This story would both evade censorship and be effective in getting the word out that risk was likely real, given the regime's low level of credibility by the 1970s.

Anti-regime opposition in the city during the 1960–1975 period – whether in the form of urbanist criticism of the Franco city, neighborhood protests, labor strikes, or clandestine political party strategies – prefigured and prepared these sectors and interests for the transition and democratic periods to come. As Franco's health became more fragile due to his Parkinson's disease and other ailments, casting doubt on the continuation of his *Movimiento* ideology and state apparatus, activists began to vigorously debate whether the future should be in the form of an evolutionary transition from Francoism or a complete demolition of Franco structures (a *ruptura democratia*, or democratic break). On November 20, 1975, after 36 years in firm control of Spain, Generalismo Franco died at the age of 82.

CONSTRUCTING DEMOCRACY'S URBAN TERRAIN: URBANISM IN THE POLITICAL TRANSITION

From 1975 to 1980, Spain and Catalonia undertook the long and ultimately successful transition from dictatorship to democracy. In the first three years alone, a dictatorial regime was transformed into a pluralistic, parliamentary democracy in a step-by-step rebuilding process that occurred without major violence. In December 1976, a major political reform package was passed that legalized political parties and trade unions, in June 1977 there were the first democratic national elections since the 1930s, in October 1977 a significant multi-sector economic and social agreement was signed, and in December 1978 a new national constitution was approved in a countrywide referendum. Locally, municipal democracy was restored with the Barcelona city council election of April 1979 and the election of socialist Narcis Serra as mayor. For Catalonia, the Generalitat was restored in 1977 after 40 years of exile. A regional autonomy statute for Catalonia giving back many powers taken away by Franco was passed December 1979, and democratic elections to the Generalitat occurred March 1980, with Jordi Pujol of the Catalan *Convergencia i Unio* winning the presidency.

An inspection of urbanism and planning during this extraordinary transition period reveals two main conclusions about the role and potential of urbanism.

1. Urban planning played a key focusing and shaping role as part of the formative processes toward democracy, changing the "prevailing logic" of how Barcelona should grow and who should benefit. The timing of planning interventions early in the transition, not after, increased planning's ability to articulate a new democratic development vision for a more equitable and livable post-Franco city. Urbanism did not wait for the formal beginning of democracy and new institutions, but was instrumental during transitional uncertainty in anticipating and implementing the basic parameters of a democratic urban terrain.
2. Planning, in its ability to articulate an alternative urban future for Barcelona, was significant in bringing together diverse strands of democratic interests around a collective project. This helped consolidate political opposition and increase leftists' ability to express new societal goals in concrete terms. The period of political transition and uncertainty created prime conditions for planning support and effectiveness. The fundamental disruption of societal relationships led many interests seeking post-Franco political power to align themselves tactically with urbanism and its "rational" face.

Spain existed from November 1975 to June 1977 in an uncertain phase that was neither dictatorship nor democracy. Three significant and simultaneous crises were occurring – the uncertain transition to a new political system, the uneven economic transformation and de-industrialization causing shocks to the social system, and the unsustainability of the current model of urban spatial development premised on the authority of the private market and speculation. It was during this amorphous early part of the transition, in 1976, that the General Metropolitan Plan [GMP] (*El Plan General Metropolitano*) for Barcelona was passed. This plan (Metropolitan Corporation of Barcelona 1976) would fundamentally change the prevailing logic of city and society. The fact that such a revolutionary document was approved so early in the long transition process shows the ability of a collective planning project to contribute as a formative agent and catalyst in a society's reorganization. Many months before national democracy became a formal part of Spain, and years before the first local and regional democratic elections, new rules for creating a city of democracy were formulated and approved.

The GMP sought to build balanced communities, alleviate the drastic urban shortages, and create new patterns of urban life. It showed that there was another way to structure cities, and asserted the "authority of the public interest over the private interest" (Joan Solans, GMP co-author, director of planning for Generalitat 1980–2001, interview). In contrast to private speculation and its corrosive effect on the community-wide interest, the plan "showed the ability to have a collective project" that could only be done with the legitimate representation of all interests in the city, not just a chosen few (Juli Esteban, director, territorial planning, Generalitat, interview). In its prescription about how to actively correct, fix, and heal the urban landscape through density and other development regulations, the GMP became an essential shaping and focusing tool for the astonishing array of urban projects that would revitalize the city in the 1980s and 1990s. GMP's focus on numerical limits and clear norms established effective benchmarks which were complemented by subsequent *Planes Sectoral* done for specific projects and *Planes Especial Reforma Interior* (PERIs) used for urban rehabilitation of existing urban landscapes. It systematized the concepts of a post-Franco, democratic city and set the terms of reference for development in Barcelona, up to and including the current time, along with influencing the plans of many other municipalities to follow. Underlying the glory years of urban project-based architectural successes in later decades is the "hidden importance" of mid-1970s large-scale planning, which articulated strategies during the pre-democratic years to correct the urban imbalances and dysfunctions of the Franco city.

Upon the densified, speculation-driven, dehumanizing, and under-serviced landscape of the Francoist city, the GMP was "a type of urban surgery that never had been done before" (Albert Serratosa, plan co-author, interview). Within Barcelona city and the 26 municipalities of the first suburban crown, it radically slashed allowable urban density levels, cutting allowable future growth by almost 50 percent. The GMP lowered the nine million population "build-out" potential that existed under the 1953 County Plan and its amendments to 4.6 million residents (the 1970 population of this area was 2.7 million). By cutting future growth, it enacted huge economic losses to real estate interests and landowners, many of whom had handsomely profited from Franco-era urban speculation. It limited the abusive heights of city residential buildings and the new floors and top-floor and rooftop apartments allowed under Franco (interviews: J. Solans, J. Esteban). In the Eixample district, largest in the city, three floors were knocked off the height limit. Meanwhile, in areas of transformation (such as from industrial to residential), the allowable amount of building mass relative to physical site was reduced from 2/1 and 2.5/1 to 1.2/1. At the same time, the plan increased the total acreage allocated for urban green zones by 500 percent, and acreage allocated for amenities by 250 percent (Serratosa 2003). The GMP also put forth a model of polycentered growth in the metropolitan area, one that would take development pressure off the center city and increase the attractiveness of urban and suburban nodes well connected by transport to each other and the center. This showed a metropolitan consciousness, an effort to plan at the multijurisdictional level.

Amazingly, work on this plan so disruptive of the Franco status quo started during the regime period. A Commission on Urban Planning and Joint Services, representing the numerous municipalities, first formulated a metropolitan plan in 1970 that would counter the growing speculation allowed by the 1953 Plan. This plan did not move forward, but a 1974 version with similar attributes was alive for discussion when Franco died. After modifications, this version was provisionally approved in April 1976, five months after Franco's death.

Final approval, with amendments, was granted three months later. Compared to many other local governments in Catalonia and Spain that waited until the institutionalization of the formal mechanisms of democracy, the Barcelona metropolitan area plan was exceptional because it was approved in the tenuous, pre-democratic years of the transition. The timing of the GMP is remarkable, producing one of the earliest concrete indicators of what the future could be like.

The formulation of the GMP was not the sole planning endeavor during the last period of Francoism leading into the early years of the transition. In the early 1970s, there began the "banking" of land by the city of Barcelona for future community and collective uses. An early assertion of a collective interest in city building occurred in the last five years of the regime when the city of Barcelona began to buy and expropriate land that would later be used for public parks and squares or would be sold to trade unions, cooperatives, or private developers to carry out specific social projects.[15] Nor was the GMP the only effort during the transition that aimed to fundamentally change the nature of urban development. In 1977 and 1978, agreements at the national level redirected a significant share of future development profits from developers and landowners to municipalities and their citizens. The *Pactos de la Moncloa* (Moncloa Pacts) in 1977 and the successful ratification in 1978 of the new Spanish Constitution both rebalanced the scale between private interests and the public interest, directing more of the *plusvalor* – the surplus value or unearned increment of profit from development – to municipal governments. This was an about-face from the status quo; "the answer of democracy to the massive speculation of the Franco years was to put into place a new system of urban economic fairness," states Joan Trullén (professor, political science, Universitat Autonoma de Barcelona, interview). With democracy ascendant, it was no longer possible for speculator appropriation of these increased profits to continue.

Collective urban planning and policymaking endeavors made great strides in shaping and focusing pre-democratic strategies for creating a more livable post-Franco city in which there would be greater equity in the distribution of urban benefits. The fortunate timing of these interventions relative to larger political dynamics explains part of their success. However, the inherent qualities of collective planning also gave it advantages during the uncertainty of the transition, namely its ability to create a platform for societal consensus.

Amidst the uncertainty, "every force that was seeking democracy tried to show that it was possible to build a community in a different way than Franco" (Joan Solans, interview). Indeed, urban progressives had spent the later Franco years reflecting on the city and the changes that would be needed. In this sense, urbanist thought was more developed than other disciplines in terms of what the post-Franco era would be like, and became one of critical consciousness's most important contributions to the transition and the restoration of democracy (Montalban 1992). After decades of disempowerment and with the foreseeable end of Franco's rule in sight, there was a natural skepticism among progressives of any project developed during the regime years and a wish to make a clean break with that era when Franco died. Thus, when the General Metropolitan Plan was being formulated and reviewed in 1974, public opinion and those political leaders who had been most vehemently opposed to, and ostracized by, Franco viewed it as a creature of the regime and opposed it outright. An overwhelming number of the 36,000 comments received on the draft plan in the 1974 public meetings were in opposition (J. Solans, interview).

However, with the Plan on hold during the last months of the dictatorship, interesting re-alignments of attitudes toward the GMP began to occur. Albert Serratosa (GMP co-author) recalls that six days before Franco died in November 1975, he was fired from the GMP planning office because he and the plan were viewed as being antagonistic to landowners and other powers linked with the regime. When the regime then came out against the GMP, what formerly were opponents came out in support of the plan (A. Serratosa, interview). By 1976, the Communist Party witnessed the negative reaction of the landowners and developers and understood that, "This plan is not of the old" (J. Solans, interview). Many social movements and neighborhood groups found agreement with the many progressive and anti-speculation attributes of the GMP and came out favoring the plan now viewed as the appropriate path away from Franco urbanism. The GMP became a magnet for consensus across a broad range of democratic interests. Over 14,000 comments in support of the plan were received during the 1976 public meetings.

A striking feature of the GMP's political dynamics during the early months of transitional uncertainty after the end of the dictatorship is the degree to which it crystallized consensus across the range of the Franco political opposition. The ability of a plan whose work began during the regime to be a catalyst to bring together the political opposition around a common project was surprising even to the plan's director (A. Serratosa, interview). In an atmosphere in which numerous sectors of society had different prescriptions about how to reform society and at what rate, the GMP appears to have provided a badly needed template for consensus. This consolidation of support was so great that multiple anti-Franco sectors of society supported a document that was indeed progressive and path breaking, but whose broad parameters had nonetheless been formulated during the Franco regime.

Planning, in normal times viewed as a technical profession that is outside the political realm and lacking in independent power, assumed in the form of the GMP significant political importance as a symbol of democratic possibilities to come. Rather than being subservient to political and organizational constraints, such urbanism constituted a form of hope and faith that better opportunities were possible in the future. Similar to the way urban criticism was allowed by the regime in the 1960s and early 1970s, the formulation of the GMP was allowed to proceed during the regime's final years. In both cases, the political implications of planning were underestimated by the regime because urbanism was viewed as not overtly political. What normally is often a handicap to planning – its lack of independent power – actually became an asset in the face of potentially repressive state power, allowing it a zone of protection not accorded other activities seen by the state as more threatening.

The early, proactive timing of urban planning discussions and interventions during transitional uncertainty portrays how planning may contribute to the transformation of basic societal goals and outcomes, as described by urban theorists such as Friedmann (1987). The case of Barcelona in the mid-1970s exemplifies a formative path wherein urban policymaking anchors and fore-shadows the broader societal changes to come. The transitional moment, and the momentum it contained, likely enabled the approval of a more radical and clear-cut plan than would have been approved during a period of more stable and institutionalized democracy. The crisis presented by transitional ambiguity appeared to have presented an opportunity for the collective planning function to shape and focus policymaking along distinct policy paths. Indeed, as described by Solans (1996, 205) the crisis in politics and transition "helped to carry through a process which, in normal circumstances, would have been impossible or very difficult." In this view, the plan was passed not despite the political uncertainty, but because of it (A. Serratosa, interview).

Planning was able to exert power during this transitional period because it filled a societal need. During times of fundamental regime change, societal relationships can become scrambled and those seeking political power look for avenues for expression. One such avenue is planning and its "rational," seemingly non-political, face. As Flyvbjerg (1998) has shown, such planning rationality can be an attractive foundation upon which to legitimize the exercise of political power. Because societal uncertainty created the need for the production of a planning template to guide society forward, the "power" of planning in Barcelona during the transition period became significantly greater than we would witness in more politically and economically stable periods when social and economic interests are more institutionalized and protective of the status quo.[16]

Albert Serratosa and Joan Solans
Principles and practices of "the two great urbanists"

Serratosa and Solans are the two great urban planners during and after the transition. They are not site-specific architects who will gain fame later during the 1980s and 1990s, but rather they are thinkers in terms of urban scale, systems, and relationships. Serratosa, a trained engineer, and Solans, an architect, both played instrumental roles and were partners in the development of the path-breaking 1976 General Metropolitan Plan. Serratosa mentored Solans as supervisor of the GMP. Both are controversial and are self-identified technician-professionals. Solans views himself as a neutral public servant able to "keep out the political noise;" Serratosa advocates sound planning concepts resistant to capture by either leftist or rightist political aspirations. And both see the assertion of the public interest to be a vital part of urbanism and of critical value in the early post-Franco years. Even in their views of appropriate spatial structure for the Barcelona area – a type of multi-nodal and connected metropolitan region – their views appear to have more similarities than differences.

Yet, I was also told repeatedly in interviews that these two men do not see eye to eye, and their disagreements are of a fundamental nature. Their differences appear to lie in their modes of operation, not their substantive beliefs. For Serratosa, the public interest is to be asserted more actively and unilaterally; for Solans (who spent over 20 years as Generalitat director of planning), the pursuit of the public interest needs to be tempered by private sector and local government realities. For Serratosa, principles and concepts are to be the guideposts; for Solans, practicalities and applications on the ground are to lead. Solans (interview) describes Serratosa as somebody who "works solely by heuristics and doesn't wish to understand that planners must work with material, land, and developers. He works with just pure spirit." Their difference in how to best achieve good planning extends to the structure of local government, too (Francesc Carbonell, Institut D'Estudis Territorials, interview). Serratosa views as essential the creation of super-municipal planning mechanisms as a way to produce a multi-nodal metropolis, viewing the politically fragmented local government landscape as a major obstacle. Solans, in contrast, worked during his years at the Generalitat within this web of local governments and would often negotiate agreements with specific municipalities involving economic development and infrastructure.

Are these two professionals necessary parts to a whole – one principle-based, one practicality-oriented? Their differences appear more rooted in methods of engagement than in substantive views of desired regional growth. Such differences notwithstanding, these two urban planners exemplify the central and significant role that planning played during a critical juncture of Spanish and Catalan change.

I have described the role of urban planning and policy in two time periods – during the political opposition years of the late 1960s and early 1970s, then during the political transition period from 1975 to the early 1980s. I have found a remarkable influence of planning in both time periods to express urban visions that countered regime powers. This type of influence is easily underestimated by the political elite and was thus able to operate in a type of protective space. The criticism of urban chaos and speculation during the opposition years was an indirect and covert criticism of Francoism and an indirect advocacy for a more democratic city and society. Then, during the political transition, the mold-breaking General Metropolitan Plan, in particular, and collective planning more generally, were early expressions and actualizers of the unfolding urban democratization and the city's physical transformation in the 1980s and 1990s.

IMPRINTING DEMOCRACY: PLANNING IN THE 1980s and 1990s

The strength we had from the transition gave us the capacity for innovation, for self-confidence, for public support, for long-term visioning.

Manuel de Solá-Morales
Professor, urbanism
Universitat Polytechnic de Catalonia

The success of planning and urbanism in affecting meaningful and positive change in Barcelona is evident in its ability first to support opposition to the Franco city as an indirect criticism of its regime, and in its capability then to change the prevailing logic of speculation and soul-less development in the early years of transition. A third test of urbanism would arise three years after the GMP's approval once local democracy was formalized in Barcelona. This concerned whether urbanism would be able in the newly democratic Spain to produce tangible on-the-ground change in the urban and human landscape in ways that would articulate and reinforce new democratic directions.

In April 1979, for the first time in over 40 years, the city population of Barcelona elected a mayor and city council. Members of the PSUC (Unified Socialist Party of Catalonia) became mayor (Narcis Serra) and comprised the majority on the council. This socialist control of Barcelona city government remains until this day. Those who were opposed to the political system were now brought into it. Architects and planners from the outside were now incorporated into the new administration along with people who worked in the old administration just before Franco's death. Jordi Borja, who during the Franco years was helping neighborhood organizations in their fight to gain resources and power, now was brought into the administration and worked to create a decentralized administrative structure for the city that would facilitate participation by ten new neighborhood district councils. Joan Solans, co-author of the General Metropolitan Plan in 1976 and who worked in the city administration during the transition to buy land for collective purposes, would remain in city government for a short time before moving onto a long-term position with the regional government. And, Oriol Bohigas, urban critic since the late 1960s, would be director of planning for the municipality from 1980 to 1984.

The momentum to engage in new urban policies after decades of political suppression was palpable. The city was ready and, beginning in 1979, Barcelona would become an urban laboratory wherein a socialist strategy would be applied for the first time since the 1930s (J. Trullén, interview). There was an acute consciousness of the urban crisis by the new administration, and "after struggling, waiting, and hoping for so long, we were in agreement as to what was needed to be done" (J. Borja, interview). Because so many people who fought with the public against the regime were now in city government, there was a remarkable urban social consensus between the new city leaders and the population at-large in terms of the need and broad outlines of how to move forward. One key participant, J. Borja (interview), describes these early years of democracy: "It was easy for us to invent public urbanism. All intelligentsia of the left were in City Hall and this facilitated an important consensus. I was in the government, at the extreme left. All questions had unanimity; it was an exceptional situation." This consensus was so great a factor that one interviewee described it as the "secret of the success we have had" (Manuel Solá-Morales, interview).

This unlocked democratic momentum and social consensus was applied to a daunting set of urban problems. The speculative Franco city had left the Old City and the Cerda Eixample district physically deteriorated and congested, and had created working class peripheral neighborhoods with vastly inadequate infrastructure and services (Borja 2001). Further, democratization came amidst an ongoing and deep economic hemorrhaging in the urban region. In the 1978–1983 period, there was a remarkable 30 percent decline in total industrial employment in Catalonia, with construction declining 49 percent (J. Trullén, interview). In the city during these five years, 100,000 jobs of all types were lost (300,000 total jobs in the city and first suburban crown, combined) [Borja 2001]. This economic crisis lowered speculative pressure on the city – which allowed the new administration some time in implementing new policies – yet created emergency social situations that city policies and programs needed to address. Simultaneously, such an economic recession restricted the amount of public money available through taxation to fund new public initiatives. Within this context of hardship and constraint, but with a readiness to act and broad citizen support to do so, what could Barcelona architects and planners do in the early years of democracy?

The democratic years are classifiable into three urban policy phases – early democracy (1979–1986), the Olympic planning years (1987–1992), and post-Olympic years (1992–2004.) The tactics, strategies, and interventions used in these phases differ in kind and degree, and responded to different circumstances and stimuli. What is provocative is how throughout this 25 year period there has been an active and innovative public sector applying the lessons from one period to subsequent ones.

In order to build democracy into the city landscape, to imprint in the minds of Barcelona residents the urban attributes and possibilities that could now exist, an intentional strategy of small-scale, project-based interventions was begun.[17] Public park improvements, central plazas, and streets could make a difference in the life of everyday Barcelonans over a relatively short time and could be done within funding constraints. These local, small-scale interventions were seen as catalysts for the overall upgrading of the city, with public investment positioned as leverage to encourage private sector interest. In his 1985 book, *Reconstruction of Barcelona*, Oriol Bohigas proposed small-scale urban projects as a strategy more useful than the abstraction of master planning. However much it was path-breaking in changing the development logic of the city, the GMP was nevertheless aimed at controlling and restricting what could be built rather than stimulating the new growth needed to counter severe economic decline and illustrate democracy's benefits. Needed now, instead, were city interventions that would be visible, more immediate, and thus influential upon the city's residents. What was needed, says Bohigas (director of planning for Barcelona, 1980–1984), was to "move from systematic but unspecific future visions to precise proposals and specific activity" (Bohigas 1996: 211). Such improvements to urban and green spaces in the early 1980s included work on urban parks, plazas and gardens, urban corridors (pedestrian and automobile improvements), large-scale parks, basic sewer and drainage services, and social, cultural, and athletic facilities (Busquets 2004). Projects were focused on enhancing the quality within specific quarters of the city, and were spatially targeted in such a way to increase the value of those parts of the urban fabric that had been lagging (Esteban 1999). Approximately 150 projects that created or rehabilitated public space were completed within the city during the 1980s (Monclús 2003). These were carried out throughout the city, with interventions having their greatest psychological effects in heretofore under-serviced working-class neighborhoods.

Thus, in the early years of democracy, the community-specific and small-scale interventions of architects and designers became more valuable in imprinting democracy upon the Barcelona landscape than the more abstract and broader scale plans of urbanists. There was an important public education function linked to these project interventions, wherein new democratic and cultural values could be translated to the population. As described by M. Solá-Morales (interview), "the recovery of public spaces in the neighborhoods, the creation of new parks, and the renovation of the central city were very pedagogical in their content." This public education and political translation function for architecture had been anticipated by Barcelona urbanists in the 1960s and 1970s, with Oriol Bohigas and Manuel Solá-Morales putting forth the idea that architecture could educate the people and translate democratic cultural values (J. Montaner, professor of urbanism, Universitat Polytechnic de Catalonia, interview). Cultural debates by urbanists in the 1970s (and covered in professional journals such as *Architecture Review* and *Architectura B*) had produced many ideas and concepts related to design and culture that were implemented and elaborated upon with the coming of democracy. Project interventions during the early democratic period, together with the new community rights to part of the surplus value (*plusvalor*) of urbanization, were able to prove that democracy was an effective way to organize collective activities and respond to urban shortages (interviews: J. Trullén, J. Solans). The master planning function that expresses city-wide order and layout and which had played such a key role earlier during the transition period took a back seat to architecture and design during these emergent years of formal democracy.

Residents could experience democratic ideals most particularly through the recuperation of neighborhood squares and buildings. Actions in design marked a sharp contrast between past and present: "there was a symbolic importance attached to this civic recuperation" (M. Solá-Morales, interview). Primary among the foci of project interventions was improvement of public and civic space. The quality of public space was of great importance to neighborhoods – public areas could facilitate mix and contact among a heretofore suppressed populous, they could facilitate and provide avenues for collective expression, and they were important to the identity of the city's working class (J. Borja, interview). In this view, public space "is not sufficient, but it is necessary for democracy in the city" (J. Borja, interview). The urban mixing that Franco repressed and contained through force and intimidation was now not only to be allowed under democracy, but was to be actively fostered by municipal government through changes in the built landscape.

The public planning and city-building function in Barcelona has shown a remarkable ability over the past 25 years to evolve in the techniques used and in the geographic scale of their application. In the mid-1980s, a significant new emphasis was added to the city's urbanist portfolio – a more strategic planning approach using larger scale interventions to modify the urban structure of the city and urban region (Monclús 2003, J. Monclús, interview). Roads, highways, and new and revitalized areas of economic development were constructed in ways to increase the economic balance between different quarters of the city (Busquets 2004). In contrast to the more spatially targeted benefits of the earlier period that utilized smaller scale targeted interventions based on architecture and design improvements, many of these projects were to have broader, even city-wide benefits (Esteban 1999). This period of strategic planning was catalyzed and made possible through one dramatic event – the approval by international organizers in 1986 of Barcelona as the site of the 1992 Olympic Games.

The Olympics event, and the many years of preparing for it, provided the city and urban area with the significant opportunity and resources to restructure major parts of the urban area.[18] The Olympics were the key intervening factor in the mid-1980s, providing substantial sums of money from the European Union and national and regional governments, along with contributions from private enterprise. As recalled by F. Muñoz (interview), "Suddenly we had all this money to develop all these projects that had been in the box for more than 30 or even 40 years." For the six-year period from the 1986 nomination to the 1992 event, "event-driven urbanism" was a major force in the city.[19] The shared goal of hosting an Olympics event brought together central, regional, and municipality governments that had heretofore not been in agreement about spending priorities or even governing ideologies.[20] The heightened level of resources now available for infrastructure in the city was used strategically by the Municipality to change the urban spatial structure for the Olympics event specifically and for the longer term. Total investment into the city during the 1987–1992 period for facilities, infrastructure, and development related to the Olympics was approximately 6.5 billion euros (about 65 percent of which was from the public sector; 35 percent from the private sector) [Ricard Frigola, Urban Development Corporation, City of Barcelona, interview, 7-2-04]. Fortuitously, Barcelona's successful Olympics bid also coincided with the start of an economic upswing in the city and country.

An urban plan or strategy was necessary to guide such significant investment and restructuring of the urban system. However, by the mid-1980s, confidence with large-scale public planning for the city had waned and numerous subarea plans (*Planes Especial Reforma Interior* or PERIs) to implement the 1976 GMP remained on paper (M. Herce, civil engineer, interview). Thus, a different type of large-scale planning – more proactive, interventionary, and catalytic – was needed than that supplied by the more blueprint-oriented GMP. In formulating such an approach, Barcelona policymakers and planners looked at their own experience in small-scale interventions and extended these tactics to the larger-scale strategies now needed (Bohigas 1996). At a smaller scale in the early 1980s, public investment was used to catalyze change through improving the physical environment and providing infrastructure and social facilities. These improvements occurred where people lived, in the meeting places and settings where collective identity is produced (Bohigas 1996). Such a catalytic role of urban improvement was now to be applied at the scale of entire subareas throughout the city, both to develop Olympic sports facilities and non-Olympic parts of the urban tissue. By enhancing accessibility, land values, and the overall attractiveness of city districts, Olympic strategic planning would stimulate the overall economic vitality of the city, counter Franco era urban disequilibria by revitalizing peripheral areas, and increasingly link the city to the larger metropolitan region. A public sector able to tactically intervene in the urban marketplace, an approach successful in proving the worth of democracy in the early 1980s, was now focused on catapulting the urban region to a more advanced level of urban connectivity and vitality.

There were three major programs during the 1987–1992 period – the creation of "new centralities" throughout the city, the building of a major road beltway system around the city, and the revitalization of the old city. The "Areas of New Centrality" (ANC) approach, first formulated in 1987, sought to implant through public investment and improvement new economic value in 12 distinct areas of the city (see Busquets 1987). Four of these – Vila Olympica, Montjuic, Hebron, and University – were to be sites of Olympic sports events and supporting residential development. The ANC program utilized city building projects to attract private capital to each area, simultaneously asserting a public interest as the overriding criterion (M. Herce, interview). ANCs, when considered as a network of revitalized nodes, would create a greater interdependence and connectivity between different parts of the city, both to each other and to the city center.[21] The program shows recognition among decision-makers that the Olympics event could be a stimulant to a longer-term restructuring of the city.[22] For example, Vila Olympica, one of the ANCs and the site of most of the housing for Olympic athletes and officials, entailed the wholesale re-creation of the heretofore dilapidated and obsolete waterfront area, creating a mixed residential-commercial-hotel district, along with a new boat harbor and beaches. A massive revitalization effort, Vila Olympica necessitated the near complete removal and relocation of the Spanish railway lines and one of Spain's oldest rail stations. There was considerable intergovernmental cooperation required for this project, and state, regional, and municipality shared the substantial infrastructure costs.[23]

A second major program implemented in the pre-Olympic years was reorganization of the road and highway network in the city, including construction of a new beltway (*cinturones de ronda*) system around the city and improvement of a secondary network of grand avenues and boulevards penetrating into the city's fabric. The 25-mile long new beltway system (comprising Ronda de Dalt, Ronda de Montana, and Ronda de Mar) opened in 1992 and was built to encircle the city and provide new points of access to avenues connecting to the center. This beltway system changed the basic calculus of access into and around the city, helped stimulate the peripheral ANCs, and because of greater mobility changed the perceptions of many residents and employers from a city to a more macro scale (M. Herce, interview).

The third major intervention begun in pre-Olympic years was the redevelopment and revitalization of the Old City (*Ciutat Vella*) area and conversion of the old port (*Port Vell*). In the historic old city – particularly in the subareas of Raval and Santa Caterina, and in adjacent Barceloneta – the congested and deteriorated urban fabric was opened up through the creation of new public spaces and pedestrian walkways and avenues (*ramblas*), façade improvement, housing and small business projects, improvement of basic public services, and creation of new cultural facilities. From the program's beginning in 1988 until 2000, Busquets (2004) states that nine billion euros (over 11 billion U.S. dollars) were invested in the Old City and Barceloneta, with 50 percent of funding for infrastructure and 25 percent for open space improvements. In the Port Vell, old and dilapidated facilities were converted into a pedestrian way for leisure and recreation, connecting it to the Old City immediately to its north and to the new Olympic Village to its east.

Each of the endeavors that enhanced the city for the 1992 Olympic Games involved major public investments. The significant effects of planning during this intense period of preparation earned the moniker of the "great transformation" (Borja 2001). This constitutes a period when large-scale urbanism and planning re-emerges in a form more proactive and catalytic than during the political transition years. With the strategic broader-scale interventions of planning during this time combined with the proven and continued effectiveness of smaller-scale, design-based interventions, Barcelona was at the top of its game in the late 1980s and into the 1990s.

The 1987–1992 period bears witness to use by the city of an event-driven urbanism as a way to reshape the urban region and to leverage private sector interest and investment. The Olympics was used as a catalyst not only to construct event-related sports facilities, infrastructure, and housing, but also to fundamentally restructure the urban landscape through the creation of new economic nodes, greater connectivity, and a revitalized historic city core. The strategic large-scale activities were necessary for the Olympic Games themselves, but also instrumental to the post-Olympic vitality and transformation of the city. Instead of creating imbalanced urban development and ephemeral spaces that would lie dormant after the Olympics, the city was effective in using Olympics-related public investment to create multifunctional spaces and amenities with long-term benefits, to catalyze private sector urban development, and to distribute new development across the urban region (F. Monclús, interview). The glamour of hosting the Olympics was more than enough to secure significant investments from national and provincial governments, which otherwise would not have been forthcoming due to competing countrywide needs and differing views of the city within Catalan nationalist politics.

The Olympics planning period also displayed the continued significant and impressive consensus and alignment between the desires of the general public and the programs and policies of the municipal government. The Games had an extraordinary capacity to mobilize dreams, economic resources, and social support. Public promoters and the general public viewed the Olympics achievement as an affirmation of the city and region, as a patriotic triumph for Barcelona and Catalonia.[24] Despite the risk of cultural conflict within the city population (between ethnic Catalans and Spanish immigrant origin), the Olympic goal was pursued with a high degree of social cohesion, organization, and participation. In many respects, to foreign observers, the city appeared as a "paradigm of civil society" (Borja 2001).

In the post-Olympic Games period (1992–2004), the pattern of major public investments used to restructure urban opportunities continued on an even broader scale aimed at opening up the metropolitan region at-large to European and global dynamics. Key interventions and investments have again been used to restructure growth and connect opportunities to broader and interjurisdictional scales (Ajuntament de Barcelona 2003; Infrastructures del Levant de Barcelona, SA 2004; Barcelona Regional, SA undated). These projects include the Llobregat Delta Plan for the area west of the city center, which aims to double the water port capacity, divert an urban river to accommodate increased logistics servicing, and double the airport capacity. In addition, to host the Universal Forum of Cultures 2004, there was the massive redevelopment of the dilapidated area near the River Besos in the eastern sector of the city, characterized by 500,000 square meters of total constructed space (including an easily identifiable 25,000-square-meter blue triangular Forum Building, a 70,000-square-meter International Convention Center, major new hotel and office buildings, and residential towers), and new shoreline, public urban spaces, and green spaces.[25] In addition, this period also is one where the city is transforming, through a program of regulatory adjustment and public investment called *22@*, an old industrial area (Poble Nou) into one that encourages new knowledge-related economic activities amidst mixed urban and residential development. Finally, in the Segrera area northeast of city center, there is the development of a large, new intermodal station to service the Spanish high-speed rail system (TAV) that will connect to high-speed lines for Europe and Madrid. In all, during a 10-year period of developing these four initiatives (each ongoing at the time of this writing), approximately 9 billion euros will be spent by public and private sectors, more than 30 percent more than during the six years of development in preparation for the 1992 Olympics (R. Frigola, interview).

When examining urbanism in the 1980s and 1990s and through the first five years of the new century – from the early years of democracy to the Olympic planning years and beyond – one is presented with a "dense succession of ideas . . . with contextual references and specific aims that have changed over time" (Esteban 1999). In the early 1980s, in the context of hard economic times and the need to implement and validate democracy, public sector interventions were small scale, targeted, design-based, and project-specific, and aimed at psychological upliftment. In the mid-1980s to early 1990s, amidst a rebounding economy and the attraction of hosting an international event, interventions were large-scale and utilized a strategic, proactive planning approach to restructure the city's spatial and economic structure. In the years after the Olympics, this strategic planning approach consolidated earlier period improvements and was moving the urban region to the next level of connectivity with European neighbors and the world. The existence is clearly evident in these 25 years of active and innovative public sector intervention at increasingly greater geographic scales. In a public learning process, urbanists learned how to intervene at small scales first and then used these lessons at intermediate and then larger scales. Project-specific architectural efforts and larger-scale urbanism shifted over time in which had the upper hand, depending upon the nature of the specific challenges facing the urban governing regime. Throughout this period, however, the two city-building approaches appear to be intertwined and mutually supportive in seeking shared goals, rather than pushing opposing views of the city.

Many observers describe a trailblazing quality to planning and design intervention in Barcelona. Ward (2002, 371) considers Barcelona "one of the most potent international models of urban planning of the late 20th Century." In 1999, a major urban analysis commissioned by the British Labour government (Rogers 1999) asserted that British design and strategic planning "are probably 20 years behind places like Amsterdam and Barcelona." Monclus (2003) positions Barcelona within a larger international planning movement, yet the city has "had distinct temporal rhythms and technical variations," being in the lead due to its unique political needs in bringing many of these ideas about public space and context-specific interventions to practice (Monclus 2003: 408). The explosion of urbanism after Franco suppression and the catalytic effects of Olympics urbanization have led to faster application of innovative planning ideas in Barcelona than elsewhere in Europe. There was a pent-up quality to creative and progressive urbanism that burst out in the late 1970s and early 1980s after the end of Francoism. The rigidity of the Franco years may have, unintentionally, produced a creative stimulus to urbanism once the regime ended. This would explain why the innovative tools and techniques used in the early years of democracy in Barcelona predated their use in other western European cities and regions. Because government policies in other western European countries had not suppressed normal urban development, the need for active and creative techniques of urban normalization and restructuring was not as immediate or compelling. In the same way, Barcelona's strategic planning in the 1980s was part of a broader international movement toward strategic urbanism and city entrepreneurialism, yet its use in the city preceded, and exceeded in scope, much European (and North American) practice. This time it was not due to the distorting effects of Franco rule, but because of the catalytic effect of the 1986 successful winning of the 1992 Olympic Games.[26]

In both approaches – design-specific interventions of the early 1980s and strategic planning in the late 1980s and early 1990s – Barcelona was ahead of its time. In the first case, a unique local historical moment of political transition and democratic recovery propelled onward the aggressive use of international planning ideas linked to context and local place. In the second case, the needs of event-driven urbanism spurred a widespread use of strategic planning in the city that was earlier and broader than most strategic interventions in Europe and North America. Amidst the ongoing flow of international planning movements, the Barcelona case displays an originality that sets it apart, and above, design and urbanism practiced in other cities during those times.

CITY AMIDST NATIONALISM

Barcelona illustrates the catalytic roles of urbanism and design in a society undergoing political transformation. It also provides a 25 year track record about whether, and through what means, urban governance and policy in a democracy can effectively accommodate cultural group-based differences, in this specific case those between Catalan nationalists and Spanish immigrants. While I have focused thus far on how urbanism works within transitional uncertainty, I come now to the second main query of this research – how planning and urbanism address issues of group identity differences based on deep historical and cultural factors. In examining the relationship between urbanism and nationalism, I explore Sassen's (2000) hypothesis that dynamic cities like Barcelona are sites where group identities can be formed and re-formed in ways that produce new types of transcendent identity and nationalism.

It appears at first glance to be a substantial challenge to maintain the vitality and utility of Catalan nationalism midst a mixed society of nativist and Spanish immigrant populations. In several important respects, Barcelona and Catalonia host two societies – one ethnically Catalan and for whom regional history and tradition matter, and one not ethnically Catalan for whom Catalonia culture does not have such resonance. Social, demographic, linguistic, and economic data support this dual society argument. Yet, due to massive Spanish immigration over the past 40 years, the society appears to be a more integrated and assimilated one than demographic data alone might suggest.

Social diversity and cohesion

> *There are many elements in Barcelona that predispose it to being polarized – different ethnic groups, histories, and political aspirations – yet this has not happened.*
>
> Oriol Nel-lo
> Former member, Catalonia Parliament
> Interview

Barcelona, and its Catalonian region, is simultaneously nationalistic and porous. Astoundingly today, approximately 58 percent of regional population is composed of first-, second-, or third-generation immigrants from elsewhere in Spain (P. Vilanova, professor of political science, Universitat de Barcelona, interview). Ethnic Catalans are in the numerical minority. Waves of Spanish immigrants searching for jobs, at the end of the nineteenth century, in the 1920s, and in the post World War II decades, has given Catalonia a history in the assimilation of newcomers that no other region in Spain has. Such has not happened without concern by some about immigration's effect on Catalan identity and the cultural cohesiveness of the urban population. A book in 1957 by Francesc Candel, *Donde la ciudad pierde su nombre* ("Where the City Loses its Name"), pointed to the immigrants' lack of shared cultural roots. And, in his 1964 *Els altres Catalans*

("The Other Catalans"), Candel worried about how the immigrants of the 1960s tended to live in independent ghettos and did not seek to establish links with Catalan culture (Montalban 1992).

Indeed, there are indications today that the region, and city, constitutes two societies. In the province of Barcelona, almost 45 percent of survey respondents felt "more Catalan than Spanish" or felt "purely Catalan," while 18.3 percent felt more Spanish than Catalan or felt purely Spanish in identity (Institut d'Estudis Regionals I Metropolitans, 2002, p. 413). There is also spatial segregation between ethnic Catalan and Spanish-immigrant families and their offspring in the urban region of Barcelona. If we use "native language spoken" as a measure of these two groups, we observe distinct differences in where these two groups live, both across the urban region and Barcelona city neighborhoods. Economically, there exists a middle and upper class in the Barcelona urban region that consists of people of Catalan ethnicity who are highly educated, speak Catalan as their primary language, and are well connected to societal networks (Institut d'Estudis Regionals I Metropolitans, 2002a). This is a cultivated, intellectual, professional, and well-bred class that has a strong internal network.

Despite the linguistic, economic, and spatial differences that contrast the ethnic Catalan and Spanish-immigrant populations, there is also a degree of hybridization between these groups. During the 1960s immigration of southern Spanish immigrants, there was a "sense that they were different, but they were cool" (F. Muñoz, professor of geography, Autonomous University of Barcelona, interview). Mixed marriages were not uncommon within the lower middle classes. Furthermore, the large industrial factories played a crucial role in social assimilation as local and immigrant workers mixed at the workplace (C. Navales, interview). In addition, having a common enemy in the repressive Franco regime further amplified the social cohesion among these groups.

Today, at the level of "on the street" functionality, the Barcelona area approximates a functionally bilingual society (see Figure 3.5). Due to Catalan linguistic training in the public schools and the growth of Catalan-speaking television, there has been significant growth over the last 20 years in the average person's knowledge level of the Catalan language. In 1985, about 25 percent of individuals in the metropolitan area replied that they could speak and write Catalan while about 40 percent stated they were unable to speak it. By 2000, nearly 50 percent of respondents said they could speak and write Catalan and only about 25 percent could now not speak it (Institut d'Estudis Regionals I Metropolitans, 2002a, 404). In addition, one-third of the surveyed population perceive a dual identity, feeling "as Catalan as they are Spanish." Likely as a way to increase their chances economically and socially, a significant number of native Spanish speaking persons have adopted a dual Catalan-Spanish identity (Institu d'Estudis Regionals I Metropolitans, 2002a). There appears to be Catalanization of first- and second-generation Spanish immigrants, both in terms of their perceived dual identity and in their knowledge of the language gained through public education (interviews: A. Serratosa; F. Muñoz). In terms of identity, an older Spanish immigrant may say, "I am not Catalan, but my

Figure 3.5   Bi-lingual Barcelona, Catholic masses in Spanish and Catalan

son is Catalan" or "I am from [neighborhood name], my son is from Catalonia" (C. Navales, interview). One effect of the growing knowledge of Catalan is that the public use of Catalan by native Spanish speakers may be creating an increasingly fluid and integrated public sphere.

While the ethnic Catalan and Spanish immigrant-origin communities are distinguishable through socio-economic and ethnographic data, there is no severe fragmentation of the social fabric. When evaluating the degree of social cohesion in a nationalistic region, Jenkins and Sofos (1996) make a useful contrast between two types of nationalist identity, one based on ethnic purity and community and one that is genuinely pluralist and based on rights of residence. The second type of nationalism, more inclusive and less ethnic, appears more consistent with goals of social cohesion and integration, especially in regions with mixed native and immigrant populations.

It appears that the inclusive and place-based, rather than exclusionary and ethnic-specific, nationalism has been ascendant in Catalonia since the beginning of Spanish democracy. Evidence supporting this argument includes the degree of adoption by Spanish speakers of a dual Spanish-Catalan identity and the increasing knowledge of the Catalan language. In addition, Catalonia has the highest degree of all Spanish regions of so-called "dual voting," where voters do not always support the same political party in a monolithic way, but rather have supported different political majorities, depending upon whether the election is for Barcelona city council (socialist support) or for the Catalonian regional parliament (Catalan nationalist party support for over 20 consecutive years after autonomy). This is evidence, says Pere Vilanova (interview), that voters view reality at least partially though a functional lens, a perception that can moderate nationalistic divisions. Political scientists view this as a sign of a healthy political system, one not bound by the rigidities and dogma of nationalism (Horowitz 1985, Nordlinger 1972). Further evidence supporting the inclusive nationalism argument is the phenomenal support by the populous for public sector interventions (seen in the most pronounced way for the Olympic Games) after Franco.

Catalan political interests appear to value and respect the need for inter-group inclusiveness and respect. My interviews with individuals inside the political world indicated a keen awareness of the need to be sensitive to differences in a nationalistic region of emotive history and passionate views. "We seek a nation that accommodates all the different political wills. As a government, we must include them all in order to avoid an endless cycle of conflict between groups," stated Domènec Orriols (Secretary of Communication, Generalitat of Catalonia, CiU, interview). The CiU party appeared to be aware of the overriding political need for inter-group compatibility in a multi-ethnic society. Joaquim Llimona (Secretary of External Relations, Generalitat de Catalonia, CiU, interview[27]) explained, "Social cohesion is very important and fundamental to us in maintaining our political coalition; there are challenges constantly that might break this solidarity." Llimona describes how CiU's leader, Pujols, was able to present a message and image that was comfortable to many types of individuals in Catalonia and he uses this approach to his own work in communicating CiU's message – "My heart says one thing, but as a member of the Catalonia government I have to take account of this complex social reality."

Catalan nationalist leaders made smart decisions during the Franco years. As far back as the 1960s, nationalists emphasized a place-based nationalism rather than one limited to ethnic identity. Jordi Pujol, in his book *Immigracio I Integracio*, asserted "anyone who lives and works in Catalonia and who wants to be a Catalan is a Catalan" (Guibernau 2004: 67). Amidst the significant immigration into Catalonia during the 1960s, this statement by CiU's future leader was significant in its emphasis on social rather than ethnic identity. Being Catalan was to be a choice, or free decision, by the immigrant. Spanish immigrants were welcome to the region, but residence here also meant a duty to respect and accept the identity of the Catalan community. If such respect was shown, the Spanish immigrant was to be fully incorporated into the Catalan project.

If Catalan nationalism had not adjusted to the demographic realities of mass Spanish immigration, and if instead nationalist leaders sought a nationalism defined strictly by Catalan ethnic origin, the nationalist political project today would likely be in jeopardy. As stated pragmatically by Catalan nationalist J. Llimona (interview), for a region of such substantial immigration, "if we had had a policy only for the 50 percent born here it would have been a disaster."

Instead, the nationalist political project has accommodated this large influx by balancing demands for Catalan cultural recognition (such as linguistic rights) with demands that greater financial resources come to the region to benefit the population at-large (both native and immigrant origin). Those from Spanish immigrant backgrounds thus would become aligned with the Catalan nationalist political project to the extent they perceived benefits to them accruing from greater Catalonian regional capacity vis-à-vis the Spanish central state. They may not feel culturally Catalan (especially the older generation), but claims for greater Catalonian resources for highways, schools, and public services would resonate with them.

I believe that substantial Spanish immigration has positively influenced the nature and contours of Catalan nationalism. Not only did Spanish immigration play a fundamental role in the social hybridization of Catalan society, but it also has led to a relatively porous and inclusive Catalan nationalism. The industrialization policies of the Franco regime in the 1950s and 1960s stimulated a flood of immigration into Catalonia, which over the long term meant that the Catalan nationalism that would re-emerge after Franco was to be, by nature of circumstances, more inclusive of immigrants.[28] Without Franco-induced immigration, Catalonia nationalism today might well be more inward looking and insular. The public consensus that has existed since democracy is also a phenomenon likely born during the authoritarian dark years. Both groups experienced the debilitating hardships of material and cultural deprivation and in the process a consensus across native and immigrant populations was forged. Both through promotion of immigration and by creating a shared enemy, the Franco regime likely created the unintended effect of promoting a broader and more balanced contemporary Catalan nationalism.

Without underestimating the deep historic cultural roots of Catalanism, the confluence of factors discussed here suggests that the nationalism of today may be one based more on sense of place (Catalonia as a place worthy of greater autonomy and resources, no matter where the resident was born) rather than one based solely on origin (wherein one's family must be from Catalonia and suffered personally the cultural repression of Franco's dark days). This place-based nationalism is a move away from the rigidities of ethnically defined nationalism and facilitates a more normal pluralism of views and demands expressed through democratic political channels. Violence – its presence or potential – is not part of the political vocabulary in Catalonia. While there is stratification and a Catalan ruling class, those with Spanish immigrant histories feel more a part of Catalan society than not, and many have assimilated (through language and identity) into what has been a dynamic, fluid society of opportunity.

Since the early 1990s, a new challenge to the social cohesion of Barcelona and Catalonia has emerged – that of immigration not from elsewhere in Spain, but from other countries like Morocco, Ecuador, Peru, Colombia, Argentina, and Pakistan. Many interviewees, while acknowledging Barcelona's and Catalonia's ability to accommodate Spanish immigrants, spoke of foreign immigration as a phenomenon that the region may not be able to handle. "New immigrants" was the most common challenge to the Barcelona model mentioned by these individuals. Policy-makers wrestle and are troubled by this issue, at national, regional, and municipal levels.

Spain's foreign immigration is not distinguished by its size (about 6% of total population, similar or below other European Union countries), but rather by its quickened pace of growth (increasing from 1.1 resident immigrants in 2001 to 2.6 million in 2003). In the seven-year period 1998–2004, the pace has been unprecedented. "We are very fast approaching the percentage of immigrants that are in France, Austria, and Belgium, but they did it in 30 years and we are doing it in ten years," states P. Vilanova (interview). This is due in part to the booming Spanish economy that has created new demand for cheap labor. Further intensifying the uncertainty of today's immigration is that a large percentage of immigrant residents may be in Spain illegally "without papers." Catalonia is the region in Spain that has the greatest number of legal foreign immigrants (in 2003, over 420,000 plus an uncertain number of illegal residents). This represents about 6 to 8 percent of total Catalonia population, but again it has been the pace of legal immigration (plus the uncertain number of immigrants "without papers") that has caught policymakers' attention. The issue has been especially salient in Barcelona city proper, where over 200,000 of the legal immigrants in the region live, comprising about 13 percent of the city's population (Rioja 2004).

Cultural difference in the city and region, heretofore seen as an addressable challenge, is now increasingly seen as a threat, especially when viewed through the lens of larger religious (Muslim–Christian) tension in the world. Whereas a common enemy (Franco) helped bond together Catalan and Spanish immigrants in the past, now foreign immigrants are being viewed more as the enemy (C. Navales, former member, commission on immigration policy, Catalonia, interview). There is now, according to F. Munoz (interview), the development of genuine foreign immigrant ghettos in Barcelona city due to linguistic and religious barriers. Whereas the middle-class largely stayed in the city throughout the years of Spanish immigration, now there is the beginning, and potential deepening, of middle-class out-migration to the suburbs due to this foreign immigration flow. There is a new element of exclusion, mainly based on religion, and spatial concentration that is lessening social contact between the city's cultures, and this trend is in danger of intensifying (F. Munoz, interview).

Non-Spanish immigration is a potential inflammatory political issue, "a very delicate one that political parties thus far have been responsible with, not using the issue against one another" (P. Vilanova, interview). Uneasiness of public opinion about immigration in Catalonia is skewed by the mistaken belief that most immigrants are from Muslim countries. In reality, those who are Muslim and from northern Africa or Pakistan make up about 30 percent of legal foreign immigration in Catalonia (Rioja 2004). Nevertheless, the pace of immigration can be overwhelming, many of the new immigrants are culturally alien to Catalan or Spanish life, and immigration has been unevenly distributed both across Spain, with Catalonia the largest receiving zone, and across the urban and metropolitan space of Barcelona. These conditions could lead to a backlash against new immigration if the processes of incorporation and assimilation are not effectively handled by the state, region, and city. This has potentially unsettling political consequences in terms of the inclusiveness and rhetoric of Catalan nationalism if such a backlash stimulates the creation of a more exclusive Catalan nationalism aimed at protecting its homeland from cultural aliens, a type of reaction akin to the emergence of Jean-Marie Le Pen in France.

Moderating this potential harsh reaction would be a coherent set of policies to effectively incorporate these new immigrants into the country. At the state level, immigrants who have been living and working in Spain for a minimum period of six months and can produce a contract signed by their employees had a three-month period in 2005 when they could legalize their situation.[29] In the future, there may likely be some greater selectivity in who can immigrate legally into the country (P. Vilanova, interview). In respect to urban and metropolitan management of new immigrants, key goals are to counter spatial and social exclusion by linking these individuals to broader economic opportunities. One characteristic of where some new immigrants are locating may be beneficial to this project. Similar to the past when children of Spanish immigrants were living in the peripheral areas of the urban region, new immigrants from northern Africa are increasingly occupying residential zones between the city and today's second crown suburbs. Due to the outward push of development to the second crown suburbs over the past decade and the existence of a good urban highway system now linking different sectors of the urban region, these formerly "peripheral areas" have become more akin to subcenters connected to metropolitan opportunities and this may provide new immigrants with an enhanced ability to assimilate (F. Muñoz, interview). Since many of these peripheral areas experienced significant rehabilitation after the emergence of democracy, they also contribute a certain quality of existence for the new immigrant, in many cases having better ratios of public space due to their high-rise residential fabric.

Notwithstanding such possibilities, challenges remain in addressing the issue of how to accommodate new immigrants within the urban fabric. Contrasting senses of history and tradition between cultural groups are little studied, hard to measure, and more often examined by anthropologists and sociologists than by urbanists and architects. Thus, principles about how to support inter-group accommodation – such as how public space use and housing needs may vary across cultural groups – are not well developed in the professional literature (Josep Montaner, professor of architecture, Polytechnic University of Catalonia, interview). The city has used a "holes in a sponge" strategy in the redevelopment of Ciutatvella (old town), an area of density, hygiene problems, little openness and light, and now increasingly a receiving zone for foreign immigrants (I. Pérez, professor, Universitat International de Catalunya, interview). This strategy is an effort to restructure the urban fabric to allow greater openness and light, much as a sponge has holes in it that allow light to pass through. In practical terms, this means the creation of more public open spaces and the taking out of some of the older and denser housing stock.

The wave of foreign immigration is presenting the city and urban region with a challenge as to how distinctly different groups in the city can coexist (*covivir*). One immediate challenge is whether Barcelona, in the face of perceived threat and difference, will be able to maintain its "architectural respect for the street" (A. Karmeinsky, architect, interview). The urban openness in Spanish and Catalan society – seen most visibly in the ever-present street café tables and the multi-generational use of public space – is so strong and embedded that one suspects it will continue and be able to accommodate new immigrants in some way. Yet, an evolution may take place wherein a psychology of openness is preserved while incorporating some aspects of greater security (for example, a public park open in the day but closed at night to provide a psychological buffer) [A. Karmeinsky, interview].[30]

There appears to be a genuine commitment to cultural diversity on the part of the public sector in the Barcelona urban region. In the 2003 Strategic Metropolitan Plan (SMP) of Barcelona, the importance of culture, coexistence, and social cohesion are consistently emphasized and explicitly discussed (SMP 2003). The plan seeks social cohesion, mutual interaction between all groups, and for Barcelona to avoid becoming a conflictive frontier city. It views cultural diversity as positive, seeing it as a "value added" component to urban life (SMP 2003: 122). It speaks of the need to "achieve the correct spatial integration of immigrant families in new and old urban areas to facilitate the maintenance of social cohesion" (SMP 2003: 127). A strategic use of urban public space and the public sphere, seen in early 1980s planning interventions to inculcate democracy into the populous, is again posited, this time for social integration purposes. In the end, an open and culturally plural Barcelona region would "promote a new type of citizenship that is related to residence and not to nationality" (SMP 2003: 129), a goal that would push further out the boundaries of Catalan nationalism.

Urbanism and nationalism: the internal tensions of regionalism as a political project

Cities are the place where two models – nationalism and social inclusion – have met. Nationalism has not been able to kill urban diversity. In Barcelona, the two models are cohabitating.

Carles Navales (Interview)
City councilor, Cornella de Llobregat, 1979–1991.
Trade unionist and activist in the 1970s

Catalonia has to be vigilant in not unbalancing itself in many aspects: culturally, economically, socially ..... The possibility of the metropolitan region of Barcelona marginalizing the rest of the country (Catalonia) is a constant danger. The question of territorial disequilibria continues to be a pending issue .....

Jordi Pujol, President of the Generalitat (1980–2004)
From: Tobaruela, P. and Tort, J. (2002: 115).

The social, linguistic, and economic landscape and dynamics of Catalonia over the past decades indicate the plausible existence of a porous, non-exclusionary form of contemporary Catalan nationalism. Underlying this congenial picture, however, is a story over the past 25 years of ideological and tactical competition between the region's two main political parties – *Convergencia I Unio* (CiU) and the Socialist Party of Catalonia (*Partit dels Socialistas de Catalunya*, or PSC). Each party has distinct political constituencies and perspectives of how best to institutionally construct a regional nationalism; an important wedge issue in this relationship involves Barcelona and its urban growth vis-à-vis the region at large. There has not been a monolithic Catalonian aspiration as expressed by, and through, political parties. Rather, inter-party competition has been a fact of life since democracy and these competitive antagonisms have obstructed the effective operationalization on the ground of Catalan nationalism. In particular, Catalan nationalism and Barcelona urbanism have been strange bedfellows. Ironically, regional autonomy and the politics of Catalan-based parties have at times obstructed the city of Barcelona's capacity to compete in the Spanish democracy.

Convergencia I Unio (*CiU*) is a political party created in the wake of the end of Francoism. From the first regional elections in 1980 all the way to 2004, CiU was in political control of the regional autonomous government, led by Jordi Pujol. CiU is a nationalist party that makes self-government for the region a central pillar of its program. Nationalism is the defining imperative for CiU; on other issues, its politics vary from center left to right (P. Vilanova, interview). Its electoral constituency areas are disproportionately in the more traditional small town and rural parts of Catalonia, less industrialized, with less social conflict and less immigration from other parts of Spain (Marcet i Morera 2000). Compared to Catalonia overall, CiU constituents are more likely to be ethnic Catalans, practicing Catholics, more knowledgeable of the Catalan language, middle class and self-employed, more connected to the social-economic elite of the region, and more likely to view themselves as "only Catalan" and "more Catalan than Spanish" (Domenec Orriols, interview; Centro de Investigaciones Sociologicas [CIS] 1992). CiU's political linkages are with foundational parts of the region, a conservative bourgeoisie/social class, small retail shop-owners, and rural interests (F. Munoz, interview). In political left-right orientation, the CiU voter views himself in the political middle, which positions him to the right of the average Catalan voter (CIS 1992).

CiU has effectively combined a nationalist stance with pragmatism and populism. We saw previously how CiU's rhetoric and positioning of its political nationalism vis-à-vis its Spanish immigrant population facilitated a Catalanism that is as much place-based as identity-based. Its main competitor, the PSC, has polled more votes throughout this period in central Spanish state elections but never yet has been able to out-poll CiU in regional elections.[31] After the battering of Catalan nationalism under Franco, the "CiU succeeded in offering the right product at the right moment" (F. Muñoz, interview). The party's leader, Jordi Pujol, came out of the "forgotten" generation that had faced strong political and cultural repression and had through his symbolic actions established himself in opposition to Franco. When the regime ended, Pujol was there as the leader of the emerging Catalan nationalism; through astute political strategizing through the years, he would remain the political leader of Catalonia for 24 years.

CiU competes with its main challenger, the socialist PSC, on the region's core political issue – Catalonia's appropriate legal and political relationship with the Spanish central state. Both parties advocate greater self-government for the region and a revision of the 1979 Spanish statute of autonomy; yet, the similarities end there. The CiU argues that the region, due to its unique history, language, and identity, deserves to be treated in a special way in its relationship to the Spanish state (Joaquim Llimona, interview). The other regions of Spain deserve greater autonomy too, but "because of our reality and our psychology," Catalonia's level of autonomy must clearly be above that of other Spanish regions. Particular needs for self-government are in the areas of financial autonomy and the ability to raise their own revenue, security and police, and international relations dealing with immigration and export business development. Financial qualms concerning the post-Franco autonomy agreement are particularly acute; "We feel that our wallet has been stolen" (Domenec Orriols, interview). The PSC socialist party also emphasizes that Catalonia has a different history and culture from the rest of Spain. But, in contrast to CiU, the PSC argues that increased autonomy for Catalonia should be advanced within a new plural, or federalist, framework for the Spanish state. "We want to be there to reconstruct Spain, not to be alone," states Maria Badia (Secretary of European and International Politics, PSC, interview). PSC asserts that Catalonia should be "autonomous and working together with the rest of Spain in terms of federalism" (M. Badia, interview).

The CiU criticizes the PSC for being too Spanish; the PSC claims that CiU thinks too selfishly of only the region's needs and aspirations. The PSC has a limited sovereignty relationship with the larger Spanish socialist party (PSOE, Partido Socialista Obrero Espanol) and has at times needed to distinguish itself as a Catalan party. To the CiU, this relationship with the PSOE handicaps the PSC; the Catalan socialist party "has different political sensitivities" due to this arrangement and this "limits their capacity to act on behalf of the Catalan cause" (J. Llimona, CiU, interview). Catalan socialists counter that CiU acts in non-constructive ways that obstruct autonomy goals. Socialists assert that CiU uses consistently an "us-them" strategy, usually in terms of Catalonia versus Spain. "It is always the object of nationalism to have an enemy. Socialists don't need this external 'other'. We don't need or want to create the enemy. CiU thinks they need to," explains Badia (PSC, interview).

The perspectives of the two political parties looking outward toward the Catalonia-Spanish state relationship is one part of the story; the other part involves how the two main parties view how to best manage and govern the Catalonia region, how development policies are linked with each party's electoral constituencies, and how this has produced a competition between Barcelona city and Catalonia region that has been largely counterproductive.

The political orientation and basis of the CiU explains much of its development policy over the 24 years of its Generalitat leadership. Subiros (1993, 48) describes: "The ideological discourse of Jordi Pujol and a good part of this political strategy is based on a historicist, essentialist, ruralist and anti-metropolitan script that looks for and uses confrontation with Barcelona and the central government to confirm and reinforce the autonomy and competencies of the Generalitat." In contrast to the more socially inclusive "city-state" notion of Catalan nationalism espoused by PSC and Pasqual Maragall (Barcelona's long-standing socialist mayor), Jordi Pujol and the CiU has used a "bourgeois regionalism" approach that downplays the city of Barcelona in favor of a regionalist stance (McNeill 1999).

To govern Catalonia and to stay in power, CiU's development tactics consolidated direct relations between individual towns (especially those having CiU constituencies) and the Generalitat, creating a set of local political leaders in these smaller towns who directly benefited from CiU's largess.[32] These tactics of localism sought to prevent the emergence or growth of extra-jurisdictional arenas of power that could compete with CiU's regional control. This CiU approach of localism appears rational from a tactical political point of view because it has rewarded its small town and rural constituencies.[33] However, it has created difficulties for Catalonia because the small scale nature of the CiU strategy is ill fitted to the contemporary needs to plan and coordinate major multi-jurisdictional transportation facilities and to address the concerns of multinational corporations (A. Ulied, environmental consultant, interview). The CiU project has led to chaos and uncoordinated urbanization because each local government tends to be atomized, with only limited relationships with its neighboring jurisdictions (E. Madeuño, interview). M. Solá-Morales (urbanist/architect, interview) describes CiU's approach as one of developmental permissiveness, an "excessive laissez-faire" perspective that regards any restriction or containment of Catalonia growth as anti-nationalist. Without articulation of regional development objectives, "those who are so adamant in their defense of Catalonia have been guilty of destroying its quality of life more than anyone else" (E. Madueño, interview).

The CiU has viewed Barcelona – because of its urban centrality and the city's propensity to vote for the Catalan socialist party – as a competitor. During its years in control of regional government, CiU thus did not seek to support and strengthen the Barcelona metropolis as the economic and cultural center of Catalonia. Given the importance of cities as engines in today's urban world, from the outside this approach of the center-right party would appear to be counter-intuitive and contrary to their own nationalist aspirations for Catalonia to be a strong and independent economic and political force in the world. In CiU's arguments against Barcelona growth, there is use of the old discourse and rhetoric of "territorial imbalance" that took hold in the 1920s and which portrayed "macrocephalic" Barcelona as encroaching and eroding the traditional moral and social order of the Catalan rural heartland. Inherent in the territorial imbalance argument is a model of the regional landscape as a "Catalonia Ciudad," a region where all cities, towns, and individuals would have the same opportunities and resources no matter where its location in the region (Joan-Anton Sanchez, interview). This implied a region of more dispersed opportunities than would be found in a region of metropolitan dominance.

Since the Franco transition, the CiU has utilized this territorial imbalance argument politically to appeal to the core non-Barcelona elements of its constituency (O. Nel-lo, secretary of territorial policy, Generalitat [PSC], interview). The idea that Catalonia was imbalanced was "very much alive" in the mid-1970s, an idea having special appeal to the Catalan nationalist movement (Nel-lo 2002). The method through which to balance the region – the Catalonia Ciudad model – was adopted by the CiU. Yet, whereas in its original version the model expressed the hope for equal opportunities across differently populated settlements, the CiU manipulated its meaning and used it to push for a more equal distribution of growth and population across Catalonia and to suppress metropolitan Barcelona development (J. Esteban, interview).

The Generalitat under CiU "always saw the city council as a counter-power against their interests" (L. Permanyer, columnist for *La Vanguardia* newspaper, interview). Accordingly, services and spending were channeled away from the city and the "red belt" of socialist-leaning suburban cities around it. The clearest manifestation of political competition between the Generalitat and the Municipality occurred in 1987 when the Catalan legislature abolished the Metropolitan Corporation of Barcelona (CMB). The Corporation, established in 1974, was involved in the approval of urban planning and the delivery of urban services of metropolitan significance for a 27-municipality area (M. Garcia 2003). Through the years, the CMB gained increased power and financial resources, and had expressed ambitions to become a fuller-fledged political institution. Jordi Pujol and the CiU saw this mounting presence as a competitive force in Catalonia. Mistrust and rivalry intensified as Pasqual Maragall, Pujol's archrival, assumed the presidency of the Corporation in the early 1980s. Against a separate identifiable layer of governance for Barcelona area cities and feeling that such mega-urbanism would surely favor the socialist cause, the Generalitat terminated the Corporation "and disaggregated it for political reasons into multiple single-purpose entities" (D. Orriols, interview). Since the abolition of the CMB, there has not been a coordinated, institutionalized governance of the Barcelona region, a circumstance that strengthened the Generalitat's hand in metropolitan development affairs.

Hemmed in by the anti-metropolitan predilections of the CiU Generalitat government, on the one hand, and the "centralist" ideology of the Spanish state, on the other, the city of Barcelona has acted creatively to break out of these constraints. Part of the motivation behind the city's use of grand and prestigious events to catalyze urban activities and investment (such as the 1992 Olympics and the 2004 Forum of Universal Cultures) undoubtedly lies in the city's desire to burst out of the political quagmire produced by the dual constraints of regional nationalism and state centralism. Through the hosting of important world events, the city has been able to obtain funding and coordination from the Generalitat and from Madrid that it otherwise would not have obtained. The city's genius and creativity appear at several times in its history to come about, and define themselves, in relation to the constraints that it has faced, whether it is Franco repression, the spatial limits of its bound geography (hemmed in by sea and mountains), or the competing and contrary visions of regional and central governments. Squeezed between the rural traditions of Catalan nationalism and the centralizing tendencies of the Madrid government, Barcelona city has nonetheless been an active and self-transforming entity through the years of post-Franco democracy.

THE RE-SCALING OF GOVERNANCE

The jockeying for 25 years between Catalan nationalists and socialists illuminates the degree to which geographic scale – nonmetropolitan, metropolitan, urban – has been connected to political motivations for power. Now, two contemporary challenges to Barcelona governance – metropolitanization and Europeanization – are stimulating a significant rescaling in the geographic scope of many urban activities and a reconsideration of the appropriate levels of governance. Contemporary Barcelona and Catalonia find themselves within a more complex and evolving framework of governance due to globalization and Europeanization. Amidst the reformulation of the nation state, there is an "explosion of spaces" and the existence of multiple and intersecting scales of political geography.[34] In Catalonia, the scales of governance most important in a global future are city, metropolitan area, subregional district (province), substate region, state, western Mediterranean region, and the European Union.

In a circumstance where political geographies are stable, political competition resembles a zero-sum game. However, when metropolitan and European economic and institutional geographies are shifting and uncertain, new opportunities exist for political interests to define themselves in relation to these new geographies, even to the point of redefining what nationalism is in a globalizing Europe. Amidst the metropolitan complexity of Barcelona and the new relationships emerging from the European integration project, there are new possibilities and obstacles in the management of urban space, as well as new senses of identity and political party maneuverings that seek to exploit them.

New metropolitan realities

Spatial diffusion and extension of the Barcelona urban region over the past thirty years has produced a complex pattern of human settlement that is transcending and eclipsing the politically-inspired rhetoric that juxtaposes the city of Barcelona vis-à-vis the Catalan countryside of small towns. From 1970 onwards, there has been substantial decentralization of population and economic activity in the urban region of Barcelona. While two out of three residents in the urban region of Barcelona lived in the city in 1960, only one in three residents lived in the city in 2004 (see Table 3.1).

There has been strong dispersion of population in the urban region over the past four decades. This is attributable to increases in income and consumption,

Table 3.1 Population of city, metropolitan area, urban region, and Catalonia[1] (in millions)

	1960	*1970*	*1981*	*1996*	*2004 (est.)*
City	1.56	1.74	1.75	1.51	1.58
1st crown	.28	.98	1.39	1.40	1.45
2nd crown	.64	.84	1.09	1.32	1.45
TOTAL urban region	2.48	3.57	4.24	4.23	4.48
Catalonia	3.89	5.11	5.96	6.06	6.81

Sources: National Institute of Statistics (various years); Ajuntament de Barcelona 2003; Institut d'Estadistica de Catalunya 2004.
Notes: [1]"Urban region" designates the city and both crowns of suburban cities; this is consistent with Catalan and Spanish use of this term.

relocation of industrial activities, and regional enhancements in the road network. The city proper peaked in population in 1970. The immediate "first crown" cities around Barcelona city are now stagnating compared to the second crown, which has experienced the bulk of new growth since the early 1980s. From 1991 to 2001, city population declined 8.5 percent while areas of the province outside the metropolitan area – meaning for the most part the second crown suburbs – grew 17.8 percent in population (National Institute of Statistics 2001). The differences in new housing units constructed between 1987 and 2000 across the three component parts of the urban region are dramatic; about 70,000 housing units built in the city, 113,000 in first crown cities, and 207,000 in second crown cities (Garcia 2003). Many suburban municipalities have been growing 5 to 6 percent per year. In terms of geographic space, the city proper is small (98 km^2) compared to the first crown (378 km^2) and the expansive second crown (1,983 km^2). And, politically, Barcelona city is one of over 160 municipalities in the urban region; 65 percent of these municipalities are small – less than 10,000 population; and 80 percent are in the second crown hinterland (Garcia 2003).

Low-density housing development and the increased building and use of "second homes" in remote country settings magnify the spatial extension of urbanization in the Barcelona region. Since 1990, more than 50 percent of housing in Barcelona province has been built at low-density urbanization levels, adopting in many cases American style suburban techniques (F. Muñoz, interview). In addition, the construction of "second homes" used during the summertime and holidays has grown significantly. Indeed, Garcia (2003) estimates that in 1991 about 30 percent of all housing in the second crown was used as a second residence. These circumstances become more problematic as Catalans today are seeking country locations to live in year-round and are increasingly converting these second homes into first homes.

Economically, there has also been strong dispersion of activities, particularly industrial production locating around the more distant industrial cities in the second crown. Barcelona city proper, meanwhile, has been transformed more toward the service sector. Throughout the urban region, there is a marked economic or functional specialization of activities where there is a concentration of specific economic activities and their support sectors in certain municipalities. Much of this specialization builds on previous patterns of industrial consolidation in the region's mature cities – such as Mataro, Terrassa, Sabadell, Matorell, Badalona, and Manresa – that have good connections by train to Barcelona. Counterbalancing the high numbers of smaller municipalities in the region are these six large municipalities outside Barcelona that have populations greater than 100,000 and thus the ability to generate and attract economic activity.

Such complex metropolitan realities do not resemble the picture of the city and region painted by the traditional nationalist rhetoric that had drawn distinct lines between Barcelona city and Catalonia countryside. Barcelona city has not taken over the Catalan countryside, as feared by territorial imbalance critics, but neither has Catalonia developed into a political region of equilibrium. The urban region of Barcelona, differentiated and complex and home to 7 of every 10 Catalans, now dominates Catalonia. However, it is an urban region not itself dominated by Barcelona city, but one whose growth has overtaken Barcelona city and first crown metropolitan area boundaries. Catalonia is increasingly becoming a single interconnected assemblage of cities and towns, a network of cities more than a region of smaller scale towns (J. Trullén, professor of political science, Universitat Autonoma de Barcelona, interview).[35] Politically, this network of urbanity likely causes difficulties to the political project of the Catalan nationalist party since the CiU developed its electoral base through relationships with smaller towns and the traditional countryside. On the other hand, the profound extension of urbanization beyond Barcelona city and its "red belt" (first crown) of politically similar suburbs points to the decreasing ability of Barcelona city to guide and shape growth in Catalonia. While the city has remained extremely active in urban restructuring *within* its borders, much growth in the urban region is now occurring beyond the reach of its jurisdiction and influence.

The city of Barcelona's tactical approach through the last 25 years of intervening at increasingly greater spatial scales – site specific design intervention in the early years of democracy to larger-scale restructuring before and after the Olympic Games – is now meeting its outer boundaries at the city's edge. As described by one insider, "urban officials realize we keep discussing what happens inside the city, but the problems are outside" (M. Herce, interview). In the years to come, efforts by the city to catalyze its regional connectivity and attractiveness (through significant investments in high speed rail and water port expansion) will mean the city will maintain some shaping ability beyond its borders. Yet, as politically fragmented suburban cities increasingly lock in lower density and regionally uncoordinated growth, and due to a minimum of multi-jurisdictional governance, the window of opportunity for effective metropolitan planning and coordination may be closing for the Barcelona urban region, and its innovative leader, the city of Barcelona.

Due to the expansive urbanization of the Barcelona region (inclusive of the second crown), the call for multi-jurisdictional regional leadership and management has intensified. To improve the spatial organization of the larger Barcelona urban region, many urbanists feel that the increasingly poly-centered and economically specialized nature of the region should be reinforced and deepened. However, the attempt to implant such a vision of the Barcelona urban region in an approved plan has been difficult. The Catalan Parliament approved a Spatial Policy Act in 1983 that required spatial plans at regional geographic levels that would guide urban development plans done by cities. As part of the territorial planning system, a "metropolitan territorial plan for Barcelona" (MTP) was to be developed. The geographic scale of this project is significant: it is an attempt at producing a spatial plan and growth vision for the larger urban region of Barcelona inclusive of the second crown. The MTP was technically completed in 1988, reviewed in draft form in 1990, arose again in the mid-1990s and languished in draft form until 1998. As of 2005, the Barcelona urban region was still lacking a comprehensive spatial plan for its development.[36]

With the coming of political power at the Catalan regional level of the Socialist Party (and its coalition) in late 2003, chances increased that metropolitan planning and governance for Barcelona would rise again. A source of emulation and inspiration for such a possibility is the strategic planning processes led by the city of Barcelona over the last 15 years. Four strategic plans have been approved, in 1990, 1994, 1999, and 2003.[37] The fourth plan in this series, the Strategic Metropolitan Plan (SMP) of Barcelona, was approved in March 2003. Although the SMP says nothing about governance options "because that topic is too political," the plan does focus on metropolitan-wide issues of economic, territorial organization, and social cohesion (M. Rubí, interview). Although the SMP was intended to have no direct political power, its ability to find consensus concerning crucial points of metropolitan influence may be building the foundation for some type of formalized metropolitan governance in the future. In 2005, the Generalitat started considering a major institutional reform for the territorial structure of Catalonia. One effort would create seven new administrative districts within the region, and one would correspond with the boundaries of the expansive Barcelona metropolitan-region scale of urbanization.

The spatial quality of Catalonia's growth is one that neither resembles the dominant Barcelona of socialist urbanists' dreams (because regional urbanization is not producing a bigger city but an interconnected network of cities) nor the territorially balanced Catalonia of nationalists' dreams (because growth is not dispersed evenly but is supporting a clear hierarchy of different size urban places). In the future, decisions regarding space in Catalonia will undoubtedly be intimately connected to historically embedded and potent emotional and political triggers.

The city in Europe

In terms of cultural identity, you need to invent, create,
and adapt to the new conditions of this widening world

Francesc Morata, Interview
Director, Institut d'Estudis Europeus,
Universitat Autonoma de Barcelona

Nationalism is not a static concept, but a political project and process that needs to evolve with changing social, political and demographic realities. We have seen how Spanish immigration into the region during the Franco years reshaped Catalan nationalism and the city of Barcelona, creating a more inclusive nationalism and bicultural city. More recently, the complex, spatially expansive, and poly-centered urbanization of the past decades is forcing the two main political parties to new understandings of how urbanization and nationalist politics intersect. A third influence, the changing institutional landscape of Europe, provides new footholds, and chasms, for those advocating a robust Catalan nationalism in the twenty-first century.

The 1992 Maastricht Treaty set off a process of political and economic integration that is fundamentally reshaping Europe and which has significant implications for region-based nationalism. Catalan nationalism, from its beginning, has constantly sought greater links with Europe as a way to counter and bypass Madrid's efforts at greater centralism (Joan Subirats, professor of political science, Universitat Autonoma de Barcelona, interview). During the Franco years, Europe was used as a resource by Catalonia in its struggle for democracy, rights, and cultural expression (F. Morata, interview). Thus, with greater European integration comes increased opportunities for Catalan nationalism to play its internationalist card. Although almost all of the 17 autonomous regions in Spain have European Union offices combining tourism and commerce functions, the fact that Catalonia (and the Basque Country) have such offices is a sensitive issue because centralists view it as presaging more formal links between the region and the EU. And, indeed, Catalonia has used the European issue in practical and tactical ways to express its identity and to undermine state dimensions (Pere Vilanova, interview).

To the Catalan nationalist, European integration and economic globalization provide new opportunity spaces. The establishment of a Committee of the Regions (COR) in the EU apparatus has brought together representatives of city councils and regions across Europe, and has put Catalan officials in frequent contact with other nationalistic regions across Europe. A globalizing world provides the ability for an area like Catalonia to differentiate itself (on the basis of culture and genuineness) in an increasingly interdependent and homogenizing world. At the same time, the maintenance and intensification of international and European linkages will likely sustain the openness of a region-based nationalism.[38] A nationalism that has found its place in a widening world could thus potentially combine the benefits of heritage and authenticity with those of openness and inter-relationships with the outside world. As described by J. Subirats (interview), if the region stays internationally connected and does not go inward to parochialism, "globalization and Europeanization create a very good scenario for an identity movement in Catalonia."[39]

Politically, within Catalonia, increased internationalization may produce a political calculus that brings the two main political parties closer together. The scale of problems and interrelationships today is overwhelming CiU's use of the territorial balance, Catalonia Ciudad, argument and leading to reconceptualization by this mainstream Catalan nationalist party of what Catalan nationalism means today. The party's traditional approach that stifled intermunicipal coordination is ill fitting in a region and in a Europe of increased integration, scale, and interrelationships. Thus, in the last years of CiU regional governance there were efforts to develop a new spatial model that would support a "cosmopolitan nationalism" facilitative of increased flows and cooperation (J-A Sanchez, interview). In an update of an old concept, Sanchez states, "To consider again Catalonia Ciudad in the 21st century calls for it to be considered one large metropolitan region within and connected to its European context."

If the mainstream nationalism of the CiU evolves toward a more open, interconnected nationalism, it will move toward the Catalanist model as espoused by its competitor socialist party in Catalonia. This party, and its leader Pasqual Maragall, has adroitly used international linkages to put Barcelona and Catalonia on the world stage, and a "cosmopolitan nationalism" would find commonality with PSC's and Maragall's view of nationalism as being city-led and metropolitan.[40] Political infighting in Catalonia will likely remain a fact of life, but it may increasingly take place within an accepted framework that emphasizes openness, European-ness, and integration across the Catalonia landscape. This scenario is illustrated by the similarities, and the differences, in two macro-regional projects initiated by CiU's Pujol and PSC's Maragall in the early 1990s. Both CiU's "Euro-Region" project and PSC's "C-6 Network" were linked with the new dynamics of Europe and sought greater cooperative and economic ventures between Spanish and French subnational governments in the Mediterranean region. Yet, whereas the "C-6 Network" was a consortium of regional capital cities and sought to develop new leadership potential for Barcelona city after the demise of the Metropolitan Corporation, the "Euro-Region" was a project of regional governments that left out cities almost entirely. The conflict between capital cities and regional governments, and in Catalonia specifically, between socialist controlled Barcelona and its then CiU controlled regional government, was reinforced as new transnational linkages emerged in the new Europe (Morata 1997). Thus, old internal fights are continued on new turf. Still, the commonality between these two efforts is informative and displays how internationalization and Europeanization are viewed as compatible with the promotion of a nationalist project; that a nationalist region can strengthen itself through the growth of international links.

CONDUIT, CROSSROADS, AND CRUCIBLE

Barcelona has been conduit, crossroads, and crucible. In each of these roles, the city has played a key role in mapping the contours of democratic and regional autonomous Catalonia in the late twenty-first century. Barcelona represents the transforming power of urbanism amidst political transition and nationalistic aspirations and in many ways its story is an uplifting one. Examining the social and political dynamics of Barcelona during and after the political transition from Franco domination shows us the power of the city to be a *conduit* or channel for political opposition and expression midst authoritarian repression. Community and political opposition that came from urban material disadvantage in the Barcelona metropolis gave rise to an articulation of an alternative, better urban future. Planning and urbanism during this period prepared the terrain for democracy and showed the populous the new capacity to engage in a collective public project after decades of authoritarianism. At the same time, the city exposes the rich internal differences within Catalan nationalist thought concerning the appropriate power of the city amidst a strong historically-based regional nationalism. In this *crossroads* role, the city is at the discursive focal point between two differing perspectives on the nexus between Catalan nationalism and urbanism. In addition, Barcelona has been a cultural *crucible* in its ability since the 1950s to absorb into its social and political life significant numbers of southern Spanish immigrants. Immigration and nationalism have coexisted in a mutually supportive relationship. Far from diluting the Catalan nationalist project politically, this synthesis of newcomers through the decades has led over time to a transformed, stronger and more inclusive nationalism better able to negotiate its future in contemporary Spanish politics.

In its capacity to navigate through societal uncertainty, transition, and democratization, and in its ability to manage productively issues of group difference related to Catalan nationalism, city leaders and urbanists have exhibited an attitude, a mood, and a vision and were able to implement these in pragmatic ways to affect change on the ground (S. Mercade, interview). A partnership between the municipality with its visionary mayor, Pasqual Maragall, the private sector, and the public at-large worked in ways that achieved significant and positive outcomes. Certainly, the dark years of Franco created a significant stimulus for the new democratic local regime to think fast and big. However, although this sense of crisis created the conditions for public sector mobilization, it did not guarantee the positive outcomes that ultimately took place. That credit should go to the city, its leadership, and its progressive democratic urbanists.

In the future, political and planning imagination and creativity will be needed again to accommodate the complexity and uncertainty of contemporary trends, including the complicated and interconnected urbanization of Barcelona and Catalonia that is confronting traditional political rhetoric, challenges to social cohesion in the form of fast-paced foreign immigration, and the ongoing political projects of European integration and Spanish state restructuring. Political leaders that recognize the need to act proactively and constructively will again be challenged to formulate innovative public sector strategies to address these contemporary uncertainties. Needed, also, will be the imagination and innovation of the prodigies and students of the pioneering urban professionals in Barcelona who contributed so significantly during the Franco transition and in the democratic years to the social and physical transformation of their beloved and authentic city. What likely is at stake is the future of Barcelona, and Catalonia, as places of genuine and positive identity amidst a multi-faceted global society.

Notes

1. Three months earlier, the Spanish Parliament had approved the new autonomy plan by a vote of 189–154.
2. The province is an expansive geographic area larger in size than the greater urban region of Barcelona. There are four provinces in Catalonia.
3. Knowledge of Catalan, in contrast to "native language," is higher. Due to Catalan linguistic training in the public schools and the growth of Catalan-speaking television, there has been significant growth over the last 20 years in the average person's knowledge level of the Catalan language.
4. So-called because of the socialist political leanings of many households with immigrant histories within this first crown of suburban cities.
5. The day in which Barcelona, and Catalonia, fell to Castilian and French troops has become the National Day or *Diada*, a national "tradition" begun in the late 19th century.
6. The county (or *comarca*) of Barcelona consists of the city and four other municipalities. There are seven comarcas in the Barcelona metropolitan region; a total of 41 in Catalonia.
7. The Eixample (or Extension) district is the most populated district in the city, with about 300,000 residents. A civil engineer interested in ideas of utopian socialism and egalitarianism, Cerda developed for this new extension of the city one of the first comprehensive city plans in history. The plan uses a grid street pattern superimposed over two large diagonal avenues, extra-large street blocks with rows of housing at the edges and inner courtyards of common open space on the inside, an urban railway system, and "chamfered" building corners that open up the built landscape at intersections.
8. Nel-lo (2002) attributes the re-emergence of the territorial imbalance criticism during these years also to anti-Francoists who saw a devolved cantonization as a good counter to the centralization of the Francoist provincial model.
9. In its intended coverage, the 1953 Plan exceeded the five municipalities within Barcelona County and thus the plan's name is somewhat of a misnomer. It's other name, *The Urbanization Plan for Barcelona and its Area of Influence*, is a truer reflection of the extent of its planning geography.
10. It was not until 1982, however, that the threat to democracy in Spain (in the form of *golpismo* or coup d'etat) appeared to have exhausted itself. As late as February 1981, a coup d'etat was attempted that would have turned out the elected national legislature, the *Cortes*.
11. There is no discipline or profession of urbanism (urban planning) in Spain per se. Rather, urbanists are a subset of professionals trained in architecture who tend more than pure architects to have a systematic view of a city and its interrelated parts.
12. Neighborhood associations were not legalized by the Franco regime until 1972 (C. Navales, labor activist, interview).
13. The role of civil society, energized over daily urban shortages and challenges, in proposing alternative urban plan-making and in connecting urban daily issues to broader political ones was also found in my research on Johannesburg urban policy during South Africa's transition from apartheid to democracy (Bollens 1999).
14. Much of the following material on labor activism and the 1974 strike is based on an interview with Mr Navales.
15. During this time of early land banking, developers linked to the regime were pushing through as much construction as possible as they anticipated potential political change ahead.
16. The societal shaping function of planning has been discussed by Etzioni (1968) and Alterman (2002), among others. My hypothesis about planning's power during uncertainty is supported by the contrasting lack of consensus over a later 1988 proposal for

the Metropolitan Territorial Plan of Barcelona, an effort that has not yet been approved as of 2004 (A. Serratosa, interview). By the late 1980s, Catalonian society had normalized to the extent that economic and social interests had become more institutionalized and turf-conscious. I thank Paul Lutzker (political consultant, interviews) for his insights on the relationship between planning and societal disruptions.

17. The strategy was initiated in late 1980 when mayor Serra appointed a five-member planning commission and subsequently created an Office of Urban Projects (Rowe 1999).

18. This was not the first time the city used an event to stimulate and foster substantial change in the urban fabric. The 1888 Universal Exposition fundamentally reshaped the Citadel Park and proximate waterfront; the 1929 International Exposition created new buildings and monumental statues, fountains and landscapes up the side of Montjuic Mountain. And, this strategy was used again more recently in developing the city's Besos area for the 2004 Forum of Universal Cultures.

19. It is more accurate to speak of a 10-year planning period because preparations for Olympic candidacy actually began in 1981. Certain urban actions were defined and some even implemented before acceptance of Barcelona's nomination in 1986 (Borja 2001).

20. Although the Socialist Party controlled both central and municipal governments during these years, the national government's objective of redistributing economic opportunity across Spanish regions would not have resulted, absent the Olympics, in the distribution of disproportionate resources to Catalonia. Differences between regional and municipal governments, meanwhile, were based on contrasting perceptions regarding the role and power of Barcelona city within the region.

21. ANCs have had varying levels of success. Besides the positive impacts of the four Olympic nodes, there has been stimulation of office development near the Terragona Street node, greater connectivity at the Meridiana node, and improvements along the Plaza Glories node (M. Herce, interview).

22. If the regional Generalitat had its way, the city may never have had this restructuring opportunity. Much less enthralled with the city, CiU nationalist politicians wanted to site the Games in the suburban city of Sant Cugat (40 kilometers from Barcelona).

23. Vila Olympica is not without its critics. The director of the project during the pre-Olympic planning stages (M. Herce, interview) recounts how the projected social and lower-income housing units for the post-Olympic period were changed to units for the wealthy and middle class.

24. This sense of patriotism was heightened by the fact that it was perceived as the completion of a dream for a Barcelona Olympics squashed in 1936 by the rise of Naziism and world political tension.

25. The Universal Forum of Cultures, from 9 May to 24 September 2004, was an international event to celebrate cultural diversity, sustainable development and a culture of peace. Organized by the city, the regional and central governments, together with the United Nations Educational, Scientific and Cultural Organization (UNESCO), it consisted of more than 40 international conferences and a range of special exhibits and artistic performances. It drew, according to official attendance figures, about 3.5 million people. It was not without criticism – of its public cost, exorbitance and vagueness of substantive goals, and ties to private property developers and advertisers.

26. This conclusion is not to suggest that Barcelona was acting in an international vacuum. Italian and French architectural conceptualizations influenced Barcelona design in the early 1980s, and the International Building Exhibition in Berlin in 1987 helped propagate ideas about recovering historic context and urban public space. There were international examples in other cities that preceded or were concurrent with Barcelona Olympic preplanning and planning stages, including Boston's

Quincy Market in the late 1970s, Baltimore Harborplace in the early 1980s, and the Docklands, London in the mid-1980s. Comparatively, Barcelona's strategic interventions have been massive, of long duration, and are ongoing.

27. Interviews took place November 2003 when CiU was still in control of the regional Generalitat.

28. It is also possible that Franco's repressive and strong-armed tactics forced upon the region a degree of inter-group stabilization during this immigration influx that may not have occurred without such repression. A Catalan population with greater political freedom may have more actively resisted such a demographic assault.

29. In late 2004, the Spanish Immigration Secretary estimated this legalization program could increase the legal immigrant total in Spain by as much as 1 million (reported in www.euroresidentes.com accessed 4/26/05).

30. As of July 2004, only one public square in all the city used security cameras – Plaza George Orwell in the Ribera district of the old town.

31. This remained true even when CiU was dislodged from regional government control in 2004. CiU received more votes than PSC but not an absolute majority. PSC constructed a coalition with two other parties to form the regional government. In the 2006 regional elections this pattern was repeated. The CiU party has benefited in post-Franco democracy from an electoral system that gives rural votes more weight than urban votes (Eugeni Madueno, editor, *La Vanguardia* newspaper, interview).

32. The CiU government created administrative units called *comarcas* to counter and bypass the four existing provincial governments in Catalonia, created by the central Spanish state and thus frowned upon by regional nationalists (Julio Ponce, lecturer of administrative law, University of Barcelona, interview). These *comarcas*, smaller than preexisting provinces, were an effective way to reinforce CiU's strategy of linking nationalism and small town localism.

33. One interviewee who worked in the CiU generalitat contrasts CiU's tactical approach to PSC's more technical orientation – "socialists do policies; CiU does politics" (Joan-Anton Sanchez, formerly, office of the presidency, Generalitat, interview).

34. Lefebvre 1979, 290; Brenner 2004.

35. Oriol Nel-lo (2001), who took over as the Generalitat's secretary of territorial policy in 2004, describes Catalonia today as a "city of cities" ("ciudad de ciudades"), an urban place of increased metropolitan integration of activity and settlement nodes.

36. There has never been an approved urban development plan or governance entity that encompasses the larger Barcelona urban region of over 160 cities. The General Metropolitan Plan of 1976, as well as the jurisdiction of the Metropolitan Corporation 1974–1987, encompassed the smaller 27-city Barcelona metropolitan area. The 1968 Provincial Plan covered the larger urban region but did not have regulatory power.

37. Strategic plans emphasize socio-economic factors and are thus different than spatial, or territorial, plans that focus on land use attributes of urbanization processes.

38. F. Moraga (interview) recalls that European linkages that existed at the time of the formulation of the Spanish Constitution were a positive influence, introducing a greater complexity into the Spanish system that ameliorated the dichotomous central versus regional government argument and created new obligations and the need to compromise on certain issues. Spain became a member of the European Community (Union) in 1986.

39. That internationalism and identity-based nationalism can coexist, even mutually support each other, goes against the fears of many globalization critiques that a pervasive homogenization is the inevitable result of increased economic globalization.

40. For example, in the late 1980s Barcelona city was actively engaged in setting up a 60-member transnational network of major European cities, called "Eurocities." And, in 1992, the Olympic Games put Barcelona on the world's front stage for 14 days and nights.

4 Sarajevo: misplacing the post-war city

Figure 4.1 Kovaci (Martyrs) cemetery, Sarajevo

Should it be further explained that such a fine and complicated totality like Sarajevo – in which the entire country of Bosnia and Herzegovina is reflected as in a mirror – must be fragile? Should it be especially mentioned how natural it is that such a totality attracts and enchants prisoners of an epic culture just as the interior of a marble attracts and enchants savages? The fundamental difference should be stressed by all means, however: an enchanted savage admires the center of a marble, but he will never break the glass to get to it because the savage is reverent; he knows that the spell and the enchantment that make it all worthwhile would then disappear. But a prisoner of an epic culture – a culture that plays its music on a single string and is almost entirely contained in it – stares at Sarajevo and circles around it, while the city eludes him as the marble's eye escapes the savage. But then the epic man shatters Sarajevo, for he has lost his reverence and his ability to enjoy enchantment, because of the illusory nature of his epic cultivation.

Dzevad Karahasan
Sarajevo, Exodus of a City (p. 16)

Dedicated to Elvir Kulin, who saw things no 18-year-old should.

"URBICIDE" AND DAYTON

The siege of Sarajevo by Bosnian Serb and Serbian militias that completely blockaded and encircled the city lasted 1,395 days (from May 2, 1992 to February 26, 1996), killed 11,000 civilians, including 1,600 children, and damaged or destroyed 60 percent of the city's buildings. Today, the political "solution" and the "peace" have an imposed feeling. The Dayton accord of 1995 institutionalized a de facto partition of Bosnia-Herzegovina. The autonomous Bosnian Serb entity of Republika Srpska created by Dayton comprises 49 percent of the country's territory and lies immediately to the east of the city, a reward for its ruthless fighting machine. The "peace" now in Sarajevo approximates what one finds when visiting a cemetery.

The story of Sarajevo illuminates the significant obstacles faced in governing group-based differences after calamitous hostilities. It points to both the tenuousness and importance of decisions made during transitional periods. During a period of great uncertainty, the international community was asked to delineate new political demarcations with momentous and durable ethnic import. The transition out of war was a fragile period when important decisions about new ethnic boundaries at a subnational level had to be made. These decisions established new political geographies to assure a degree of inter-group stability in the short and medium term. However, this redrawing of political space may create long-term conditions that handicap the emergence of a genuine multicultural democracy in the future. The "misplacing" of the city in the state's new political geography foregoes a major opportunity for Sarajevo to constitute a multicultural center in an otherwise fragmenting state.

Much like Barcelona, Sarajevo has experienced a period of societal transition in which fundamental parameters of governance and urban development have been scrutinized. Such transitional uncertainty had to contend with how contentious group identity conflicts should be managed, now and in the future. The Sarajevo experience exposes difficult moral and ethical decisions about whether the social composition of the pre-war city should be the goal of international programs and policies, or whether the new ethnically sorted city should be supported as a means toward inter-group stability. The lessened ethnic diversity of post-war Sarajevo and reduced likelihood that nationalism and group-based claims will obstruct urban policymaking may facilitate the city's redevelopment. Yet, this possibility puts international overseers in a difficult space – should the international community, in the name of urban stability and redevelopment, accept the relocation of Serbs and Croats who moved out of the city? Or, should international organizations continue with the morally acceptable stance of encouraging all possible returns of war-displaced people and thus sustain hope for a multicultural Sarajevo again, even if this might hinder urban redevelopment?

Bosnian urban policymakers and the international community have faced fundamental practical and ethical challenges in reconstructing Sarajevo physically, economically, and socially. Three transitional processes in Bosnia and Sarajevo are being undertaken concurrently: post-war reconstruction, democratization, and movement to a free market economic system (Pejanovic 2002). Sarajevo and other cities in Bosnia provide important potential foundations upon which to start to rebuild a multiculturalism killed by the war, and from which to develop viable democratic governance and economic interdependences that could normalize the country. Yet, the Dayton redrawing of political space in Bosnia into two ethnic entities, and on the Muslim-Croat Federation side, into ten cantons, leaves little room for cities like Sarajevo to act as societal transformation agents in the future. Ironically, while the war damaged but did not eradicate the multicultural spirit of Sarajevo, the political boundaries drawn to stop the war may over time slowly deplete any ability of Bosnia and international policymakers to resurrect the country's, and the city's, integrative and tolerant capacity. As Lovrenovic (2001: 207–208) describes, the Dayton agreement "saved Bosnia from further war but not from exhausting political contradictions and tensions." Emergent was a "new, criminally-based, ethnically-structured Bosnia" supposedly created in the interests of the three nationalist groups, Bosniaks (Muslims), Serbs, and Croats (Lovrenovic 2001: 208).[1] Burg and Shoup (1999: 415) assert that the Dayton accord "resembled an armistice between warring states more than a social compact for the rebuilding of Bosnia."

Bosnia imploded into war twelve years after the death of Marshall Josip Tito, who ruled Yugoslavia from 1943 until 1980. Tito combined authoritarianism with elements of a market socialism, seeking to carve out a third way midst Cold War dualities. Through appeals to "unity and brotherhood" and outright repression of ethnic identity, Tito was able to make Yugoslavia into a seeming model of multi-ethnic coexistence for decades. After his death in 1980, Yugoslavia went through an uncertain transition period, both politically and economically.[2] The future of Yugoslavia, a federation of the republics of Serbia, Croatia, Bosnia-Herzegovina, Slovenia, Montenegro, and Macedonia (see Figure 4.2), had become a subject of intense debate and argument. Serbia pushed in the 1980s for changes that would put more power in the hands of the federal, central government, while Croatia and Slovenia by 1986 were advocating a loose confederation among the republics (Burg and Shoup 1999). When the electorate in most of the republics opted politically to put nationalist parties into office in 1990, this started a period of "permanent crisis" (Burg and Shoup 1999: 69). Croatia and Slovenia began to call for their sovereignty, a process that led to their secession

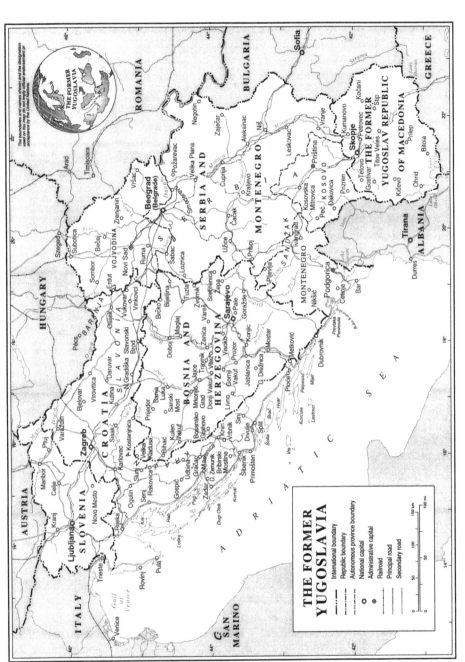

Figure 4.2 Former Yugoslavia and its Republics

from Yugoslavia in June 1991 and war in Croatia. Slobodan Milosevic, meanwhile, gained power in Serbia by adroitly exploiting Serbs' uncertainties and fears regarding their economic well-being and cultural cohesiveness.

As Yugoslavia began to unravel as a state, Bosnian politics and society became polarized along ethnic lines as three main nationalist political parties formed. More mixed ethnically than the other Yugoslav republics, it was only a matter of time before the inter-ethnic splintering that was tearing Yugoslavia apart set its sights on the Bosnian republic and its 4.3 million residents.[3] Bosnian Serb interests in remaining within Yugoslavia became aligned with Serbian leader Slobodan Milosevic's interest in creating a greater Serbia through territorial extensions. Bosniak Muslim political leaders felt their options to remain part of Yugoslavia narrowing as Serb or Croat nationalists claimed substantial parts of the republic.

The city of Sarajevo, meanwhile, stood as a potential bulwark against the ethnic fragmentation of Bosnia. Not only were the three ethnic groups in close contact with each other in the urban setting, but there also was a lesser propensity among residents to identify themselves ethnically. Instead, there was a greater propensity to adopt a more multi-ethnic, "Yugoslav" identity.[4] Nonetheless, a deadly dynamic in Bosnia had begun which would in the end consume Sarajevo. The Yugoslav People's Army (JNA), under Serbian control, began to transfer arms to Bosnian Serbs, a Croatian paramilitary force began arming itself in the Herzegovina part of Bosnia, and the Bosnian Muslim Green Berets were organized in the fall of 1991. After a March 1992 referendum in which 62 percent of Bosnian voters opted for independence (with Bosnian Serbs abstaining), the Serbs on the next day erected barricades in Sarajevo and called for the end of debate about Bosnian independence. On April 6, the EU recognized Bosnian independence, with the U.S. following the next day. On April 6, fighting began in Sarajevo and on April 8 the Yugoslav army entered the fight. The horrific Bosnian war was to go on for over four years.

Sarajevo, Bosnia-Herzegovina has always been a special, transcendent place. Although a political boundary created through brutal war is now within its urban sphere, Sarajevo is not now a divided city with partitions and checkpoints. Nevertheless, it is a traumatized one that must deal with issues of group difference and identity. Like the cities of Johannesburg in South Africa and Beirut in Lebanon, how this city is physically reconstructed and socially reconstituted will play a key role in whether the larger region and nation will bear witness to a sustainable and productive peace or an unstable and prolonged period of uncertainty that is not war but is not peace either.

Sarajevo is now a different city. In 1991, it was a mixed ethnicity city of over 500,000 people, was composed of ten urban boroughs, and was about 50 percent Bosnian Muslims. In 2002, it was an ethnically sorted city of about 300,000 people within four urban boroughs and had an over 80 percent Bosnian Muslim majority.[5] A substantial part of the ethnic demographic shift is due to in-migration of Muslims during the war and out-migration of Serbs since the war. An important underlying influence is the demarcation by Dayton of new ethnic "entity" boundaries that placed over 35 percent of the land in the urban area outside of the official "city" and within Republika Srpska. Although many Bosnian Serbs stayed in the city during the war in defense of the bombarded concept of multi-ethnicity, substantial numbers fled after Dayton, fearing retaliation and possibly encouraged by Bosnian Serb leaders. During the war, Muslim refugees from ethnically cleansed Eastern Bosnia (now Republika Srpska) inhabited shelled and burned-out flats in the city's worst war-torn neighborhoods, and many have remained in Sarajevo.

For the young state of Bosnia-Herzegovina overall, its economic and political condition ten years after the war locates it somewhere between a "post-conflict" situation in which it must deal with the aftermath of war and a "transition" agenda in which it would concentrate on fuller individual freedom, economic well-being, and institutional self-sustainability (Commission of the European Communities 2003a; World Bank Group 2004). The war decimated its physical capital and contracted its economy by over 90 percent (World Bank Group 2004). Its economic situation seven years after war remained precarious – estimated gross national product in 2002 was less than half the pre-war level, the average net salary in BiH in August 2002 was about 3,000 euros per year,[6] and the official unemployment rate in 2002 was about 40 percent (although the World Bank estimates the true rate to be about half that number, due to employment in the unofficial, "grey" economy) [Commission of the European Communities 2003a]. In the period 1991 to 2002, the European Commission had sent approximately 2.4 billion euros in assistance to BiH; from 1996 through 2001, European Union member states contributed an additional 1.2 billion euros. International aid has declined since 2001, and is now linked to BiH's progress toward possible future European Union membership.[7]

Snapshots from a war city[8]

"Urbicide" – the attempted killing of a city – was part of a secret plan called RAM designed in Belgrade (Bublin 1999). In smaller towns of BiH having strategic value to Serbia's territorial ambitions, local territorial forces of Bosnian Serbs were to provoke an incident; then paramilitary groups from Serbia would make raids and "cleanse" the territory of non-Serbian residents. The Army aligned with Serbia would then roll in its armor and create a buffer zone around the conquered territory. In larger urban centers such as Sarajevo, neutralization actions included artillery bombardment and sniper fire, shelling of non-Serbian residential areas, blockades of traffic, propaganda campaigns, surprise attacks, and the destruction of vital urban structures. The result of these military tactics was the besieging and holding hostage of the city for almost four years.The lines of confrontation in the Sarajevo urban area between Serb and Bosnian government forces are shown in Figure 4.3.

Sarajevo, like Jerusalem since 1948, is now a frontier city between opposing political territories. The boundaries between the Dayton-created Muslim-Croat

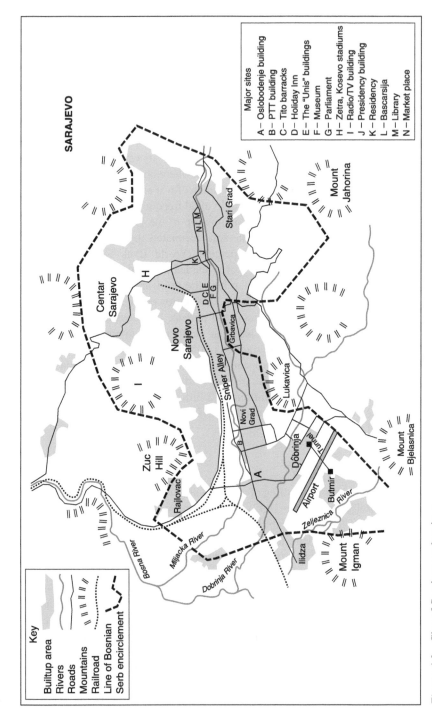

SARAJEVO

Major sites

A – Oslobodenje building
B – PTT building
C – Tito barracks
D – Holiday Inn
E – The "Unis" buildings
F – Museum
G – Parliament
H – Zetra, Kosevo stadiums
I – Radio/TV building
J – Presidency building
K – Residency
L – Bascarsija
M – Library
N – Market place

Key

Builtup area
Rivers
Roads
Mountains
Railroad
Line of Bosnian
Serb encirclement

Figure 4.3 City of Sarajevo under siege

Federation and Republika Srpska entities (in international speak, Inter-Entity Boundary Lines or IEBLs) are within the southeastern part of the urban area and contain no checkpoints and no visible signs of differentiation, except for the Cyrillic written alphabet present in the Bosnian Serb entity. Indeed, in an affront to the logic of aggression, by 1999 there was new road building to connect the two entities and the creation of "universal license plates" to facilitate automobile travel from one part to the other. Was not land and its control what the heinous 1992–1995 Bosnia war all about? Another crossing nearby, although also without a checkpoint, reveals who is sponsoring this reconnection. Electronic monitoring and transportation vehicles of NATO's Stabilization Force (SFOR) were obvious and busy, contradicting the otherwise intended normalcy of the unmarked crossing. Do those now seeking connection believe that which was torn apart by war can be normalized four short years after slaughter? And is the absence of an armed border four years after war a good sign or a bad sign? Should its absence be treated as a sign of mutual tolerance or an indicator of an artificially imposed peace? By 2003, the argument that functional and joint interests across the Dayton-created entities may help reestablish cross-ethnic links was gaining some credence within the international community. Yet, can joint economic development initiatives help put back together what was lost in the war and move a society forward midst contorted political space created out of war?

> *During the day there is progress due to the international community. But at night, it deconstructs in people's minds.*
>
> Activist and writer (confidential)
Interview 11/19/03

Sarajevo is the scene of a crime, a rape, and devastation. It is an affront to humanity and rationality. Blown off limbs, punctured heads, humiliation, playgrounds, and soccer fields turned into cemeteries because these were areas that hillside snipers couldn't see, the famous ice rink from the 1984 Winter Olympics shelled and afire, building after building shattered and burnt. Zeljko Komsic, president of Sarajevo's Municipal Council in 1999, states that, "Divided city is a difficult term for us; we are the cruel victims of history." Political graffiti is surprisingly limited in Sarajevo; when it is present it frequently asserts SDA, the Muslim's main Party of Democratic Action. The town of Pale, the headquarters of Bosnian Serb war leader Radovan Karadzic, stands ten miles to the east of Sarajevo in the strange new political geography of Bosnia-Herzegovina. New canton (county) borders have been drawn on Sarajevo's side to accommodate the new jigsaw-like boundaries.

In the aftermath of war, there are heroes that provide hope. Jovan Divjak is a square-jawed, grey-haired man, solidly built, with a face etched in war. He is a retired general in the Bosnian Army and was in command of the forces defending the city of Sarajevo against Bosnian Serb militias and Serb paramilitaries. He is a believer in a multi-ethnic Bosnia and advocates a return to the more ethnically mixed Sarajevo of pre-1992. Most amazing about the man is this startling fact – he is a Bosnian Serb. We hear so often of those who play the ethnic card and manipulate identity to divide and conquer. Divjak, in contrast, is a living and vital example of someone who embodies the spirit of inter-ethnic tolerance. Divjak is steadfast and determined in leading a tour of the city in 1999 for political and community leaders from Belfast, Beirut, Jerusalem, Nicosia, and Barcelona. He spends time to clearly describe the logistics of the war and to show us the hubris left from the siege in the form of shot-out buildings and overflowing cemeteries. At the beginning of the siege in 1992, the aggressor forces surrounded Sarajevo with 260 tanks, 120 mortars, and vast numbers of rocket launchers, anti-aircraft machine guns, snipers, and machine guns. In contrast, the city's defenders were left with minimal arms for protection. Every day the city was hit by some 4,000 shells; among the targets were hospitals, schools, mosques, churches, synagogues, maternity hospitals, libraries, museums, open-air and sheltered food markets, and any place where people stood in line for limited supplies of food, bread, and water.

Divjak is firm and unemotional in his recall. He is accompanied by a female translator who has frequently been by his side as he has described the war over the past three years to all those who are interested. At a military cemetery that used to be a playground where he took his grandchildren, Divjak breaks from our group to hug and console a mother remembering her son. As we walk through the city, many residents on the streets embrace him. They want to touch this man, to thank him. It is clear that he represents a valuable part of them that they struggle to preserve as they battle the emotional exhaustion and pain of ethnic hatred. The retired general is heavy-hearted about the future of Bosnia. He dismisses the sustainability of the new internal boundaries of Bosnia-Herzegovina negotiated in Dayton. The 49 percent of Bosnian land that is now the Bosnian Serb autonomous zone is indicative of a victorious campaign of war and ethnic cleansing. Divjak is concerned with the translation of this war and its meaning to today's youth; he asks, "Are parents capable of excluding children from these manipulative mechanisms we play?"

We grew up during the war, but we don't know when.
Jasmina Resulovic (interview)

Jasmina Resulovic and Arnan Velic are in their middle to late 20s. Jasmina is a short, round-faced, bespectacled young woman with contemporary flair. Arnan is a lean man, dark-featured and handsome. Jasmina says, "I guess by our parents' birth we are Muslim." Both are architecture students at University of Sarajevo. They both stayed in the city during the four years of war, Arnan fighting in the Bosnian Army for five months and Jasmina mired with her parents and other family in a high-rise flat near the front lines of hand-to-hand fighting. During the war, they attended abbreviated "war school" in lieu of high school. A few years after the war, they and a few other students led tours of the historic and war-affected city. I spent one and a half days with Jasmina and Arnan as they guided me around the city and I queried them about the "indescribable." They were

15 and 14 years old when the war started. Four years after the war, in 1999, they are now kids with the wisdom, sadness, and perspective of adults. We stand for many quiet moments at the Vraca Monument on the hills overlooking new town Sarajevo. It is a remembrance of the power of brotherhood in the communist partisans' successful crusade against fascism in World War II. Arnan finally speaks: "It's unreal, it is like that war never took place; we learned nothing." Their long stares at this monument likely owes to this cruel fact; it was from within that monument celebrating inter-ethnic unity that the heavy guns of the Serb militias were first fired from the hills at the Grbavica neighborhood of the city below.

It is a different life now. "Everyone was equal during the war," says Jasmina, "now money follows money." And in a cruel irony, Arnan painfully describes how "we are looked down on now by those who left during the war and now are back with new cars and clothes. Sometimes I just want to strangle them." Jasmina's mother is a teacher and now makes about one-third of her pre-war wages. Her underemployed father now makes less than her mother does. When Jasmina was able to work as a translator for seven days, she was embarrassed to take the wages back to her household because it was as much as her mother makes in one month.

One of the few ways to leave the city of Sarajevo during the siege was through the Dobrinja-Butmir tunnel. This was a passage two feet wide, less than four feet high, and running for about 830 feet under the Sarajevo airport, formally controlled by the U.N. for most of the war but de facto controlled by Serbian militia snipers. Although the tunnel was built for military purposes, intended solely for getting Bosnian troops in and out of the city, it became a passage for people dragging bags of potatoes and eggs into the city. Nedzad Brankovic, the army engineer who built the tunnel, responds this way to the idea of turning the tunnel into a war monument: "It should not be a monument to war, but to the indifference from Europe and the international community to the atrocity of Sarajevo as we approach the millennium."

The governance of Dayton

Bosnia-Herzegovina collapsed under the strains of war soon after its creation as an independent country in 1992.[9] International involvement continues in all aspects of the peace-building process. The Office of the High Representative, under the authority of the United Nations Security Council, has repeatedly made binding decisions on recalcitrant local officials. A NATO-led peacekeeping force (the Stabilization Force, or SFOR), originally 60,000 strong, was replaced in late 2004 by a European Union-directed peacekeeping force of 7,000 troops. Bosnia today, with strong United Nations oversight, is a loose confederation of two "entities" whose boundaries were created largely through war and ethnic cleansing. After three years and nine months of fighting in Bosnia that killed over 200,000 people and expelled over one-half of Bosnia's 4.3 million people through ethnic cleansing, the Dayton Accord (General Framework Agreement for Peace in Bosnia and Herzegovina) was signed in 1995. The accord provided for the continuity of Bosnia and Herzegovina as a state, creating two constituent entities – the Federation of Bosnia and Herzegovina (with a post-war Bosniak [Muslim] – and Croat majority) on 51 percent of the land, and Republika Srpska (mostly populated by Bosnian Serbs) on 49 percent of the land. The estimated Bosnian population in 2001 was about 3.36 million (approximately 1 million less than before the war); about 2.3 million people are estimated to live in the Federation; about 1.07 million in Republika Srpska (United Nations Development Programme 2002b).[10] The accord institutionalizes a de facto partition of Bosnia-Herzegovina based on the final locations of the warring parties. The state and Federation capital is Sarajevo; the capital of Republika Srpska is the city of Banja Luka (see Figure 2.3, page 27).

Central institutions for the state are weak.[11] The Parliamentary Assembly has two chambers – the House of Peoples and the House of Representatives. The former has 15 members, five from each sub-population group (Croat, Bosniak Muslim, and Serb), nominated from the Bosniak-Croat Federation and Republika Srpska legislative chambers. The House of Representatives has 42 members, 28 elected from the Bosniak-Croat Federation and 14 from Republika Srpska. A majority of those present in both chambers is the basic requirement for decisions in the Parliamentary Assembly, but each constituent people has the right to declare any prospective decision "destructive of a vital interest," in which case concurrent majorities within each of the ethnic group representatives is needed. There is also a three-member Presidency composed of a directly elected leader from each ethnic group. A majority decision is possible; however, decisions "destructive of a vital interest" can be referred to either the Bosniak or Croat members of the Federation House of Peoples, or to the Republika Srpska Assembly, where a vote of two-thirds of the relevant group renders the decision null and void. The Presidency appoints the government, or Council of Ministers, of which no more than two-thirds of Ministers can come from the Federation, and deputy ministers may not be of the same constituent people as the minister. All the central mechanisms for the state require broad agreement and consensus to function. Given the existing animosity and absence of trust, such consensus has not existed.

For all intents and purposes, the Bosnia state is de facto a two-state structure loosely and minimally held together by a dysfunctional central government apparatus. The Bosnian state constitution[12] devolves most powers to the entity level and below in order to provide as much ethnic self-rule as possible (Jokay 2001). Given the continued animosity of the three sides, a strong centralized state would have been unrealistic. The two entities have their own military forces (with international monitoring), while the state cannot have an army. All powers not the exclusive domain of the state[13] are possessed by the entities. Each entity has its own independent budget, and the state is reliant on budgetary contributions of the two entities. The Federation and Republika Srpska are responsible for police protection, environmental policy, social policy, refugees, reconstruction, justice, and taxation. Each constituent entity has its own legislative and administrative structures. The Bosniak-Croat Federation has a mixed system with a president and a parliament that must approve the president's choice of a prime minister. Republika Srpska has a parliament-president system.

In the Federation, ten cantons were created as a layer of government between the entity and local governments; no such layer exists in Republika Srpska.[14] Eight of these cantons have clear ethnic majorities (five Muslim majorities, including Sarajevo Canton, and three Croat majorities); two cantons are mixed Muslim-Croat (see Figure 4.4). Similar in motivation to the establishment of ethnically-based entities, Dayton structured and empowered the Federation's cantonal governments in order to facilitate ethnic self-rule. This also means that cantonal organization reinforces in most of the Federation an ethnic compartmentalization. Cantons have all those powers not explicitly granted to the Federation, and are particularly influential in their authority over local government, land use

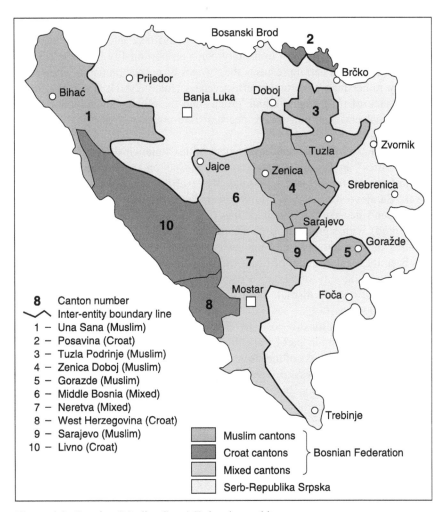

Figure 4.4   Bosnian (Muslim-Croat) Federation and its cantons

planning, local business development, and local economic development (Jokay 2001). The cantons, through their taxation and regulatory powers, can organize the Federation's municipalities and cities as they see fit; such is their influence that Jokay (2001) states that in the Federation there are basically ten systems of local government with significant variability across cantons.

Local governments in Bosnia are municipalities and cities. In 1991, the state had 109 municipalities. Ten years later, that number had increased to 145 due to the creation of numerous new municipalities, many along the Inter-Entity Boundary Line. In numerous cases, including Sarajevo, there has been the split of pre-war municipalities into separate ethnic municipalities. Power-sharing requirements for local governments in the Federation and Republika Srpska have sought to allocate municipal council seats proportionate to 1991 ethnic distribution, yet post-war local demographics call into question the sustainability and value of such standards.[15] Further, appointments to administrative positions and hiring in public employment should take into consideration the registered ethnic distribution in the city. Such proportionality in public employment constitutes in certain respects the "wages of peace," aimed at inter-group stability rather than government effectiveness (Jokay 2001). Given the fundamental lack of trust among ethnic groups and the ethnic fragmentation of the professional class, administrative decisions are commonly distorted by ethnic considerations.

The primary form of local government is the "municipality;" there are only three "cities" in Bosnia – Sarajevo and Mostar on the Federation side and Banja Luka on the RS side – and the city level on both sides is institutionally weak. Sarajevo has a "city" government, but within the city borders on the Federation side exist four municipalities (Stari grad, Centar, Novo Sarajevo, and Novi grad) with significant local autonomy. There are also five municipality suburbs outside Sarajevo city on the Federation side. On the RS side, there has been the establishment of East ("Serb") Sarajevo city just across the IEBL. What power the city of Sarajevo does have comes from its control over the business and historic district, and its ownership of some residential and commercial properties.

The political realities of post-war Bosnia and the continuing antagonisms of the three major ethnic parties are dispiriting. Muhidin Hamamdžić (Sarajevo mayor, interview) speaks of the strong fragmenting forces present in the country: "The goal of the war was to separate the country. We haven't finished the war. We stopped the killing, but the desire to move Bosnia toward the goal of separation is still present." For most of the ten years after the war, nationalist parties for each of the three sides have been able to dominate politics through patronage and the propagation of ethnic insecurity. An international election observation mission on the 2004 municipal elections documented the "continuing primary of ethnicity as the central underlying issue in politics in BiH and in this electoral campaign" (Council of Europe 2004). In those elections, the Bosniak (Muslim) nationalist Party for Democratic Action (SDA), nationalist Croat Democratic Union (HDZ), and nationalist Serb Democratic Party (SDS) won some 100 of the 122 municipalities. This domination of local councils by ethnic parties came after the previous round of local elections, in 2000, had presented some hope that nationalist voting was on the decline (Drazenovic 2000). Another troubling aspect of the 2004 local elections was the low turnout of registered voters (45.5 percent), the lowest participation in any election since the Dayton Accord. Particularly disinterested were individuals 30 years and younger, of whom about 10 percent voted (Human Rights House Foundation 2004).

At state and entity levels also, nationalist parties continue to enjoy widespread support. The division of Bosnia into separate ethnic voting blocs has remained remarkably stable if one compares 1990 to 2000 (European Stability Initiative 2001). There has been some nationalist fragmentation (particularly among Bosniak and Serb segments) and an increase in support for non-nationalist parties;[16] yet, "the nationalist parties do not show any signs of disappearing. They are likely to remain a feature of the political landscape ... for the foreseeable future" (European Stability Initiative 2001). Finally, in the Federation, the ten cantonal elections have tended to reflect ethnic population distributions and the fact that eight of the ten cantons have clear ethnic-nationalist majorities.

Morning Glory Sarajevo

Ferida Durakovic is a poet, born in 1957 in the Bosnian village of Olovo, approximately 25 miles north of Sarajevo. Many of the poems she wrote before the war focused on Sarajevo; in the 1980s it was not a feeling of bad things coming, but she did write about divisions within and between people. She describes the city as a morning glory, a flower that is lazy and opens when the sun comes. Sarajevans were content, living slowly and enjoying life, with the cross-cultural history and traditions of the city. Yet, the city was overwhelmed by history in the end.

She is a self-described secular Muslim – "I call myself a Bosniak now – a political decision – but I have more important things to do than think about ethnicity and religion." She cannot make sense of what happened: "I saw too many evil things; even today I don't feel comfortable living here." During the siege in 1993, she once described to a visitor that she never felt life was so wonderful as then. Amidst hell and deprived of all of life's material things, "to wake up and see blue skies and see that you are alive and meet your loved ones and to know that they are alive – this feeling is not to be described." She continues, "When you find yourself at the edge of something – an existential moment – it's the edge between two worlds. You look to the other side, the abyss, and then turn around and see the other side and see that it is so wonderful."

Her grandmother lived through the Bosnian war, dying in 1995 at the age of 85. It was her third war, worse than the World Wars because in this one neighbors were killing each other. Her mother, a small girl during WWII, lived through two wars. Ferida has now lived through just one. By this logic, she concludes and hopes, her 8-year-old daughter will live only through peace.

THE REDRAWING OF POLITICAL SPACE

The City and Canton of Sarajevo today is of markedly less ethnic diversity than before the war. For the Canton of Sarajevo, about 45,000 Serbs lived within its borders in 2002, compared to about 139,000 Serbs who resided in the greater urban area of Sarajevo before the war.[17] This means that substantial numbers of Bosnian Serbs are no longer considered residents of the "official" post-war Sarajevo city. Part of this decrease in Serb population in Sarajevo is due to Dayton's redrawing of political lines and partly to an overall lesser presence of Serbs in the greater urban area, irrespective of Dayton boundaries. In the city itself, today's population is about 80 percent Bosniak Muslim and only about 12 percent Bosnian Serb (compared to approximately 50 percent and 30 percent, respectively, before the war).[18] A look at the data for one of the districts within the City, the Centar municipality, displays a microcosm of city-wide trends. In this administrative and business center of the city, the estimated number of Bosnian Serb residents in 2003 was almost 10,000 less than in 1991 and Bosniak Muslim residents 12,000 more than in 1991. Having a pre-war population of 79,000, it is estimated that during the war about 7,000 residents were killed or wounded, about 27,000 residents abandoned the district, and over 60 percent of housing was destroyed or seriously damaged. The 2003 population was estimated to be 67,000 residents (Centar Municipality 2003).

The "city" in societal reconstruction

Sarajevo, a target during the war, was for Bosniak and Bosnian Serb political leaders a prize to hang onto after the war, and for the international community, an ideal of multiculturalism to uphold. Multiple pressures acted at cross-purposes, in the end producing new city and subnational boundaries and a resident population of strong Muslim majority. Before the war, the city of Sarajevo contained ten municipalities and a far wider geographic area (including the municipality of Pale to the east) than it does today. The city boundaries encompassed its natural economic region in the then-Yugoslav Republic of Bosnia-Herzegovina, unencumbered by today's cantonal or entity borders[19] (Morris Power, Sarajevo Economic Region Development Agency, interview). The blending of Turkish and Austro-Hungarian history, culture, and architecture, of Orient and Occident, produced a special multicultural quality of Sarajevo. Inter-group differences were tolerated, even to the point of lacking salience. One interviewee[20] recalls, "It was completely unknown to me people's religions during communism. With the regime and in daily life, everyone was equal. To have talked about religion or ethnicity would have been like fiction – why talk about that? We weren't multicultural, we were one nation." In terms of residential location and activity patterns, there was both mixing and separation. The city had a mixed ethnic population of approximately 50 percent Muslims and 50 percent Bosnian Serbs, Bosnian Croats, "Yugoslavs," and "other" in 1991. Karahasan (1994) likens the *mahalas* (or neighborhood districts) of the city to rays spread out from the common focal center district of Charshiya. The Muslim mahala of Vratnik existed on one side, the Catholic Latinluk on another side, the Eastern Orthodox mahala of Tashlihan on a third side, and on the fourth side the Jewish mahala of Byelave. Before the war, upon leaving the center of the city, "all Sarajevans retreat from human universality into the particularity of their cultures" (Karahasan 1994). Religious neighborhoods had developed into a pattern of "tigers' spots" – together but separated (Said Jamaković, director, Sarajevo Canton Institute for Development Planning, interview).

Lovrenovic (2001: 209) speaks of the essence of "composite integration" at the core of Sarajevo's cultural identity, the parallel existence of three separate traditions. He distinguishes between the sphere of "high culture" (religion and civilization) marked by isolation between the three cultures and the sphere of "folk culture" (cultural traditions and practices that "live in the shadow of the towers and their clocks") which was where the three groups blended and integrated (Lovrenovic 2001: 222). In the 1980s, in the years leading up to societal breakage and transition, Lovrenovic observed a new affirmation of ethnic and religious traditions, linked at that time not to fixed ethnic ideologies and boundaries but to an integral cultural ambience of multiple perspectives and traditions. In a psychiatric analysis of Bosnia life before the war, Heine (1999: 14) focuses on the concept of *merhamet* as a central cultural value in its multicultural life. Of Muslim origin, *merhamet* emphasizes an empathy and kindness to others and a collective, inclusive identity amidst differences – an ability to get along with others in a humanistic way.

Remarkably, during the siege from May 2, 1992 to February 26, 1996, the urban area held together as a multicultural entity, albeit with ethnic groups spatially sorted due to the realities of urban war and the control of urban territory by warring parties. The siege formed a front line that divided the city's inner neighborhoods from its outer ones; Muslims fled inward from areas such as Ilidza, Ilijas, Hadzici, Grbavica, and Vogosca, leaving these neighborhoods almost 100 percent Serb. At the same time, many, though not all, Serbs in the inner sections of Sarajevo left for Serb-controlled territory in the urban area (Lippman 2000). During the war and before Dayton, there were several diplomatic efforts to preserve the city's multicultural quality. In March 1993, the Vance-Owen Plan would have divided the state ethnically through the creation of ten ethnically structured provinces. Although it would have placed Sarajevo within a Muslim-majority province, it also stated that all three groups would participate on an equal basis in the governing of Sarajevo province (Burg and Shoup 1999). Later that year, in August, an "Owen-Stoltenberg Proposal for Partition" of Bosnia-Herzegovina separated the state into three ethnic zones, but proposed for Sarajevo a demilitarized zone that would be administered by the United Nations or European Community. In the "Washington Agreement" of March 1994 and in a subsequent July 1994 "Contact Group" proposal, Sarajevo was again suggested as a United Nations administered district. By structuring the governance of Sarajevo to de-ethnicize its management, these proposals aspired to hold all three groups physically inside the city's administrative boundaries.

The Dayton Accord of December 1995 set off processes that unraveled efforts to create Sarajevo as a multicultural space within a fracturing state. Amidst increased NATO military actions against Bosnian Serb locations, Serbian leader Melosevic conceded the city and portions of surrounding hills, transferring to Muslim control those territories that had been key to Muslim-Serb fighting in Sarajevo and which had been under Serb control since the start of the siege (Burg and Shoup 1999). In this "reunification of Sarajevo," there would be the transfer over a three-month period of the districts and suburbs of Grbavica, Ilidza, Hadzici, Vogosca, and Ilijas – home to about 60,000 Bosnian Serbs – to the Federation (see Figure 4.5). Although able to stay in the urban area throughout the war, this planned transfer awakened fear of intimidation and retribution on the part of Sarajevo Serbs. In addition, there is evidence that Serb military and paramilitaries, and RS officials, sent forceful messages that Serbs should leave the area to increase the concentration of Serb population elsewhere, such as in the municipalities of Brcko, Srebrenica, and Bratunac (Kumar 1997; Lippman 2000). In efforts to maintain Serbs in a "reunified" Sarajevo, NATO Stabilization Force (SFOR) personnel were used to provide security for these to-be-transferred Serb areas. Despite these international efforts, and whether by choice or by force, what resulted was a mass exodus in early 1996 of some 62,000 Sarajevo Serbs from inside what would be the Dayton borders of Sarajevo city and its suburbs within the Federation (Internal Displacement Monitoring Centre 1996).[21] The result would be an increasingly mono-ethnic city. Ironically, the Dayton peace process had catalyzed an ethic purification of Sarajevo; as stated by Kumar (1997: 114), "Sarajevo, which had so proudly resisted ethnic divide during the war and occupation, was being driven to it by reunification under the peace agreement." Dayton also did not create an international administrative zone for Sarajevo, but instead incorporated most of the city's population within Federation territory, and in addition accepted its eventual location within a strong Muslim majority canton. The submersion of this once multicultural city within an ethnically prescribed canton led then-mayor of Sarajevo, Tarik Kupusovic, to resign, forecasting that such a decision "has pronounced a death sentence on Sarajevo as an urban environment" (reported in Kumar 1997: 115).

The ethnic homogenization of Sarajevo occurred despite heightened awareness and aspirations by the international community of the significance of the urban area to state peace-building. Just before the large Serb out-migration from the city, "The Rome Statement," signed by Bosniak, Croat, and Serb leaders, February 1996, asserted that Sarajevo "will be a united city" and appealed for all to stay in the city as a means toward "reconciliation and peaceful living together." Eight months later, in October, an agreement under the auspices of the U.N. Office of the High Representative[22] accepted the creation of what would be a Bosniak-majority Sarajevo Canton in the Federation, composed of nine municipalities including Sarajevo. However, it also sought to create, albeit unsuccessfully, a "State District" within Sarajevo Canton, consisting of government institutions, that would not be subject to the jurisdiction of either entity, but rather governed by Bosnia State institutions directly. The agreement sought ethnic

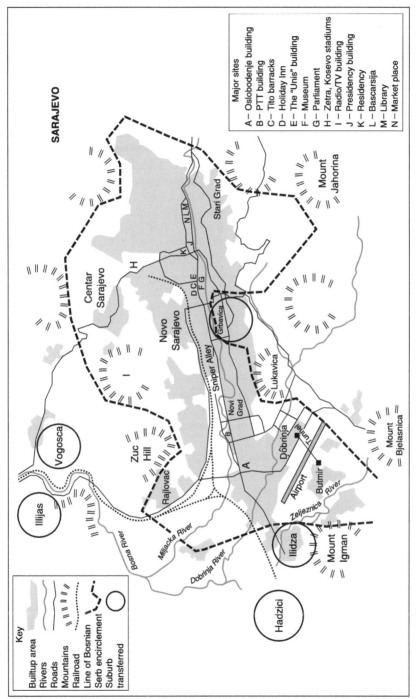

Figure 4.5   Sarajevo's suburbs transferred in post-Dayton transition

power sharing in Sarajevo City: at least 20 percent of City Council seats were each guaranteed to go to Bosniaks, Croats, and to "Others." In March 1997, the city constitution was amended to assure this multi-ethnicity on the 28-member City Council.

Over two years after the Dayton agreement, in early 1998, the international community expressed continued concern in the "Sarajevo Declaration" and established the tangible goal of enabling the return of at least 20,000 "minority" (i.e. Serbian and Croatian) pre-war residents to Sarajevo Canton in 1998.[23] It spoke of Sarajevo as a "model of co-existence and tolerance for the rest of the country."[24] A report by International Crisis Group (1998a, 1) published earlier that month stated that success in Sarajevo would "portend the success of minority returns in general and Bosnia's ability to defy the goals of ethnic cleansing." Despite the 20,000 minority return goal, official estimates were that 9,400 minority returnees came back to Sarajevo in 1998; of those, only 4,400 returnees registered and thus documented their physical relocation (U.N. High Commissioner for Refugees [UNHCR]).

Sarajevo for beginners

Ozren Kebo is editor-in-chief of a Sarajevo newspaper, *Start*. He says, "I am confused when I think about where Bosnian society is today. It seems at this moment that we have at least three societies. Sarajevo was a very cosmopolitan, modern city with strong energy. This creative energy remains today but it is no longer a modern city. In all aspects of life, it is a wounded city." The book, *Sarajevo for Beginners*, is a compilation of newspaper columns and published materials that Ozren wrote during the siege. "It is about people in war and our ordinary intentions and goals." Although Sarajevo today is now majority Bosniak Muslim, it "has a normal atmosphere and you don't have to speak about 10–15 years from now, it is that way now." The problem, he believes, is not Sarajevo but the rest of Bosnia. Regarding the Dayton agreement, "they made a terrible compromise with fascistic forces; would you give Saddam Hussein 49 percent of land? Dayton created an impossible state." He calls himself a "moderate optimist," but he also is "afraid this could be a brave position." Regarding future reconciliation of Muslims, Serbs, and Croats, Ozren describes an enigma: "Conversations can be normal and very pleasant with Serbian and Croat counterparts, but between political leaders, or what counterparts might say when they are not with me, oh that is the problem."

Misplacing the city

They could kill you at any moment at any place. There was no way to confront them. You had to live a normal life, which was the only and strong way to oppose them. You couldn't control the siege; if you are a loser, such as in Sarajevo, Ok, you change your tactic.

Writer, member of nongovernmental organization
Confidential Interview (11/19/03)

The endeavor to reconstruct the political geography of Sarajevo city and region since the war shows the tension between efforts to rebuild and maintain the city's multiculturalism (mainly by the international community) and ethnic group-based responses to the realities of nationalist animosities during and after the war. At the national level, the concentration of power at the entity level and, within the Federation, at the cantonal level, contributes significant impediments to Bosnia's possible future transition to an ethnically shared democracy. And, at the local level, in its ethnic circumscription of territory in and around Sarajevo city, the Dayton Accord substantially restrains the city's ability to play a peace-building role in Bosnia's future. In an electoral environment where group identity remains the voting foundation, the redrawing of political space in and around Sarajevo city in the months after the war established an ethnically circumscribed playing field upon which a fundamental game of rigid ethnic competition has taken place. In containing Sarajevo behind negotiated political boundaries built upon ethnic differences, international negotiators allowed for the "misplacing" of the City in post-war Bosnia. An alternative placing of the City, one that would have established the spatial, demographic, and political framework to encourage multiculturalism over the medium to long term, was not pursued amidst lingering ethnic antagonisms.

Negotiations and bargains engaged in during and at the end of the war by Serb, Muslim, and Croat ethnic leaders and the international community over borders and jurisdictions produced a set of ethnically-demarcated containers within which the different ethnic groups could continue to exert power after the war. Rather than a unitary state that would encompass mixed ethnic populations within single jurisdictional geographies, the Washington Agreement of early 1994 (pertaining to what would become the Muslim-Croat Federation) and the Dayton Accord of late 1995 (pertaining to all the state of Bosnia) pursued political strategies that divided the state into cantons (on the Federation side) and entities having, in almost all cases, clear ethnic majorities. International acquiescence in this strategy established an "institutional frame" that supported a key plank – the establishment of ethnic territories and constituencies – of Croat and Serb ethnic leaders' programs of "divide and rule" (Kumar 1997: 97). While wartime goals of division were largely accommodated in diplomatic agreements, there was little done to counter the negative consequences of ethnic compartmentalization and autonomy – the entrenchment of ethnic identities territorially, the compromising of fundamental values of the state, and the obstructing of compromises necessary for the success of such a state (Ghai 2000).

The effect of such ethnic demarcation and gerrymandering was to tighten the screws on Sarajevo city's ability to act as an opportunity space for multiculturalism in the future. The Dayton Accord and predecessor agreements gave little space to Sarajevo to integrate and assimilate different ethnic populations over time. Sarajevo city has been submerged and quartered through four main ways, each of which I examine in detail. Decisions influenced by ethnic imperatives have (1) overshadowed efforts to neutralize and share governance of the urban area, (2) drawn Serb entity boundaries within the city's functional sphere,

(3) created Bosniak-majority cantonal boundaries that engulf the "official" city, and (4) restricted the city's spatial reach and functional scope. Divided further by four municipalities that make up the city and frequently have greater powers than the city administration itself, Sarajevo city faces strangulation both from outside and within.

Governing the city

An initial strategy during early diplomatic efforts to counter possible ethnic claims on Sarajevo was to create a special status as a district under United Nations or European Community administration. This "corpus separatum" strategy resembled the unsuccessful proposal in 1947 to protect the city of Jerusalem through United Nations oversight, and was premised on Sarajevo being different and special in Bosnia-Herzegovina. The urban region was to be preserved as a multi-ethnic capital for all the state, and a place where collective rights would be protected (Dragan Ivanovic, interview). The Owen-Stoltenberg proposal of August 1993 recommended that Sarajevo city, inclusive of all ten pre-war municipalities except for Serbian stronghold Pale, be under United Nations governance. One year later, the UN neutral zone idea still appeared alive in diplomatic discussions; the Contact Group Plan of July 1994 recommended UN administration of a spatially expansive Sarajevo district. However, by the time of the Dayton Accord near the end of 1995 that ended the war, the idea for international governance or oversight of the city had been overtaken by the give-and-take negotiations of ethnic leaders.

Absent international oversight, an alternative approach to preserving the city's multiculturalism would have been to construct a cross ethnic power-sharing agreement for city governance. For the first several years after Dayton, the city had a joint governance and administrative structure that sought fair representation by minority groups. A mayor and two deputy mayors had been selected in a way that each of the main nationalistic groups – Bosniak (Muslim), Croat, and Serb – had one of these positions. From 1997 to 2000, the allocation of city council seats was also engineered to assure multi-ethnic representation. In the run-up to the first municipal elections in September 1997, there was the detailed management of ethnic balance for the city by the international community. The Protocol on the Organization of Sarajevo (OHR 1996) specified that at least 20 percent of city council seats go to Bosniaks, Croats, and to "other." In March 1997, the constitution of the city was amended to reflect these ethnic quotas. The intent was that the city would be governed on a shared basis and this would show a way forward for the Bosnian state. With the 2001 Election Law for Bosnia, however, this ethnically engineered electoral system for Sarajevo was replaced by an electoral system where council seats are allocated proportionately to popular votes garnered.[25] As of 2004, without minimum representation quotas, an ethnic party representing Croats or Serbs held only one of the 24 city council seats. The days of ethnically shared governance of Sarajevo city appear over.[26]

Municipal boundary drawing

A way to facilitate the maintenance of minority Serb and Croat groups in the city would have been to create municipal borders that intentionally bridged the Dayton autonomous entity boundaries. Before the war, Sarajevo city was a geographically expansive city that spanned well into what today is Serb Republic territory; in 1991, the population of this greater Sarajevo was over 500,000. If re-established after the war, such expansive entity-spanning city boundaries would have provided increased space necessary for the three groups to live in the city, over the short term "together separately"[27] and, over the longer term, in ways that would revitalize the urban relationships and processes of the pre-war city. Instead, under Dayton, Sarajevo was jurisdictionally located within the Muslim-Croat Federation, with the urban part of the city in the Federation and the rural part in the Serb Republic. If this rural part had been included within larger, entity-spanning city boundaries, the Serbs who had since relocated there would likely feel part of Sarajevo rather than ostracized. Instead, today living outside the city borders and functionally disconnected from the urban system, "those on the Serb Republic side do not have a future under division" (Muhidin Hamamdžić, Sarajevo Mayor, interview). Soon after the war, the area became known as "Serb Sarajevo," indicating a psychological and territorial claim to that part of the former city that will likely strengthen over time.

   To the credit of Dayton negotiators, there was a strategy to "reunify" post-war Sarajevo and not let the urban area become ethnically fragmented. As described earlier, the transfer of the districts and suburbs of Grbavica, Ilidza, Hadzici, Vogosca, and Ilijas into the Federation sought to re-integrate and unify the Sarajevo urban area. Yet, the "reunification" of Sarajevo was not a proposition without ethnic salience because the Serbs to be reunified within the urba area were simultaneously incorporated into the Muslim-Croat Federation. This psychological factor spawned the substantial out-movement of Bosnian Serb population from the transferred districts and suburbs to nearby Serb Republic land and to other places in that republic. If, instead, the "reunified" city would have been a spatially expansive zone that spanned entity boundaries and thus was not fully contained within the Federation, more Serbs would have probably stayed in these neighborhoods.[28] With expansive and spanning boundaries that marked the city as neither Federation nor Serb Republic (a "D.C. Sarajevo" model[29]), "reunification" of the city may have occurred without the Serbian out-movement that diluted the city's multiculturalism.

Cantonization

The vision of a multi-ethnic Sarajevo has also been compromised by larger state restructuring decisions that have shaped strongly the distribution of decision-making power in the Federation and the Sarajevo greater area. Such state restructuring did not support the city's multiculturalism as the way forward, but rather perceived it as a direct threat to its vision of a Bosnian state. The idea to create Cantons for the Federation – a layer of government between entity and local governments[30] – became an important pivot around which international negotiators, and Muslim and Croat leaders, converged.[31] "The whole focus was on cantons because that was the point where power was distributed between the negotiators," observes Dragon Ivanovic (interview). International negotiators saw it as a way to devolve and disperse ethnic tension and give the Federation (and thus the state) a chance to endure. Muslim and Croat negotiators saw it as a way to carve up territory and protect their people, amidst and subsequent to the Croat-Muslim "war within a war" in 1993 and 1994 centered on the city of Mostar.

The idea to organize Bosnia into cantons began in early 1993 with the unsuccessful Vance-Owen Plan to create ten ethnic provinces or cantons that would cover all of Bosnia. Negotiator David Owen (1995: 59) recounts the logic behind cantonization, that it "promised the most stable form of government, since much of the predicated intercommunal friction could be kept from the central government by giving the provinces competence over the most divisive issues, for instance, police, education, health and culture." Cantonal organization was subsequently given substance by the "Washington Agreement" of 1 March 1994 that outlined the principles of a Muslim-Croat Federation, stating that it would be a single entity composed of a number of cantons yet to be determined[32] (Burg and Shoup 1999). This agreement went into effect nominally, midst war, with its adoption later that month by the rump parliament of the old Republic of Bosnia-Herzegovina. Then, after the Dayton Accord of December 1995, the exact boundaries of the cantons in the Federation were eventually established in May 1996.

Back in 1993, the Vance-Owen negotiations had proposed a special power-sharing arrangement for Sarajevo Canton, similar to what would be established later in the city. Unique among the ten proposed provinces,[33] the distribution of seats in the interim Sarajevo area government would be equally allocated to the three ethnic/nationalist groups – Bosniak, Serb, and Croat. However, the Muslim-led Bosnian government objected to the ethnic sharing of this important region, stating that what would become Sarajevo Canton should have the "same status" as other provinces, and thus due to its demographic proportions be a Muslim majority province (Burg and Shoup 1999: 225). The leading Bosniak political party, the Party of Democratic Action (SDA), did not want a model of multicultural inclusion and power sharing to spread too widely in the urban region (D. Ivanovic, interview). A firewall was thus created, in the form of a Sarajevo Canton without "special status" that would make decisions in line with the wishes of its Bosniak majority.

City space and function

The Bosniak SDA party has worked successfully to limit the political space of the city through re-creating Sarajevo city (with its multicultural enterprise) as a spatially smaller jurisdiction with powers subordinate to the canton. As cantons emerged after the war as the important subentity power structure, cities as institutions in Bosnia "basically disappeared after Dayton" (Dragon Ivanovic, interview). When the City of Sarajevo was eventually re-established two years later, it was not at its pre-war expansive size containing ten municipalities; it was now to contain just four municipalities and be approximately 60 percent of its original size (International Crisis Group 1998a). This smaller area provided Sarajevo Canton with land and population resources now separated from the "city" and dampened the city's ability to compete with the Canton. Further tightening the noose around Sarajevo city was the institutionalization of a significant power imbalance between cantons and local governments in post-war Bosnia. In the new Bosnia, cantons have substantial legislative power to formulate policy on vital urban issues such as education, culture, housing, public service provision, and local land use regulation. They have substantial own-source revenue, compared to other levels, and also have significant power to decide on how revenue coming from the Federation level is to be spent. In this context, local governments are clearly subordinate (Dragan Ivanovic, interview). The powers of local governments are ambiguous and often without force, and efforts to establish local self-government principles have languished.

The cantonal level of government, although lodged with significant powers of urban management in post-war Bosnia, is in reality more a creature of the entity government and thus one step removed from directly dealing with urban issues on a daily basis. In contrast, local governments, the level at which common citizens most frequently interact with public authority, are given more limited resources and competencies. This Dayton structuring of local and cantonal government authority has obstructed long-range urban planning and development regulation in the urban area, resulting in a lack of a cohesive city-wide vision that could shape in spatially rational ways the large amounts of international and private funding of post-war reconstruction. There has been chaos and confusion in terms of post-war planning and regulation authority (S. Jamaković, interview). All of Sarajevo Canton is guided by a "spatial plan" done in 1988 that broadly classifies expected future land use types for the entire region. For local governments, there is the development of "urban plans" and "regulation plans," the latter being the equivalent of zoning plans in the American system. Whereas before the war municipalities held competencies for planning and regulation, Sarajevo Canton took the lead in these areas during the war. Dayton then gave considerable post-war competencies in planning to the cantons. Sarajevo City, in theory, has authority of planning and regulation in the four municipalities within its borders while the Canton has authority for areas outside the City. In reality, however, the Canton has bypassed city authority and has engaged in development regulation throughout the canton (D. Ivanovic, interview). An important institutional actor is the Sarajevo Canton Institute of Planning, a paragovernmental, semiautonomous bureau that is part of Cantonal government. In cases where urban plans and regulations are out of date (which many of them are, some dating to the 1970s), this Institute frequently makes decisions regarding project approvals or denials. This has decreased the transparency and openness of what should be public decisions and has made corruption and payouts common (D. Ivanovic, interview).

Lacking a unifying vision for growth and development for the city of Sarajevo, political connections drive much private development since the war rather than sound planning principles (D. Ivanovic, interview). In addition, the chaotic state of planning has allowed for the continuation of illegal building, a prevalent practice in Sarajevo before the war. One estimate puts the percentage of housing built that is inconsistent with local plans at close to 90 percent (Said Jamakovic, interview).

Powers and capacities of the City of Sarajevo are further circumscribed by the more expansive local powers given to the four "municipality" governments within its borders – Centar, Stari Grad, Novi Grad, and Novo Sarajevo. The "city" is not a common concept in Bosnia and suffered further during the post-war transition. The Sarajevo Mayor is more a symbolic position, with the mayors of the four city municipalities actually having greater administrative power and budgetary resources (Gerd Wochein, interview). In addition, councilors for the city are elected by municipality according to their populations, so that the most populous municipality (Novi Grad) receives the most seats on the council. In such a system, municipal interests of city councilors will many times take precedence over city-wide concerns. The division of the City of Sarajevo into multiple municipalities also existed before the war. Indeed, at that time there were ten municipalities within the city. There were also difficulties with lack of coordination, and revenue competition, between city and municipalities, exacerbated by parastate and semi-independent entities with separate competences. At times, such as in the preparation of the 1984 Olympic Games, these local governments, along with professional and business communities, pulled together impressively.[34] Significantly, however, while local political fragmentation before the war lessened normal government functions and efficiencies, now it is also obstructing a process of inter-group urban accommodation that could occur at the broader geographic scale of the city and that is essential for Bosnia in its societal reconstruction. Due to the increased burden and opportunity for the City of Sarajevo in post-war Bosnia, there should have been greater attention to how multiple types of government could be structured in ways that support and encourage this common societal goal.

Who could have devised a strategy such as this?

A writer and nongovernmental activist[35] recalls how the war came. "We in Sarajevo did not believe it could happen to us. There was denial. It was almost invisible. Suddenly, it happened. It was not possible for us to think that it would be that type of war. We think how could they do it, but they did." Regarding ethnic leaders, "a leader such as Milosevic connects with people in some deep way. There were not real reasons, but you will find the reasons in the past myths. Facts are then no longer facts." In surviving the siege of Sarajevo, the aggressors were "invisible enemies in Sarajevo. You could never see them. But they could kill you from the hills at any moment and at any place. There was no way to confront them, no way to control the siege, so you changed your tactic and lived a normal life as much as you could. This was a strong way to be opposite to all the hostility and hatred." Regarding the future, "nationalist leaders today and their followers are simply using time before they assert their power again; who could have devised a strategy such as this, no one?"

In summary, the redrawing of political space during Bosnia's transition from wartime to diplomatic agreement has significant implications for future Bosniak-Serb-Croat relations. Entity boundaries, the creation and empowerment of cantons on the Federation side, and limitations of the spatial size and functional competencies of Sarajevo city government, mean that Sarajevo's multiculturalism is endangered and in many ways illusionary. Political redrawing in and around the city has disempowered it as a functioning level of government and as a contributor to the social reconstruction of Bosnia. In a situation where ethnic difference has been accommodated and reinforced through the drawing of political boundaries, efforts by the international community to build a democratically shared Bosnia lack the local foundational level of democracy to build from.

The diplomatic re-creation of the Bosnian state has "misplaced" the city in Bosnia's reconstruction, suppressing its innate ability to be a symbol for all and to re-establish over time inter-ethnic relations on the ground, in the marketplaces, and in the neighborhoods. Aspirations for the city to be a multi-ethnic anchor for Bosnia will be unfulfilled due to the city's genuine lack of authority and its subordination, both spatially and functionally, to more empowered and ethnically-based subnational governments. It is remarkable that the multi-ethnic spirit of the city of Sarajevo has survived through the onslaught of ethnic war, evidenced by strong voting support for multi-ethnic (SDP) or soft-nationalist (SbiH) parties in the 2000 and 2004 local elections. This certainly provides some room for optimism that the multi-ethnic soul of the city remains alive; the special, transcendent nature of Sarajevo may still be present. Yet, this spirit is being expressed not so much through a genuine lived process of Serbs, Bosniaks, and Croats existing in productive coexistence side by side in the city, but rather is dependent upon the sentiments of what is a strongly Bosniak majority. A more proper "placing" of the city midst Bosnia's peace-building would have provided for the development of a more genuine, existential multiculturalism. Due to Sarajevo's location at the conjuncture of interspersed pre-war ethnic geographies, a Sarajevo of special government status and genuine multicultural and shared power would have helped over the long term to rebuild centripetal forces necessary for societal reconstruction. Instead, the "ethnocentric autism" (Lovrenovic 2001: 213) of the war was strong enough to influence the drawing of post-war political geographies in ways that accept and will continue to reinforce centrifugal forces and separate futures.

When deciding on the spatial scale of political organization and decision-making, Sibley (1995: 14) asserts, "spatial boundaries are in part moral boundaries." And, morally, there may be strong reasons for allowing Bosniaks to be in political control of Sarajevo. After all, Muslims were the explicit target in the city for almost four years and clearly the war's victims. In this view, the fact that the Sarajevo urban setting today is primarily controlled by Bosniak Muslims is a moral outcome of an immoral war. This claim to the city is based on the fact that there were ethnic aggressors and victims in the conflict. Certainly, the international community – with its emphasis on human rights protection – is empathetic to such a moral claim. However, there exists another type of claim on the city, one that utilizes its unique ability to extend moral boundaries and spatial relationships in ways to counter the tribalism and sectarianism that destroy common responsibility (Tronto 1993: 13, 59; D. Smith 2000). Only in the city can there be the daily experience of difference and diversity that enables people "to see beyond their own partiality and to be held responsible for this larger domain . . ." (Sack 1997: 257). In this view, there are no winners and losers, no aggressors and victims, and the demarcation of post-war city space focuses on the urban setting as an essential starting platform and organizing framework for rebuilding and reconstructing the multicultural basis of a traumatized society. The war itself, despite its concentrated efforts, could not kill the city of Sarajevo; "the soul is still here; even though such strong energy was used to try to destroy it, they couldn't do it" (M. Hamamdžić, interview). Diplomatic agreements to end such a war should have sought to resurrect and perpetuate these special qualities, not take actions that have eroded urban capacities that are essential for the building of peace.

DIVISIONS AND DISLOCATIONS

The ethnically delineated political containers created by Dayton and other negotiated agreements, although rational responses to the need to stop war and politically steer Bosnia through transitional uncertainty, nonetheless have significant and adverse implications for Bosnia's multicultural future. Post-war Sarajevo urban area is in many respects a divided city socially, demographically, and psychologically. It constitutes an ethnically compartmentalized setting that obstructs the recovery of the city's, and Bosnia's, multicultural heritage.

The invisible wall

> *For Sarajevo, the boundary line is invisible.*
>
> Vesna Karadzic
> Assistant Minister, Ministry of Physical Planning
> Federation of Bosnia and Herzegovina
> Interview 11/24/03

The Inter-Entity Boundary Line (IEBL) just outside Sarajevo city boundaries has no checkpoints, no partitions or obstructions of any type. The sign stating entrance into Republika Srpska and the Cyrillic written alphabet evident on the Republika Srpska side of the boundary does distinguish one side from the other. But, with those exceptions, the boundary line demarcating political territory, the core prize fought for by the antagonists for over four years, is indeed not explicit and easily missed by someone from the outside.

Sarajevo is certainly not a physically partitioned or divided city such as Belfast or Nicosia. It is also not a city today terrorized by violence such as contemporary Jerusalem nor a city where war antagonists remain firmly entrenched within the urban sphere. Nonetheless, Sarajevo is traumatized and must deal with issues of ethnic, nationalist, and religious identity that were not salient until the early 1990s. The urban area houses three peoples who need the buffer of space and time to deal with the trauma of war. The political boundary created through brutal war that is within Sarajevo's urban sphere lacks a physical or intimidating presence; however, it is a line of psychological separation within an urban system and it has already influenced, and will continue to do so in the future, where people live and how and where they choose to interact. As noted, the location of the IEBL was an explicit factor considered by Serbian households when they left Sarajevo city after the war. In his work on returns and reconstruction, Morris Power (interview) discovered that many displaced households relocated not far from their original homes, but consciously chose to cross the IEBL and stay there, perhaps 15–20 miles from their original home. He observes, "displacees were moving aware of those lines and zones." Whereas before the war households were locating due to economic and functional reasons, now they were residing to be in tune with larger politics. In this way, the abstract IEBL was becoming real and taking on a life of its own – the reification of a politically potent ethnic boundary.

The Dayton Accord assigned about 35 percent of the Sarajevo functional urban area to Republika Srpska. This part of the Sarajevo agglomeration, including four municipalities created out of larger pre-war municipalities now split by the IEBL, has traditionally been more rural but now faces urban pressure as Bosnian Serbs stimulate development and strengthen territorial claims.[36] Basically a small rural village before the war, East ("Serb") Sarajevo is now recipient of significant investment from the RS government to turn it into a fuller city (Gerd Wochein, interview). In one of the municipalities in East Sarajevo, it is estimated that its population of about 10,000 people could grow to approximately 30,000 (Said Jamakovic, interview). Unconfirmed estimates in 2005 put the overall population of East Sarajevo as high as 100,000. Although development is ongoing there since the war and it has impacts on highways and the airport, Bosnian Serb and Federation officials are not in contact with each other regarding its planning and service delivery requirements (Said Jamakovic, interview). The sewage system in East Sarajevo is connected across the IEBL to the Federation system and many people in "Serb" Sarajevo work in Federation Sarajevo, yet, "the intent of Bosnian Serb officials is for these areas and these people to be separated" (Said Jamakovic, interview).

The placement of the IEBL has created hotspots of ethnic contentiousness that have needed, at times, international arbitration many years after the end of the war. In the Dobrinja area near the Sarajevo airport in the southern part of the city, five large apartment complexes had been built before and immediately after the

1984 Olympic Games. This area suffered greatly during the war, located as it was near the confrontation. The initial Dayton line gave 62 percent of the units in the five Dobrinjas to the Federation. Yet, the line drawn at large scale turned out to divide one of the buildings and even the apartments within it. In addition, it appeared to put on the RS side numerous apartments formerly resided in by households on the Federation side. Finally, in April 2001, an international arbitrator appointed by the High Representative made a decision about where the IEBL should be and awarded the contested units to the Federation side.[37] Language contained within the arbitration award is telling in terms of the realities and frustrations present when addressing post-conflict local geographies. The arbitrator reflects upon the larger context of this decision:

> *The pre-war population of Sarajevo consisted of Bosniaks, Serbs, Croats, and other nationalities. It was an international city and a very beautiful one at that. I find it very difficult to come to terms with the phrases "Serb Sarajevo," "Bosniak Sarajevo," "Croat Sarajevo" or any other appellation such as those. In my view, the city of Sarajevo is a single unit comprising the ethnic groups that I have mentioned.*

The arbitrator also declared his hope that the IEBL that was his focus would vanish in the future:

> *It would be heartening, if I were to be able to expect that in the future the line would disappear for all practical everyday purposes and that the communities generally, whilst of different ethnic origin, would come together as neighbors and would be welded by time, into a community enjoying the fruits of a rebuilt Dobrinja.*

Despite these hopes, the increased post-war salience of ethnic microgeographies and territoriality have been a fact of life for urban and international managers in the ten years since the war's end. At times, the community cohesiveness of an ethnic minority in the city becomes an issue. In the southwestern Stup area of Sarajevo has resided a significant Bosnian Croat community (pre-war percentage estimated at 70 percent).[38] Close to the first line of combat, the district was severely damaged during the war and the population was halved due to displacement. While the demographic percentages of the downsized post-war population stayed roughly the same, post-war pressures to meet the needs of the growing Bosniak Muslim population displaced into the city has created tension. When the municipality built new apartments in the area, the local Croat community perceived this public action as an intentional effort to change the ethnic balance in the area. As a result, efforts to create a regulation (land use) plan for the area stalled.

Reintegrating traumatized neighborhoods requires a difficult balancing act and sustained effort on the part of urbanists. Efforts to reconstruct damaged and destroyed community areas run head-on, out of necessity, into issues of displaced and refugee households now living in the neighborhood, the legal and illegal occupation of war-torn housing, and efforts to return original occupants to their reconstructed units. In the mid-1990s, one of the first efforts by the international community (IC) to reconstruct a settlement in an ethnically integrated manner was the Dobrinja "airport settlement" (Gerd Wochein, architect, Office of High Representative, interview). For the IC, this was viewed as an important exemplar of inter-ethnic reconstruction due to its location, straddling the IEBL between the Federation and the Republic. If successful, this effort would show how physical reconstruction could contribute to stability and reconciliation between groups. An informal urban master plan established a de-mining program, incorporated displacees and refugees into the rebuilding process, and dealt with forceful evictions and the return of property to pre-war occupants. In addition, the effort included social programs and support. The result was "a small paradise growing up in the middle of the war ruins" (G. Wochein, interview).

Despite the best efforts of the international community, the airport settlement project exhibits the limits of reconstruction as a means of reconciliation in post-war Bosnia. Five years after their establishment, most social infrastructure components of the reconstruction failed because public authorities that took over maintenance of the project from the IC were not interested in the social side of reconstruction. In addition, the quality of the actual ethnic integration that took place in the area is questionable. The demographic "mixing" of Bosniaks and Serbs at the level of the larger community is, in actuality, more a case of parallel communities that straddle the IEBL, the Bosniaks on one side and the Serbs on the other (Javier Mier, Office of High Representative, interview). The "mixing" is more a political geographic artifact than an example of genuine integrated neighborhoods and streets.

The influence of ethnic boundary lines in Bosnia on households' and political leaders' decision matrices will probably outlive by many years the original conditions which produced these boundary lines. The city will likely not again be ethnically mixed in the next twenty years as it was pre-war, a perhaps not surprising prognosis after such horrific societal trauma. As one interviewee stated (Jayson Taylor, Office of the High Representative, interview), "There are so many things that will need to happen in the next 10–15 years for there to be ethnic integration that it is hard to envision; it will be a long struggle to get to that point."

Displaced populations and minority returns

Whether displaced and refugee populations return to their pre-war locations or continue to live in their relocated post-war locations influence the extent to which war-created ethnic geographies are overcome. The international community estimates that two million of Bosnia's pre-war population of 4.3 million were displaced to some other location within Bosnia or became refugees and left Bosnia entirely. This is disruption of civilian life by military coercion and/or voluntary decision on a vast and traumatizing scale. Internal displacement due to the war ran in several different, but identifiable, directions: Bosniaks from what is now

Eastern Republika Srpska to central Bosnia, especially Sarajevo, and to Gorazde and Mostar areas; Bosnian Serbs from central Bosnia to areas close to Serbia and Montenegro borders; and Bosnian Croats from central Bosnia to the western Herzegovina region of Bosnia close to the border with Croatia (OHR 2002a; Gerd Wochein, interview). At the end of 1994, the Bosnian Institute of Public Health estimated that one of every two persons living in Federation territory had been displaced from Serb-held areas (Burg and Shoup 1999). Internal displacement was disproportionately a rural to urban movement because cities were perceived as providing greater safety to fleeing residents. Main refugee flows out of Bosnia were Bosnian Serbs to Serbia, Bosnian Croats to Croatia, and Bosniak Muslims to northern European countries such as Austria, Germany, and Scandinavian countries.

Major axes of movement during the war were into Sarajevo from towns such as Srebrenica (the site of over 7000 Muslim deaths, the largest massacre of human life since World War II), Bratunac, Zvornik, Visegrad and Brcko. Within the Sarajevo urban area, the UN Refugee Agency (UNHCR) estimated in late 2004 that there were approximately 58,500 displaced, primarily Bosniak Muslim, persons in Sarajevo city who had lived in what is now eastern RS before the war. In terms of persons displaced from the Sarajevo urban area, significant Serb and Croat emigration occurred from the five districts and suburbs transferred to Federation control in 1996 – Ilidza, Hadzici, Vogosca, Ilijas, and Grbavica. Data from Ilijas municipality are illustrative of the level of disruption. In this suburb, a "very considerable part of its population" remained displaced throughout eastern RS (particularly in the Srebenica and Bratunac areas) and in the former republics of Serbia and Montenegro, while about 60 percent of the municipal population consisted of Bosniak Muslims displaced from eastern RS (Office of the High Commissioner 2002b: 15).[39] The Sarajevo suburb of Hadzici, meanwhile, has a large pre-war population now displaced into Bratunac in eastern RS. And, in Vogosca, some 6700 pre-war residents remained displaced into eastern RS and some 7500 current residents were displaced into Vogosca from eastern RS.

The return to pre-war homes of displaced persons and refugees has been a primary goal of the international community since the start. The Dayton Agreement (Annex 7, article 1, paragraph 1) reads: "All refugees and displaced persons have the right freely to return to their homes of origin." The obligation of each of the belligerent parties is spelled out further in article 11, paragraph 1: "The parties undertake to create in their territories the political, economic and social conditions conducive to the voluntary return and harmonious integration of refugees and displaced person, without preference for any particular group." The return of displaced persons and refugees has been one of the three primary objectives of the OHR in their post-war engagement. The main implementing unit within the IC related to returns is the Reconstruction and Return Task Force (RRTF), a consortium of international entities that facilitate returns through initiatives aimed at providing returnees with housing, security, and sustainable socio-economic conditions. The RRTF also coordinates the reconstruction of housing and oversees the implementation of the Property Law

Implementation Plan (PLIP), under which all refugees and displaced persons are entitled to repossess their pre-war homes. Based on the significant involvement of the IC in facilitating returns, it is clear that it is cognizant of the importance of returns to the reconstruction of Bosnian society.

The process of encouraging returns by the IC involves numerous aspects that must be coordinated, including the identification and encouragement of potential returnees, legal repossession of pre-war property, reconstruction and rehabilitation of residential units, and provision of necessary support services (including security) to support returnees in their pre-war neighborhoods. A significant source of resistance to returnees is those displacees from the war who have been living in returnees' residential units (D. Ivanovic, interview). The property implementation law allows returnees the right to possess their old units, so in order to avoid sending the wartime occupants out into the streets, one of two avenues must be pursued by the IC: (1) the opening up of wartime occupants' housing in pre-war areas through PLIP and, if necessary, reconstruction of that unit, so that the evictee is able to return; or (2) the provision of alternative accommodation for the evictee near the household's post-war location. With some other accommodation available, usually through reconstruction, evictions of wartime occupants becomes more doable; one estimate is that about 80 percent of evictions in BiH have been without force while 20 percent have required police support (G. Wochein, interview).

The overall statistics on returns in Bosnia show significant progress from Dayton through mid-2005. Of the 2.2 million displaced persons and refugees, over one million had returned to their pre-war homes and municipalities.[40] Crucial to the question of whether returns are helping to bring back an ethnically mixed country is which areas and municipalities these returnees are moving back to – recipient areas in which returnees are in the numeric majority (so-called "majority returns") or where they are in the numeric minority (so-called "minority returns"). Majority returns reinforce ethnic separation; minority returns increase ethnic mixing. Significantly, over 450,000 of the returnees are "minority returns." Minority returns are more likely into the Federation (about 270,000 returnees, primarily Bosnian Serb and Croat) than into Republika Srpska (about 160,000 returnees, primarily Bosniak Muslim). The return of properties through the PLIP process, meanwhile, has been a spectacular success, with an overall 93 percent implementation rate by the close of 2004 (UNHCR 2005a). Individuals who were displaced from their properties during the war have largely been able to take back legal possession of those units.

With nearly one-half of those who fled during wartime having returned, with almost one-half million of these returns contributing to ethnic mixing, and with property assets being returned to pre-war owners, important progress has been made in re-creating the pre-war Bosnia. Yet, satisfaction with these numbers must be tempered by other realities. Ten years after war, the return process may be near its culmination. There have been decreasing annual rates of total returns from 2002 to 2005. Minority returns, in particular, peaked in the 2000–2002 period. Thus, the UNHCR (2005b: 1) was assuming by early 2005 that most of those who have not returned will not do so in the future. In addition, general progress on return may be hiding continued problems with the support and sustainability of these returns. For a certain percentage of people returned (especially to places where they are not in the majority), the long-term prospects of staying there may be tenuous. If those minority returnees do not stay in place, progress in post-war ethnic mixing starts to unravel.

UNHCR (2001) reports several types of obstructions to a sustainable return, exacerbated when the returnee is in the minority, including access to employment, health care, pensions, utilities, an unbiased education system, and to reconstruction assistance. Efforts by the IC to harmonize public school curricula across the IEBL have been problematic. As of 2004, parallel education systems were still in place; indeed, UNCHR (2004: 1) in its own guidelines for potential returnees to Sarajevo accommodates this reality and suggests that "those who wish their children to learn according to RS curriculum should enroll their children in nearby schools in RS," meaning in East ("Serb") Sarajevo. UNHCR (2005b: 10) reported that it is "contacted on a daily basis by very disturbing cases of individuals who are unable to access health care despite urgent needs, have no housing or income whatsoever, and are thus forced to make a living as best they can through begging." Security remains an important concern for minority returnees in BiH and continues to constitute an obstacle to return (UNHCR 2005b). The incidence of violence against minorities has been twice as high in RS as compared to the Federation; particularly high rates of violent incidents occur in RS municipalities with higher rates of Bosniak Muslim returns.

The next generation?

Dragan Ivanovic is a Bosnian Croat born in Sarajevo in 1970. With a background in computer science and in his 20s during the war, he became involved in politics because he opposed the war and the ethnic political parties that controlled post-Dayton Bosnia. He is a member of the Social Democratic Party (SDP), one of the non-nationalist parties in Bosnia that seeks to span and transcend the three nationality groups. A former municipal councilor in Sarajevo from 1997–1999, he was in 2003 deputy speaker of the Sarajevo Canton Assembly and member of the Federation Chamber of Peoples legislative assembly, which represents the cantons. He sees in Bosnia in 2003 the same vulnerability that existed before the war – an ethnic and structural compartmentalization of politics and uncertain economic times. The way forward is to get younger people involved in politics and decision-making; in a circumstance where "our politicians are very much contaminating the political system today, we need to see a generation change."

A specific examination of returns in the Sarajevo urban area reveals additional realities and hardships. As in Bosnia overall, it is evident that progress has been made in Sarajevo regarding minority returns. The highly publicized Sarajevo Declaration of early 1998[41] called for 20,000 minority returns to be facilitated in one year, but it took until early 2000 for this goal to be reached. Some momentum occurred in 2001, with about 17,900 Serbian returns registered that year (Internal Displacement Monitoring Centre 2002). Due to its sheer size relative to other towns in Bosnia, Sarajevo and its suburbs constitute one of the largest receiving zones for minority returns in the country. Still, Sarajevo also illuminates the limitations of the return process. OHR (2002a: 3) notes the exceptional situation there, stating, "While in central Bosnia, with a few exceptions, the majority of return-related issues are resolved, *Sarajevo and Eastern Republika Srpska continue to present a serious problem*" (italics added by author). Many Bosnian Serbs are not returning to Federation Sarajevo. It is increasingly clear to the IC that "there are a very significant number of people who have decided not to return to Sarajevo. Those wishing to return to Sarajevo have likely done so at this point" (Richard Ots, formerly RRTF, Sarajevo, interview.)

Primary among the difficulties of the return process is the real story behind the successful property repossession numbers. The implementation of PLIP legislation was nearing full completion by mid-2004, with local housing agencies actualizing the repossession of well over 200,000 of the total 216,904 claims throughout BiH. Dispossession of property is one of several humiliating and dispiriting consequences of war; bitterness over non-return of property can build in burning resentments for generations. Much to the credit of the IC, properties were legally returned to their pre-war residents so successfully that the IC has verified 77 BiH municipalities as having substantially completed their private property claims.[42] As one IC official described it, "Repossession of property is an important step and you would be hard-pressed to find a similar example anywhere else in the world that has been this successful in returning properties" (J. Taylor, interview).

However, the effort at returning property to refugees and displacees in Sarajevo is not restoring the multicultural residential fabric of the city as much as property repossession figures would imply because PLIP is returning properties more successfully than it is people. Property repossession only produces a real return if the former occupant physically returns to his pre-war unit. In reality, what commonly happens is that the person repossesses the unit and then sells or rents. In these cases, the return is on paper, but not on the ground. The number of repossessed units in Sarajevo that are subsequently sold is estimated to be high, likely more than 60 percent of all repossessed apartments and real property (Ombudsman Institution of the Federation of BiH 2004). OHR (2002a: 6) reports "repossession of property in the urban areas of Sarajevo, Eastern Republika Srpska and in particular, towns of Ilijas, Vogosca and Hadzici (all in the Sarajevo urban area) is often followed by sale or exchange." Substantial returns of pre-war households are not taking place despite near completion of property repossessions.

The selling of repossessed units is likely a particular feature of potential minority returns as these households decide to stay relocated amidst their own ethnic group rather than return to more ethnically precarious environments. Such a pattern of selling of PLIP repossessed units, in aggregate, would then be cementing the ethnic sorting of post-war BiH more than facilitating its ethnic mixing. In Sarajevo and elsewhere in Bosnia, despite the return of properties, the changed ethnic structure of many areas compared to what existed before the war is significant and glaring (Ombudsman Institution 2004). Indeed, the Federation's Ombudsman Institution estimates that when a count of real, physical returns is calculated that only about 30 percent of displaced persons and refugees have physically returned to pre-war locations in BiH.[43]

The overall magnitude and pattern of returns in Bosnia and Sarajevo inform us about the extent to which there has been ethnic reintegration since the war. One million persons have returned to their pre-war locations and over 400,000 of these moves have increased the ethnic mix of the recipient area. These are significant achievements. As of mid-2005, about 270,000 non-Bosniak Muslims have returned to the Federation, and about 160,000 non-Serbs have returned to Republika Srpska (UNCHR 2005b). One expert who has studied the return process in Bosnia estimates that between 11 and 16 percent of the RS population in 2004 were non-Serbs (Toal 2005).

Despite these achievements, the Dayton objective of reversing ethnic cleansing has not been achieved. Approximately 1 million Bosnia citizens remain displaced or are refugees. Barring fundamental improvements in the economic and political circumstance of Bosnia, many of the displaced and refugees will likely remain in their place of relocation. This means that one of four pre-war Bosnia residents will never come back home. For many internally displaced persons, they have built up a new life over the last ten years where they are, and they may now be linked to a social support system (in terms of health, education, and pensions) that may not be present should they cross the IEBL and return to the pre-war home. For refugees, they may have access to job opportunities in their host country that are not available in Bosnia. The magnitude of this potentially permanent population of relocated pre-war residents makes it difficult to imagine a genuinely ethnically integrated Bosnia in the future.

At the same time, the fragility of life for minority returnees who have come back home means that the ethnic integration that has occurred is tenuous and susceptible to relapse. Physical return of a minority household represents an important step in the process of ethnic reintegration, yet it is only a first step that can falter should adequate education, job opportunities, social benefits, and services not be available. Another contributor to the tenuousness of the future ethnic reintegration in Bosnia built upon minority returns is that it is older family members who have been more likely to return to the family home, with younger persons more likely to connect to opportunities in their post-war location and thus to remain relocated (S. Jamaković, interview; World Bank 2002; Toal 2005). As these older family members reach the ends of their lives, any ethnic reintegration gained due to their return will relapse as younger family members remain in more ethnically homogeneous areas. Younger persons who served in the military during the war, in particular, have been most hesitant to return to areas where they would be an ethnic minority for fear of disclosure of their wartime activities and retribution. Demographically, it is this age segment that would have been most crucial to reintegrate ethnically because they were in the child-rearing years, would have boosted housing construction, and would have had school-age children to educate.

After such traumatic conflict, Toal (2005) points out that "it is unrealistic to assume that Bosnia's demographic structure in 1991 can be restored." It is more appropriate to ask whether the current level and trajectory of ethnic reintegration (in the form of minority returns) has sufficient momentum to contribute to the reconstitution of Bosnia society, or whether Bosnia is now at a point of ethnic reintegration which will decay over time as minority returnees go back into displacement due to lack of supportive infrastructure, and as older minority returnees reach the end of their lives. I conclude that Bosnia's ethnic reintegration is not of sufficient strength or momentum. Absent significant economic and political improvement in Bosnia in the future, the extent of ethnic reintegration that has occurred in the first ten post-war years, despite extraordinary efforts by the IC, is not sustainable over the next generation.

The war, willfully, and the Dayton Accord, unintentionally, have ethnically sorted Bosnia. Despite efforts of the IC, the partitioning of BiH into separate ethno-national spaces has been largely achieved. One million pre-war residents have been cleared out of their places of origin. Those who returned and find themselves ethnic minorities are in vulnerable and tenuous conditions. The political system has been carved up and compartmentalized to align with nationalist ethnic interests, which for the most part remain in firm political control within their respective ethnic worlds. Burdened with difficult practical and moral dilemmas concerning Bosnia's future viability is the set of international organizations that have been the de facto government of the country since the end of hostilities.

THE INTERNATIONAL COMMUNITY AND ETHNIC REINTEGRATION

> *They want a normal life, but do not pretend to create a Yugoslavia again where they are all Yugoslavs. They do not believe in that, and that is what the international community is trying to do.*
>
> Javier Mier
> Office of the High Representative, Sarajevo
> (In BiH since 1994). Interview, 11/21/03

The international community has been extensively involved in the management and rebuilding of Bosnia since Dayton. The United Nations' Office of the High Representative was created in the Dayton Peace Agreement to oversee implementation of the civilian aspects of Dayton. Annex 10 of Dayton lays out the mandate of the High Representative and declares him the final authority to interpret the civilian implementation aspects of Dayton.[44] The staff and responsibilities of the OHR have grown over time, increasing from a staff of 60 soon after Dayton to the employment of 681 staff and 18 offices throughout BiH in 2000. One of OHR's governing principles calls on officials and citizens of Bosnia to take ownership of the peace process so that it can be self-sustaining. At the same time, such local ownership can threaten progress on peace building, and at these times the IC is able to make decisive judgments and impose them in top-down ways.[45] In 1997, the Peace Implementation Council (PIC)[46] that advises the High Representative gave him significant powers to remove obstructive local public officials from office or to prevent them from running for office. These so-called "Bonn powers" have been used frequently to impose IC decisions on Bosnian legislative and executive branches.

The OHR is supported by numerous other international organizations and agencies. The UN High Commissioner for Refugees facilitates the return and reintegration of refugees and displaced persons. The European Commission, in the form of the Delegation of the EU to Bosnia and Herzegovina, has played a key role in implementing external assistance to the country. The Organization for Security and Cooperation in Europe (OSCE) promoted and monitored democratic elections and processes, and the protection of human rights. The UN Development Program (UNDP) has supported local development and capacity building. The World Bank funded from 1996–1999 significant emergency reconstruction work, including infrastructure, de-mining, and housing; since then, it has emphasized fundamental economic structural reforms linked to a market-based economy.

For most of its existence in Bosnia, the OHR has focused on three priority goals: creating and strengthening effective governance institutions, facilitating return of displaced persons and refugees, and reforming the economy. Other important issues are judicial and legal reform (i.e. the "rule of law"), protection of human rights, education reform, media reform, and integration into the European Union. There was the recognition and hope by the international community by 2003 that Bosnia's recovery may have been entering a new phase focused less on the aftermath of war and more on the economic issues of a conventional transition country (OHR 2003). There is the recognition of the need for Bosnia to move toward a self-sustaining peace less dependent on international community post-war support systems and interventions, a trajectory that would put the country on the road to EU membership (OHR 2003; Commission of the European Communities 2003a; World Bank 2004).

The relationship between the IC and local politicians has been difficult and contentious at times. Andy Bearpark,[47] with over 25 years foreign experience dealing with places like Rwanda, Northern Iraq and Somalia, explains in frustration that the international community cannot mandate change in Bosnia; yet, the UN is increasingly forced, due to local obstructions and disagreements, to remove ornery local politicians from power and to impose new civilian laws pertaining to electoral representation, property ownership, economic reconstruction, and return of displaced persons. Thus, while advocating democracy, the UN increasingly has acted in authoritarian ways. This reality inadvertently allows local officials to escape responsibility and retreat to tribalism.

The Dayton accord, although critically beneficial from a military perspective, may have created a faulty foundation from which to politically rebuild Bosnia (Jaume Saura, Bosnian election monitor, professor of international law, University of Barcelona, interview). The IC's push for early elections meant that they often took place midst feelings of threat and mistrust; in such a circumstance, extremists and war profiteers were most likely put into office in the early years (J. Mier, interview). These "forces for separation," once incorporated into the state's structures of governance, would logically seek to obstruct and separate the country. One OHR interviewee suggested that the IC should have come in with greater power, acting as a protectorate, rather than accommodating to local ownership and the cooperation of internal leaders[48] (J. Mier, interview). Under this scenario, after a number of years of social and economic stability midst stronger international intervention, the electoral process would possibly have had a greater chance of producing a more balanced assortment of local politicians.

Reconstruction, returns, and relocation

Policies by the IC regarding the physical reconstruction of infrastructure and housing, the return of the displaced to their pre-war places of origin, and the relocation of those displaced who do not wish to return to their places of origin, constitute important levers shaping future Bosnian society. The challenge of physically reconstructing Bosnia and Sarajevo was significant. The war led to wide-ranging destruction of physical capital and housing stock. Over 1,400 monuments of culture were either destroyed or damaged, of which 440 were totally razed to the ground (Bublin 1999). Libraries, museums, institutes, schools, government buildings, and hospitals were systematically destroyed. In the Sarajevo urban area specifically, wartime damage and destruction was substantial. Wholly or partly destroyed were 66 public buildings, 25 cultural and leisure amenities, and 60 historical monuments. In the Canton overall, Federation Ministry of Housing[49] estimates that about 84 percent of housing units had incurred some damage (about 119,000 units of the more than 141,000 units cantonwide). About 23 percent (approximately 33,000 units) of all units in the Canton experienced structural damage of 40 percent or more. Within Sarajevo's four municipalities, damage ranged from 67 percent of housing units in Centar to 92 percent of units in Novi Grad.

Sarajevo has certain fortunate characteristics that have encouraged rebuilding. It is the most populated city and is the center of Bosnia in terms of history and culture. It is internationally connected and symbolic, both due to its hosting of the 1984 Olympic Games and its being the institutional home of most of the international organizations that have occupied the country since the war. It is important to many, both to the IC and to Bosniaks, that the city be physically reconstructed, and the IC and donor organizations have supplied much funding to make sure this happens. Substantial money from the EU, and from member states such as Holland, USA, Japan, and Germany, were dedicated for reconstruction. Between 1991 and 2002, BiH received 2.5 billion euros in assistance from the EC, plus another 1.2 billion euros from EU member states between 1996 and 2001 (Commission of the European Communities 2003a). Such investment has led to extensive and visible progress in physically redeveloping the city. Estimates of how much has been reconstructed in Sarajevo vary, but one interviewee within the IC estimates it as 80 percent (G. Wochein, interview). Even those who otherwise criticize IC strategies concede that reconstruction has been the most successful aspect of international intervention and that fewer traces of the war are evident with each passing month (M. Hamamdžić, interview). Sarajevo's gains may be at the expense of others' losses, however. "The huge amount of reconstruction in Sarajevo," explains G. Wochein (interview), "is because everybody is here – embassies, international community, non-governmental organizations. It is much easier for an agency official to travel 10 minutes to a reconstruction site in the city than it is 8 hours travel to north Bosnia. So, Sarajevo is substantially reconstructed, but 20 minutes outside the city and elsewhere in the Federation housing is still destroyed."

During my field research in the city in 1999, 2002, and then in 2003 there were discernable improvements in the physical fabric of the city, both in terms of the redevelopment of key symbolic buildings (such as the *Oslobodjenje* newspaper building, the Zetra sports stadium, and Unis skyscrapers destroyed during the war) and improvements in front-line residential neighborhoods such as Grbavica. At the same time, physical scars from the war, including the decimated Parliament Building and the destroyed National Library Building, are daily reminders of the siege (see Figure 4.6). And, in the redeveloped front-line neighborhoods, new

Figure 4.6   Parliament Building, Sarajevo

residential blocks and international organization buildings still coexisted as of 2003 alongside the bombed-out remains of adjacent apartment buildings.

The reconstruction process first focused on emergency infrastructure and service needs created by the war and then subsequently sought to support the returns process by prioritizing the reconstruction of those units having identified returnees/beneficiaries. The Reconstruction and Return Task Force within the IC put forth the overall plan and prioritization for housing reconstruction efforts in Bosnia. Each year, the main axes of returns were identified and this is where housing reconstruction was targeted. The European Commission would then solicit funding proposals from nongovernmental organizations and individual donor countries for designated return areas.[50]

Despite the achievements in the physical reconstruction of Sarajevo, the changed ethnic complexion of the city today indicates that the IC is physically rebuilding a predominantly ethnically sorted city. As of 2004, a "great gap between return and reconstruction remained for Bosnia," states Sanja Alikalfic (regional director, United Nations High Commissioner for Refugees, interview). Although linking reconstruction to likely returnee/beneficiaries (rebuilding where a return is likely) increases the efficiency of the effort, earlier and more extensive reconstruction (without waiting for identified beneficiaries) may have been able to send a stronger signal to displacees and refugees about their future viability in their places of origin. Further, extensive international intervention in rebuilding has focused on physical reconstruction rather than social rehabilitation. This means that there has been less attention paid to those neighborhood attributes that are critical to sustainability of minority returns – job availability, satisfactory education for children, security, and non-discriminatory access to urban services and social support. Such a physical emphasis implies that cities like Sarajevo may be increasingly physically rebuilt, but the population will be ethnically sorted and ethnic minorities will be socially isolated and traumatized. In addition, domestic authority resistance to returns has been a fact of life in both the Federation and RS. A Bosnian human rights institution has reported for many years "conscious obstructions by the authorities at all levels, including steps they took or failed to take, all targeted against real return" so that any returns "would not endanger ethnic domination in the part of the Federation under their control" (Ombudsmen Institution of the FbiH 2004). In Republika Srpska, delays and obstructions in housing reconstruction in eastern RS have for some time clogged the process of encouraging Serbian returns to Sarajevo because war-displaced Bosniak Muslims living in repossessed Sarajevo units have limited options about where to go.

The close involvement by the international community in the returns process in Bosnia confronts it with a difficult moral quandary. Encouraging returns to pre-war locations has moral weight behind it because it may reanimate pre-war multi-ethnic integration. Yet, it also may stimulate inter-group tension and conflict and cause hardship on returnees if they are disconnected from social and economic opportunities in the pre-war location. On the other hand, accepting non-return of a substantial number of persons who now are abroad or internally displaced within Bosnia is a pragmatic response midst the difficulty of seeking to re-create pre-war demographic geographies and may produce over the medium term a more stable inter-ethnic situation. Yet, such ethnic partition based on relocation likely has long-term negative consequences on inter-group under-standing and tolerance. This moral quandary continues to trouble the IC. As stated by J. Mier (interview), who has been part of Bosnia's international effort since 1994: "If we accept what this terrible war has produced, then what is our sense of being here? Why the intervention, then? If acceptance, it would mean a total fiasco of international policy in Bosnia." The priority of the IC has always been for minority returns; if the person or household decides to remain where they are (to relocate permanently), no international assistance for housing has been offered and the displacee must seek domestic financial support instead (S. Alikalfic, UNCHR, interview). Yet, it is also clear that the IC is cognizant that pragmatic realities obstruct its moral stance. Some observers take issue with the IC's reintegration strategy as unnecessarily standing in the way of an inevitable, even desirous, homogenization. Snyder (2000: 326) asserts that "the international community's insistence on maintaining the Dayton Accord's de jure fiction of political integration seems almost perversely designed to prevent the acceptance of an inevitable equilibrium." There is some tacit acceptance by the IC of relocation. "The OHR says that property return implies actual return," says one source,[51] "but this is often not the truth. At first, the selling of repossessed property was not accepted by the IC as a return, but now we accept it." The selling of repossessed properties flies in the face of ethnic reintegration goals and "discussion within the IC has been heated more than once," states Richard Ots (OHR, interview), "yet the law calls for choice by people where to live and OHR will not force people where to live. The property repossession law gives people a fair opportunity to live where they want."[52] In a similar vein, Jayson Taylor (deputy head of RRTF, interview) defends the IC's stance: "From our legal perspective, the important thing is that they have had the reasonable opportunity to exercise their choice of return." In particular, the property repossession is viewed as playing a key role in this provision of choice. The ability to possess and, if the person desires, to then sell his residential unit creates opportunities for the repossessor to buy a home in whatever location he wants.

The distinct possibility of an ethnically sorted Sarajevo and Bosnia in the future is in line with the political self-interests of both Federation and RS entities and the nationalist parties that have controlled their destinies. In the short term, this ethnic relocation and sorting may well create some normalcy of daily life and heightens urban and group security (J. Mier, interview). Each of the sides is able to heal wounds and to build strength and confidence within their own territories. There are indeed strong human and political impulses toward ethnic consolidation after the trauma of war. Relocation has been the preferred solution of Bosnia's nationalist parties (International Crisis Group 1998b). These political parties, in the case of the Bosniak Croats with assistance by Croatia, have provided financial support for relocation, constructed new accommodations, and facilitated property exchanges that result in ethnic consolidation. In other cases, the sheer size of displaced populations creates demand for the provision of emergency accommodations in their displaced locations; in Sarajevo Canton, for instance, 25,000 displaced persons may have the right to emergency accommodation in accordance with Cantonal laws (OHR 2002a: 3).

Amidst these realities, some suspect that the IC may be implicitly accommodating de facto demographic sorting as a means toward inter-group stability. Kumar (1997) notes how the idea of ethnic partition as a means toward addressing ethnic conflict gained a new legitimacy in Dayton. One source[53] within the IC suggests that the OHR now, off record, accepts the three-way ethnic sorting of Bosnia as a means toward stability while publicly maintaining support for the reintegration of the country. Although morally repugnant because it accepts the goals of the war makers, acceptance of relocation and sorting may indeed over the next decade aid social stability due to its ethnic consolidation effect. However, it will likely impose enormous costs in the longer term on Bosnia's degree of multi-ethnic tolerance. Any de facto legitimization of partition and sorting for the sake of security and normalcy likely facilitates in the longer term increased inter-group rigidity, lack of interaction and tolerance, and greater ability of external and internal forces to radicalize populations (J. Mier, interview). For the Bosniak Muslim population specifically, partition and segregation may lead to feelings among the moderate Muslim population that they are isolated in a Christian Europe, thus opening them to the influence of radical messages (interview).[54]

The ability to effectuate a genuine ethnic reintegration of Bosnia has been a conscious priority of the IC since the start, and significant credit is due. It is a difficult process to manage (especially when it is dependent on domestic authorities for implementation), and appears at its numerical limits ten years after the war, susceptible to lack of sustainability and relapse. Despite its significant efforts, the IC is faced with a state that is ethnically gerrymandered and sorted to a substantial degree. Any permanent relocation of persons displaced by the war, as warned in International Crisis Group (1998b: ii), "risks leaving a frustrated, hate-filled and despairing population ... and abandoning entirely the concept of multi-ethnicity in Bosnia." In the end, the foundation that Dayton constructed, based on the partitioning of political space to accommodate ethnic difference, is more likely to suspend and prolong Bosnia's ethnic conflict than to solve it.

Transcending Dayton

Ten years after the end of the Bosnia war, Bosniak, Serb, and Croat leaders signed an agreement to seek a strong and more unified Bosnian government.[55] It envisions a single state president and possibly a strong prime minister and strong parliament. Looking ahead to state reform, a US official described Dayton as having "established a state with internal Berlin walls" and asserted, "these internal walls must now be torn down" (Knowlton 2005).[56] By the time of this diplomatic promise, the international community had become increasingly aware of the constraining effects of Dayton's geography, and had already given some attention to practical strategies that could transcend Dayton's partitioned political space. These practical techniques by the IC, emphasizing economic reform and revitalization, do not attempt to explicitly overturn or rewrite Dayton's political lines. Rather, they seek new functional partnerships and alignments that, it is hoped, will minimize the independent importance of these lines in determining household and political party behavior. These techniques constitute potentially important contributors to peace building. Because they have the ability to positively affect people's daily lives, these interventions hold promise for complementing and enlivening any future advancements that are achieved at the higher, diplomatic level regarding reform of Bosnia state structures.

One approach establishes economic regions within BiH that connect Federation and RS municipalities around collective interests, and focuses on local governments as significant agents in bottom-up peace building and state building. A second approach seeks to overcome the fragmented political geography by encouraging Bosnia to join, with its neighboring countries, a collective development path that would bring it into the European Union.

Regarding the Sarajevo urban area, there is an emerging hope among some in the international community that functional linkages could increasingly transcend and de facto erase the Dayton boundary lines. This would increase the possibilities for trust building within the urban region, and the IC hopes that these local relationships could be an important building block in the overall state building project. After years of discussion within the IC, there was the establishment of an economic region for Sarajevo and a development agency, Sarajevo Economic Region Development Agency (SERDA).[57] The region is defined geographically and functionally and intentionally spans the Dayton IEBL. A significant accomplishment is that its members include Canton Sarajevo and the Township East ("Serb") Sarajevo, likely the first time these two political bodies have agreed to work together. With about 700,000 population and covering over 8,000 square kilometers, the regional partnership includes eighteen municipalities from the Federation and 13 municipalities from Republika Srpska (European Union and SERDA, 2004). Encompassing all the local jurisdictions that historically have economically gravitated toward Sarajevo, the region constitutes the biggest market in the country. SERDA's main goals are to re-establish the economic base and potential of the region, increase business linkages within the region, as well as better connect and market the region to others in BiH and Europe. Although the SERDA effort is premised on solely voluntary and functional cooperation across Dayton boundaries, its eventual aspiration is one that is more potentially significant for Bosnia's future. "Ultimately, and I don't think this is a hidden agenda," says R. Ots (interview), "we foresee these economic regions as forming new administrative regions of BiH." In other words, as functional and economic links become pre-eminent, these multi-ethnic functional regions would replace Dayton-conceived local boundaries as the main means of policymaking and organizing local programs and activities.

An interesting aspect of the SERDA approach is that, whereas RS national politicians have consistently been most resistant to IC programs, local political leaders on the RS side of the Sarajevo urban area "have been the first to grasp the SERDA idea and the benefits of regional economic zones" (M. Power, interview). In the post-war environment, many of these local political leaders in eastern RS felt cut-off and marginalized as Bosnian Serb political power consolidated in the city of Banja Luka to the north. Bosnian Serb local leaders outside Federation Sarajevo increasingly were confronted with numerous urban problems – lack of infrastructure, unemployment, revenue constraints – but had very few tools to deal with them. Thus, SERDA offers to local politicians and businesspersons in RS a degree of cooperation with Federation Sarajevo localities and provides links to the economic and business opportunities there (M. Power, interview).

Among the outcomes of the SERDA effort, still in its infancy as of 2004, has been some normalization of business decisions with some enterprises re-establishing their pre-war locations by moving across the IEBL and re-creating economic linkages that span the divide. There has also been some development of regional infrastructure, such as roads, to facilitate cross-boundary economic activity. In addition, the ability of functional relationships to alleviate inter-group tensions was illuminated during the Dobrinja dispute in which the location of the final inter-entity boundary line had to be eventually decided by an international arbitrator. In that case, a regional forum facilitated constructive discussion about the future of the functional relationships affected by the boundary realignment. The IC hopes this is an indicator of a future in which trust and relationship building across the Dayton line can make war-created partitioned geographies increasingly obsolete.

Based on such experiences, the OHR has become increasingly forthright in internal, programmatic reports about the value it attaches to local governance in the Sarajevo area in the national peace-building project. It notes the continuing weaknesses of state and entity cooperation in Bosnia and highlights the Sarajevo functional region, spanning the IEBL, as a good fit with international goals (OHR 2003a). The OHR indicates that its activities in the Sarajevo region "will focus solely on the region and its municipalities" instead of on the entities or the state. OHR efforts include: "straightening out local self-governance, promoting return, reconnecting divided municipalities, linking up the mayors through joint interests and the Economic Region initiative." Provocatively, the OHR envisions this endeavor as a "bottom-up approach" designed to create a solid foundation for State-building at grassroots governance levels (municipalities and cantons). Such language within its planning documents, and also the SERDA strategy, indicate the faith the IC now has in local governance, innovation, and cooperation to shake up the moribund and ethnically-burdened superstructure of entity and state government.[58] OHR clearly indicates its intentions: "preserving local self-governance, keeping an appropriate amount of authority at the municipal level, gathering the mayors together on common issues and problems using 'economic regions' as forums, *while disregarding the IEBL*, and generally giving freedom to the municipalities to operate, is an essential step in the process of state building" (italics added by author) [OHR 2003b: 2]. Such local and urban peace building, in its ability to create urban models of coexistence and new forms of inter-ethnic cooperation, may provide escape hatches from the deeply ethnicized political geographies of Dayton through which policy and business entrepreneurs can begin the thorny task of resurrecting multi-ethnic Sarajevo and Bosnia.

The other approach to transcending Dayton's partitioned political space aims not internally at Bosnian local governance but externally at European governance. One of the perceived benefits of European integration is that it may be able to more effectively address nationalism and inter-group ethnic differences (Dunkerley et al. 2002). In the case of the Balkans, the IC feels that the incentive of potential integration with the European Union will help advance Bosnian democratic and judicial institutions and revitalize economic linkages and structures destroyed by war. The EU views this incentive as able to promote stability not just in Bosnia but also throughout the "Western Balkans" region, defined to include Albania, Bosnia and Herzegovina, Croatia, Macedonia, and Serbia and Montenegro. The purpose of such a Western Balkans regionalism is to encourage economic transition, cooperation, and harmonization across boundaries, and new market linkages both across Bosnian entities and among neighboring countries. In mid-2000, the European Council confirmed the objective of the fullest possible integration of the Western Balkans countries into the European mainstream and recognized the countries as potential candidates for EU membership. In 2003, the European Council stated "the future of the western Balkans is within the EU." Later that year, a report by the Commission of the European Communities (2003b: 2) stated that the "unification of Europe will not be complete until these countries join the European Union."[59] The intent and hopes of the EU regarding the Western Balkans are clear.

The region's states entered into a "Stabilization and Association Process" (SAP) that is designed to help the five countries with their reforms and to bring them closer to EU membership. This process is intended to consolidate democracy, increase interstate regional cooperation and shared agendas, enhance stability, and promote economic development and transition. Western Balkans regionalism is viewed as the appropriate scale in efforts "to promote social cohesion, ethnic and religious tolerance, multiculturalism, return of refugees and internally displaced persons and combating regressive nationalism" (European Union 2003: 2).

For Bosnia, where the state and entity governments have been slow to advance needed economic and political reform, this EU process seeks to stimulate at a broader, Western Balkans scale more genuine democratic processes and a more market-based economy and vibrant private sector. Thus, the aspiration is that Bosnia may achieve through an extra-state regionalist EU agenda what it otherwise would be unable to do if left on its own.[60] Intra-state problems are to be transcended by reference to a larger scale. With interstate cooperation and the shared goal of EU integration, there is the push for harmonization of economic, social, and human rights policies and instruments across the borders of the five states. And, significantly for Bosnia, to join in this shared EU agenda, such cooperation and harmonization would need to occur across Federation and Republika Srpska entity borders.[61]

Bosnia's progress from war reconstruction to a "full transition agenda" that would move it along a path toward integration into the EU has been difficult (Commission of the European Communities 2003a). The first step of this "stabilization and association process" (that included 18 prescriptive political and economic steps deemed "substantially completed" September 2002) "took too long and its full implementation requires continued attention" (Commission of the European Communities 2003a: 4). As of mid-2003, BiH was behind most of its neighboring countries in this process, and the EU noted that Bosnia must accelerate reform and develop truly self-sustaining political and economic structures (Commission 2003a). Although observing that Bosnia has moved away politically from the chronic instability of the early post-war period, evidence that domestic institutions fully embrace reform remained limited. In assessing the overall implementation progress of the stabilization and association process, the Commission (2003a, 21) concluded, "There is little proof that BiH has used ... EU requirements to dynamize reform." Issues related to the establishment of the "rule of law," human rights, the protection of minorities (especially in RS), fiscal stability and economic reform were noted as problematic.[62]

Along with political reform, the restructuring of Bosnia's economy is of primary importance to the EU. The turmoil of 1990's Bosnia was certainly a political one influenced by nationalism; yet, also occurring in the early 1990s was a crisis of Yugoslavia's economy that helped fuel nationalist agendas (D. Ivanovic, interview). What the war then did to Bosnia was to place it many years behind central and Eastern European countries that began their economic reform paths away from their planned economies in the early- to late-1990s, and which are now members of the EU. Thus, post-war economic policy in Bosnia is not about resurrecting pre-war economic structures and recovering lost jobs, but involves fundamentally restructuring an economic system from one that was partly centralized before the war to a more market-based economy viewed by international overseers as better able to function in today's global economy.

The old Yugoslavia represented a "third way" economically, which might be called a "humane socialist model" (R. Ots, interview). Lindblom (1977) highlighted the "market socialism" of Yugoslavia as a viable alternative to market economies and communism. This old system combined elements of the free market system with social ownership of enterprises, coordinating worker control (in the form of "self management units") through the use of market mechanisms. The crisis of this economic system, coupled with the devastation of the war years and Dayton's ethnicized political geography, has left "the third way" in tatters. Ten years after the war, BiH still faced a heavy reform agenda on its path toward a more market-based economy (Commission 2003a). With the entities more in charge of economic policy than the state, there is no "single economic space" and the national market has been ethnically fragmented and distorted, causing substantial losses and inefficiencies. Inter-entity capital and labor mobility remained limited as of 2003. The average salary is one of the lowest in the region, the business climate, foreign investment and free trade potential are hobbled by ambiguous regulation and the lack of a single national market, and strategic privatization of large-scale companies has been slow. Further, despite the IC's crucial and positive role in Bosnia, international intervention has generated a "beneficiary mentality" that obstructs independent and sustainable political and economic development (Joint Declaration 2003: 2).

The provision of jobs and economic opportunity in Bosnia is viewed as a key component in many international and domestic efforts at peace consolidation. The availability of jobs and access to them on a nondiscriminatory basis are viewed as key in sustaining minority returns and bringing back to Bosnia the many refugees currently in foreign countries. Without Sarajevo's economic recovery, there will be a vacuum that will drag down efforts at national economic recovery. An expanding economic base would encourage an ethnic mixing at workplaces that could provide initial steps at regaining some multi-ethnic tolerance in a country otherwise segregated residentially.[63] One interviewee (R. Ots, OHR) portrayed how a typical Bosnian resident might feel: "Don't put me in some reconciliation program, just put me in a factory alongside my neighbor and we'll become friends."

There is no doubting the important role of economic growth in Bosnia's future and the fact that shared economic gain can bring people together. Yet, economic development based upon shaky and fragmented political institutions is likely not sustainable. What likely must come first is a fuller extension of Bosnia's democratization process so that there exist viable and genuine democratic institutions that allow for the expression of political voice and the protection of individual, especially minority, rights. With the ability of Bosnia's political structures to move the society forward still in doubt ten years after the war, economic development will likely lag significantly.[64]

Ironically, years after Dayton misplaced and contained Sarajevo and other municipalities as key assets in multicultural peace building, the international community has a newfound belief in the capabilities of local and urban governments to move Bosnian society forward. There is optimism that functional links and commonalities at the urban level can blur internal Bosnian political boundaries, while European aspirations might advance Bosnian democracy and Western Balkans cooperation. However, Dayton-created Bosnian boundaries have reified ethnic geographies and may be too hard to surmount. After years of ethnic circumscription of Bosnia's formerly interspersed cultural geographies, it may be too late. Dayton's boundary drawing established a playing field upon which ethnic development impulses have aligned and upon which ethnic political interests have congealed. In retrospect, the inability to position Sarajevo and other localities during the formative years of peace building as constituent parts in the projects of multi-ethnic reconstruction and reconciliation was a huge missed opportunity for this terribly wounded country. In peace accords that necessarily emphasize conditions that stop a shooting war, peacemakers must also be cognizant of the transformative potential of urban areas to help, with time, to reconstitute and rediscover multi-ethnic tolerance.

Notes

1. "Bosniak" is a post-war identifier that can connote religiosity (Muslim) or personal support for a multi-ethnic Bosnia. I use the term in its first meaning.
2. In contrast to Spain's transitional period after Franco (1975–1980), the period after Tito's death in 1980 until the Dayton accord in 1995 is marked by a longer duration of uncertainty, no clear trajectory toward a sustainable democracy, and then the intrusion of catastrophic warfare starting in 1991. I delimit my study of transitional democratization in Bosnia from Dayton (1995) onwards, recognizing that there are factors present in the Bosnian case not found in the Spanish case.
3. In the then Republic of Bosnia, before the war, the 1991 Census identified 44 percent of the population as "Muslim Slavs," 31 percent as "Orthodox Serbs," 17 percent as "Catholic Croats," and 7 percent as "Yugoslav" and "other." Geographically, areas of ethnic dominance in Bosnia tended to fall in isolated pockets rather than in contiguous areas. Of the approximately 100 administrative districts (*opstine*) in pre-war Bosnia, in about one-third of them no ethnic group had a strong majority or numerical advantage (Burg and Shoup 1999). One July 2005 estimate of post-war Bosnia population puts the ethnic distribution about 48 percent Bosniak Muslim, 37 percent Serb, 14 percent Croat, and 1 percent other (Central Intelligence Agency 2005).
4. In three of the major districts within Sarajevo city, between 11 and 17 percent of respondents identified themselves as "Yugoslav" on the 1991 Census. These are significantly higher percentages than for Bosnia overall.
5. The ethnic distribution of the pre-war city was approximately 49 percent Muslim, 30 percent Serb, and 7 percent Croat (1991 Census). No census has been taken since the war due to worry that it may statistically hold in place the current locations of displaced persons. Thus, post-war population figures are estimates based on best available data.
6. Equal to approximately $3,600 per year (at June 2005 exchange rates).
7. The EC aid program since 2001 is CARDS – "Community Assistance for Reconstruction, Development and Stabilization;" it provided 72 million euros in 2002 and 63 million euros in 2003, well below earlier aid levels,
8. Observations in this section come from October 1999, when the author participated in a peace-building conference in Sarajevo entitled "Breaking Walls" (Adjuntament de Barcelona 1999). Participants included political and community leaders from the cities of Belfast, Jerusalem, Nicosia, Beirut, as well as from Mostar and Sarajevo.
9. Rotberg (2004: 5, 9) describes "collapsed states" as those with a vacuum of authority; they are extreme versions of "failed states" that are deeply conflicted and contested bitterly by warring factions. Other collapsed states include Somalia in the 1980s, Lebanon, and Afghanistan.
10. A more recent estimate (July 2005) puts Bosnia's total population at 4.03 million (Central Intelligence Agency 2005).
11. During the writing of this book, Bosniak, Serb, and Croat leaders signed in November 2005 an agreement to seek a strong and more unified Bosnian state government, envisioning a process that would create a single-state president and perhaps a strong prime minister position and strengthened parliament.
12. The state constitution is a constituent part of the Dayton Peace Agreement.
13. State powers include foreign policy, customs and trade, monetary policy, international immigration, transport and communication; progress on these policy areas are often held hostage to gridlock at state government level.
14. The idea to cantonize the RS side of BiH was nixed by the IC because it was felt it could facilitate partitioning of RS and the annexation of eastern RS by Serbia (Morris Power, Sarajevo Economic Region Development Agency, interview).

15. Massive war-caused displacement such as on the scale of Bosnia-Herzegovina's causes substantial impediments to the conduct of municipal elections, tied as they are to specific geographies. Between 40–50 percent of Bosnia population no longer lived where they were registered for the 1991 Census and were provided the right to vote in 1997 municipal elections in their place of 1991 residence. The ability of displacees to vote in their former municipality meant that many city councils represented "virtual populations" of ethnic members who voted but had not physically relocated back to their former place of residence (Jokay 2001).

16. The Party for Bosnia and Herzegovina (SbiH) is a moderate, soft Bosniak nationalist, pro-state party and was in coalition for a short time with the SDA. In addition, a multi-ethnic "Alliance for Change" political coalition, openly committed to international goals and the Bosnian state, helped form governments at state and Federation levels after the 2000 general election. At the core of the Alliance was the Bosniak Social Democratic Party (SDP), a moderate, non-nationalist party that draws its electorate primarily from urban areas. The October 2002 general elections, however, brought nationalist parties back to power, halting the progress of the SDP and other reform oriented parties.

17. Sources: Sarajevo Canton government (2004), Federation Ministry of Displaced Persons and Refugees (2003). Today's Canton in the Federation excludes some areas that used to be part of Sarajevo urban area; specifically, not counted in Sarajevo Canton demographic figures are areas in Republika Srpska and populated by Bosnian Serbs. However, even holding boundaries constant pre- vs. post-war, the Bosnian Serb presence in the urban region would likely be much lessened due to emigration to other parts of RS and to Serbia.

18. See note 5; pre-war vs. post-war "city" boundaries also do not correspond. War and military actions often result in changes to municipal boundaries to accommodate post-conflict ethnic realities or aspirations (Jerusalem is a prime example [see Bollens 2000])

19. After Dayton, the part of what once was considered Sarajevo that is in the Federation constituted about 61 percent of its former territory (International Crisis Group 1998a).

20. Identity of interviewee withheld upon request.

21. It is estimated that only about 8,000 Serbs stayed in these five suburbs after transfer (Internal Displacement Monitoring Center 1996).

22. Protocol on the Organization of Sarajevo, OHR, October 25, 1996.

23. Estimates of returning minorities to Sarajevo Canton from the end of war to the end of 1997 were 13,200 Croats and 5,600 Serbs (U.N. High Commissioner for Refugees). Compare this to the approximately 76,000 Bosniaks that returned or resettled to the Canton during the same period, and to the mass outflow of Serbs in early 1996.

24. Sarajevo Declaration, Office of the High Representative, February 3, 1998. The document was the result of a UN conference on returns to Sarajevo.

25. Election Law of BiH (Official Gazette no. 23/01 of 19 September 2001, as amended in 2002).

26. On a positive note, however, the end of guaranteed representation of Croats and Serbs on the city council has not thus far meant dominance of Sarajevo city politics by the Bosniak nationalist SDA. In 2000 and 2004 city elections, Sarajevo voters opted more for the multi-ethnic Social Democratic Party and the moderate, Bosniak "soft-nationalist" Party for Bosnia and Herzegovina than they did for the SDA.

27. This concept emphasizes living within the same urban system but with inter-group psychological distance (Romann and Weingrod 1991). It is a likely scenario for a city after the trauma of inter-group conflict and war.

28. It is also possible that Serb and Bosnian Serb leaders who influenced Serbian out-movement from neighborhoods incorporated into Federation Sarajevo may, under

this alternative scenario, have supported continued Bosnian Serb residential claims to a Sarajevo that spanned entity boundaries.

29. This is a reference to Washington D.C., which is a special government district not part of any state (M. Power, interview). Such status for Sarajevo would mean it could be a capital that means something for everyone, without ethnic/nationalist coloring.

30. The parallel to cantons in the US governance system would be county governments.

31. Bosnian Serb and Serbian leaders favored a more centralized structure for what became Republika Srpska, composed of only the entity government and local governments. Thus, cantons were not a feature of negotiations over Bosnian Serb political organization.

32. This "Washington Agreement" also anticipated that the Federation at some point would enter into a confederative relationship with the country of Croatia.

33. Vance-Owen called these regional jurisdictions provinces rather than cantons.

34. The parallel with Barcelona should be noted: Sarajevo city's ability to use the Olympic Games process to strategically catalyze power within an intergovernmental web of relationships.

35. Identity kept confidential at request of interviewee.

36. This stimulation of urban growth outside an imposed political boundary bears similarity to Palestinian growth in the area of A-Ram, just outside the Jerusalem municipal boundaries imposed by Israel in 1967. In that case, Israeli restrictions on Palestinian development within municipal boundaries helped stimulate A-Ram development.

37. Arbitration Award for Dobrinja I and IV, 17 April 2001, by Diarmuid P. Sheridan, Arbitrator for Dobrinja IEBL. Office of the High Representative, Sarajevo.

38. Officially, the Stup district is within Ilidza municipality, now outside post-war Sarajevo city boundaries.

39. In terms of refugees in Serbia and Montenegro, RRTF (OHR 2002b) estimates that one-half of applications for return seek migration back to Canton Sarajevo. The Federal Republic of Yugoslavia was restructured in 2003 and renamed Serbia and Montenegro; then, in May 2006, voters in Montenegro voted in support of independence from Serbia.

40. United Nations High Commissioner for Refugees (2005a and 2005b).

41. At the time of the Sarajevo conference that produced the Declaration, it was estimated that about 4,000 non-Bosniak Muslim returns had occurred since Dayton.

42. PLIP data reported in OHR, "Statistics: Implementation of the Property Laws in Bosnia and Herzegovina" (various years).

43. This estimate of real returns should be contrasted with the approximately 46 percent rate of return reported by UNHCR.

44. Since December 1995, there have been five High Representatives.

45. There are some analyses, such as by Chandler (1999), that are highly critical of the IC's external regulation of Bosnia, asserting such regulation does more harm than good in democratizing a country, and is more responsive to an international agenda than the genuine needs of Bosnian society. Kumar (1997) views the IC's approach in Bosnia as a problematic revival of ethnic partition as an acceptable means toward ethnic conflict. My own view, developed from interviews with personnel within the IC, is that the IC is in a difficult, no-win situation in Bosnia and that it knows this. My opinion is not as critical of the IC's motivations as Chandler's; at the same time, I agree with Kumar that the IC's involvement in Bosnia has produced numerous unresolved issues related to outside intervention in peace-building.

46. This council is a group of 55 countries and international organizations that sponsor the peace implementation process.

47. Presentation at "Breaking Walls" peace-building conference, Sarajevo, October 1999. At the time, Mr. Bearpark was head of the Reconstruction and Return Task Force for Bosnia.
48. An example of such a protectorate role for the IC is the Brcko district in northeast Bosnia. To overcome deadlocks and deliberate obstructions, in 1999 the whole pre-war municipality of Brcko was placed by the IC under strong international supervision and declared a unified and neutral District. In Mostar, Bosnia, a more authoritative and intervening model of international urban management has been used than in Sarajevo, although with problems.
49. Ministry of Housing figures are reported in Sarajevo Canton (2000).
50. At times, the specific priorities of donor countries or their insistence on high individual project visibility ran counter to the IC's programmatic goals (OHR 2002c). In the immediate years after Dayton, donor countries would tend to fund reconstruction in areas where their NATO armies were located, equating physical reconstruction with greater safety for their troops (G. Wochein, interview).
51. Identity of source withheld on request of interviewee.
52. When urban ethnic animosities are high, governments often limit their explicit efforts at residential integration and instead emphasize freedom of choice for households, hoping that some households will aspire toward more integrated environments. Belfast, Northern Ireland illustrates this public stance (see Bollens 1999).
53. Identity of interviewee confidential by request.
54. Ibid.
55. Agreement signed November 22, 2005 upon 10th anniversary of Dayton Accord.
56. R. Nicholas Burns, Under Secretary for Political Affairs, US Department of State.
57. Five economic regions and development agencies have been created, with IC sponsorship, throughout BiH.
58. The OHR is not alone in this view. The World Bank (2002: 4) views "local-level institutions as critical actors in conflict resolution, effective public service delivery, and sustainable development."
59. Indeed, one view is that the EU wants the region more than the region may want the EU. Without the Western Balkans, there will be an "island in the middle of EU member states" (Bashkim Shehu, Albanian writer, interview.) Further, in a 2003 proposal to amend the Dayton Accord, a communiqué representing 24 European political officials noted that the IC and Europe "need Bosnia and Herzegovina as a proof of successful peace policies in recognition of variety and diversity."
60. This moving up of the geographic scale of policy influence, from within Bosnia to a broader extra-state regionalism, is similar to the "rescaling" process as described by Brenner (2004). However, whereas Brenner concentrates on economic competition as the driving force behind rescaling, I would add here an additional consideration – that rescaling to broader geographies may open up opportunities for addressing intrastate inter-group conflict that would otherwise be obstructed by intrastate paralysis.
61. This effort to dampen internal conflict as a means toward EU membership bears certain similarities with the failed effort to link an accord between Greek Cypriots and Turkish Cypriots on the island of Cyprus to the EU ascension process.
62. The November 2005 agreement to strengthen and unify state governance was viewed as an important step in reconnecting Bosnia to the EU ascension process.
63. Public employment ethnic hiring requirements are of importance here. By law, the ethnic composition of the workforce for all public bodies in Bosnia should be equivalent to the 1991 ethnic distribution in that local area. Municipalities are required to hire to get to this ethnic distribution. Although there are no such ethnic quota requirements for private sector employment, a private company may seek a seal of approval from the Organization for Security and Cooperation in

Europe (OSCE) for abiding by a set of fair employment criteria, including antidiscrimination.

64. Economic growth has lagged in countries that have undergone political transformations, such as South Africa in the 1990s and Spain in the 1970s. An alternative view is that economic reforms that introduce freedoms can precede and catalyze political reforms; some observers apply this view to contemporary China and other southeast Asian countries.

5 Bilbao, San Sebastián, Vitoria (Basque Country): urban dynamism amidst democratic disability

Figure 5.1 "Old" Basque Country – support for political prisoners, Vitoria

Basque Country ("Pais Vasco") in northern Spain is one of the most economically advanced regions in the country, has the greatest amount of financial and regional autonomy vis-à-vis the Spanish state, and for the last 15 years has experienced a profound physical revitalization of Bilbao, its largest and most important city. It is also a locale where militant violence in pursuit of Basque independence has been a part of the political dynamics for 30 years after the death of Franco.[1] The conflict is not between two communities living side-by-side (such as Protestant unionists and Catholic republicans in Belfast), but rather between radical Basque nationalists and a Spanish state viewed with contempt as an

Figure 5.2 "New" Basque Country – Guggenheim Museum, Bilbao

unwanted occupying force. Caught in the middle have been moderate nationalists who support greater Basque independence but reject militant violence.

Compared to Catalonia, the nationalist conflict here has been more radicalized, and the political transition to a workable democracy has been significantly more prolonged. In many respects, Basque Country has experienced two transitions – the formal and largely successful one from Franco authoritarianism to a functional model of regional autonomy, and a second ongoing one: from a regional democracy hamstrung by radical nationalism to one where nationalist grievances may one day be expressed politically rather than through violence.

Strong regional nationalism in the 1970s successfully created during the transition period the governance and financial foundation upon which Basque urban policies and programs have had significant impacts over the past 25 years. The transition from Franco, although not without complications in the Basque Country compared to other Spanish regions, secured for the region a degree of autonomy and a level of financial resources unparalleled in Spain. Basque nationalistic aspirations during post-Franco negotiations over the new Spanish state helped to create a space of self-government within which proactive urban interventions have occurred for over 20 years. In addition, nationalists' negotiations with the post-Franco Spanish state produced a historic economic agreement with Madrid, and this arrangement subsequently has enabled and catalyzed impressive urban revitalization in Pais Vasco, the city of Bilbao in particular.

Yet, existing alongside the functional level of urbanism and local governance in the Basque County has been a regional politics distorted by radical nationalists' use and threat of violence. Effective Pais Vasco urbanism has occurred within a "socially traumatic" context (Victor Urrutia, Professor of Sociology, University of Pais Vasco, Bilbao, interview). With consensus building regarding regional political issues often obstructed by radical nationalists' acceptance of violence as an appropriate path, the building of a normalized polity has remained out of reach (Pedro Arias, *Gesto por la* paz, interview). A "spiral of silence" has existed in Basque Country as political voice is squashed due to intimidation and as the Basque general population increasingly views societal violence with alarm (Mata 2004; EuskoBarometro 2005).

We saw earlier how urbanism played a key role in Barcelona during the transition from Franco in constructing an urban terrain that actualized democracy. In contrast, despite significant urban revitalization in Basque Country since regional autonomy, there has persisted a hardening of political positions about nationalism, a democratic disability, and separatist violence. Compared to a pragmatic approach to Catalan nationalism that advocates its cause within the boundaries of Spanish state sovereignty, there has been in the Basque Country a radicalized nationalist strategy more fundamentalist in approach that plays outside state institutions (P. Vilanova, interview). In such an atmosphere, there have been two parallel tracks here with semi-autonomous trajectories – a productive urban and local governance track and a destructive regional political track. The Basque Country case shows how urbanism and city development, and the ability to change and improve the quality of urban life and opportunities, can shuffle the decks in a region that otherwise would remain obstructed by political gridlock and societal violence. The effect of such deck shuffling on normalizing a society takes a long time in the face of the self-perpetuating dynamic of violence and thus at times appears irrelevant. Yet, urban enhancement represents the most visible form of the benefits of regional self-government and creates facts on the ground able to produce momentum toward social and political normalization.[2]

EUSKAL HIRIA, NATIONALISM, AND VIOLENCE

Glossary: Etxea – the Basque house; Euskal Hiria – Basque city

Basque urban and social structure

The Basque Country[3] has a population of over 2 million people (as of 2001) and is about 2,800 square miles in size (about one-fourth that of Catalonia). It lies at the western end of the Pyrenees in northeastern Spain and contains stunning mountainous and coastal terrain, together with obsolete industrialized and highly dense urban fabric. It consists of three distinguishable political components – the provinces of Vizcaya (*Bizkaia*), Guipuzcoa (*Gipuzkoa*), and Alava (*Araba*)[4] [see Figure 5.3]. There are three major cities in Basque Country, each the capital of a province. The city of Bilbao (*Bilbo*) [2001 population of about 350,000], in Vizcaya province, is the largest city in Basque

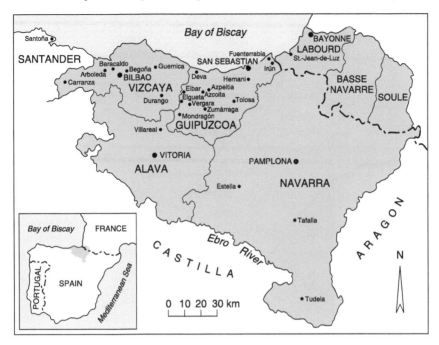

Figure 5.3   The Basque Country

Country, and its metropolitan area (population over 900,000) has for decades been the center of iron and steel manufacturing in the Iberian peninsula. The city of San Sebastian (*Donostia*), with a population of approximately 180,000, is located in Guipuzcoa province in northeast Basque Country a few miles from the French border and is well known for its physical setting and tourism. The city of Vitoria, both the capital of Alava province and the Basque regional government (Gobierno Vasco), has a population of over 200,000 (Gobierno Vasco 2002).

The urban and economic structure of the Basque Country is important to understand in examining the role of urbanism in this nationalistic region. The three major cities of Basque Country, and their urban catchment areas, comprise a polycentric system of urbanization. Each urban area is distinguishable by physical characteristics, local history, and increasingly by political constituencies; the cities are about a one hour drive from each other. On the other hand, the cities are also increasingly connected functionally and economically and create the potential for development of the Basque Country into a single integrated "city system." Interspersed throughout the region, especially to the north, is a set of medium-sized cities, many of industrial heritage. In addition, there are numerous small towns throughout the region; indeed, more than 80 percent of the 250 municipalities in Basque Country are smaller than 9,000 in population (Llera 1999). The provinces of Vizcaya and Guipuzcoa are heavily industrialized

Figure 5.4   Metropolitan Bilbao

and with high population densities, with Alava of lesser density and more recent industrialization.

The Bilbao metropolitan area, because of its size and economic importance, is of special note. Comprised of thirty municipalities, the urban region contains close to 45 percent of the population of the Basque Country with six of its cities over 50,000 population (see Figure 5.4). The gross domestic product of metropolitan Bilbao represents one-half of Basque Country economic output. It is a traditional port city that has been, and remains, an industrial heartland. Its base has been in heavy manufacturing. There has been severe manufacturing employment decline, and, along with physical dereliction and pollution, the city had become an archetype of a declining industrial area (Rodriguez, Martinez, and Guenaga 2001).

Similar to Catalonia, the Basque Country experienced significant in-migration during the Franco years as the regime's strategy of planned industrialization from 1963 onward artificially bolstered the Basque economy through import protection and other means (Clark 1979). Non-Basque Spaniards filled thousands of new steel, shipbuilding, and manufacturing jobs; by 1975, more than 40 percent of Basque Country population had no Basque parent (Kurlansky 1999). Such strong immigration during the Franco years, together with repression by the regime of the Basque language due to its link to nationalist aspirations, means that those who use the Basque language are in a clear minority. Only about 25 percent of Basque persons consider Euskera their mother tongue or consider both Euskera and Castellano languages as mother tongues (Eustat 2001).[5] In terms of nationalistic identity, surveys show the existence of a strong plural Spanish/Basque identity more than two clearly identifiable, exclusive identity communities (Francisco Llera, professor, University of Pais Vasco, interview). Sixty percent of respondents identified themselves as some hybrid mix of Basque and Spanish nationalities, whereas 34 percent of respondents identified themselves as "solely Basque" and 4 percent identified as "solely Spanish" (Euskobarometro 2005).[6] Although there is this plurality of identity, there also is a propensity to feel more Basque than Spanish; 54% feel "solely Basque" or "more Basque than Spanish" compared to only 10 percent who feel "more Spanish than Basque" or "solely Spanish."

Basque nationalism, economics and politics

Autonomy and self-government have deep roots in the history of the region. Codified local customs (*Fueros*) in the Basque Country were first written into the legal code in the twelfth century, although they had been used many centuries before (Kurlansky 1999).[7] These so-called Foral rights of self-determination, like the Basque language, are an essential part of Basque identity. A Basque assembly existed which legislated on Foral law and its early meetings were held underneath an oak tree in the town of Guernica (a place that would later become a bombing target of Francoist nationalist forces during the Spanish Civil War). When modern-day Spain was created from the fusing of the Kingdoms of Castile and Aragon in the fifteenth and sixteenth centuries, Fueros assured exemption from direct taxation by Castile. Instead, Castile and the Basque Country negotiated the amount of revenue needed for the Basque government, which would then raise the agreed sum from its own population. Foral administrations within each of the provinces had considerable self-rule powers (to this day, in Euskera the provincial governments are called *Foru Aldundia*). At times, this relationship between Basque Country and Spain has been more a relation between two sovereigns than one between state and subordinate.

The roots of Basque/Spanish conflict and violence come from long before its anti-Franco and contemporary manifestations (Francisco Llera, interview). Multiple civil wars through the past two centuries have resulted from divisions between traditionalists/conservatives, liberals, and monarchists (Carlists). Each of these groups in Madrid and the Basque Country held fundamentally different views of society and polity. Traditionalists in Basque Country believed in regional autonomy and self-rule by the Fueros. In direct contrast, Liberals believed in a strong centralized administration in Madrid as a preferred alternative to the Spanish monarchy, were anti-Church, and wished to abolish the Fueros. In Basque Country, liberal supporters came from the urban middle classes and the commercial class. The Carlists believed in the continuation of a strong monarchy and the power of the Catholic Church as the best way forward, and allowed sufficient Basque autonomy to garner the support of many Basque traditionalists, especially the clergy, peasants, and aristocrats.

When industrialization came to the Basque Country in the nineteenth century, the struggle over modernization hardened cultural cleavages between Basque traditionalists and those advocating a liberal (in this case, meaning centralized) administration of the Spanish state. Kurlansky (1999) describes this fissure as that between the *Baserritarrak* (people of the farm) who felt besieged by modernization and the *kaletarrak* (people of the street) eager to embrace a new type of society. During the second half of the nineteenth century, Basque cities were becoming cosmopolitan while rural areas were remaining traditional. The population of Bilbao grew from 20,000 in 1850 to almost 100,000 at the close of the century, with more than half of its residents in 1900 born outside the province (Kurlansky 1999). Strong industrialism and capitalism were in force during this time, with Vizcaya producing 77 percent of Spain's cast iron and 87 percent of its steel; along with such production came a Basque bourgeoisie and banking community more aligned with liberal than traditionalist camps (Kurlansky 1999).

Political Basque nationalism arose during this time as a conservative, traditionalist reaction to industrialization, socialism, and immigration. In 1893, Sabino Arana, ironically the son of a wealthy industrialist, led the first open public demonstration declaring Basque nationalism. Fundamentalist in religious meaning, this nationalism garnered the support of persons in small and medium-size towns. Although initiated by sons of Vizcaya industrial enterprises, this movement explicitly made its appeals based on the pure values exemplified by rural home life and heritage. In 1895, Arana founded an underground independence movement, the Basque Nationalist Party (Partido Nacionalisto Vasco, or PNV), a group that has been the political vanguard of mainstream Basque nationalism since.

Believing that Basque regional autonomy would be better served in a republic form of Spanish government than a central state, Basque nationalists aligned with republicans during the short-lived Second Republic in the early 1930s, and at least in Vizcaya and Guipuzcoa provinces, fought against Franco and Spanish centralists during the Spanish Civil War. German and Italian modern air forces aligned with Franco inflicted an indelible mark on April 26, 1937, when they bombed the Basque town of Guernica for three hours during a prime market day and time, killing hundreds if not thousands of innocent people. The victorious Franco regime imposed a state authoritarianism antithetical to Basque cultural and political autonomy, and the years under Franco were as dark in Basque Country as they were in Catalonia. The provinces of Vizcaya and Guipuzcoa were specifically punished by the regime for siding with the republicans during the civil war and they lost all rights of self-rule. Cultural expression was suffocated under Franco, with the Euskera language outlawed and with the symbolism and local history of the region squashed under the heavy weight of Spanish nationalism. The former government of the Basque Country left Spain and became a government-in-exile, focusing on international diplomacy from afar for the next four decades.

The other side of the story about the Franco years in Basque Country, particularly from the 1960s onward, was Franco's dependence upon, and utilization of, the region as an economic asset. In need of economic modernization to hold Spain together, the industrialization program launched by the regime in the 1960s made the three most advanced regions – the Basque Country, Madrid, and Catalonia – developmental focal points as a way to spawn overall national economic growth and to spread capitalism throughout the country (Clark 1979).[8] By 1969, the provinces of Guipuzcoa, Vizcaya, and Alava were the three highest ranked provinces in Spain in per capita income (Clark 1979). During these years, economic integration and linkages between Madrid and Basque Country were common; indeed, many Spanish industrialists were of Basque origin and a strong Basque bank sector controlled substantial amounts of investment in Spain. Ironically, at the same time that Basque nationalism was to become radicalized, the Basque industrial elite was enjoying significant benefits from being part of Franco's "false economy" based on anti-competition and anti-labor planks.

The Franco regime's economic program benefited the Basque Country not out of good will, but because Basque development advanced the regime's own nation-building interests. This is clearly brought out by examining the exploitative fiscal relationship between Madrid and Basque Country during this time. The per capita tax burden in the four Spanish Basque provinces (including Navarra) was about twice as heavy as in Spain overall (Clark 1979). And, while Madrid in 1970 withdrew from the provinces about 30,000 million pesetas in tax revenue, it returned to the provinces in the form of expenditures a little more than 8,500 million pesetas. In other words, less than 30 percent of tax revenue came back into the region. And, as Clark (1979) points out, what was returned went usually not for social and physical improvements, but often to operate central state administrative organs and to maintain public order in what was becoming an increasingly violent place. Along with industrial growth in the Basque Country came unbalanced and deteriorating living conditions. Population growth in Vizcaya, Guipuzcoa, and Alava provinces was more than triple the Spanish average between 1940 and 1970.[9] This resulted in greater Bilbao having one of the highest population densities of any city in Europe. Such high population growth and significant immigration rates were wholly inconsistent with urban planning and social objectives (Ibon Areso, Vice Mayor, City of Bilbao, interview). With local taxation powers limited and tax revenue redirected to service the central state, acute shortages developed of essential social services such as hospitals, schools, and parks.

Amidst the Franco cultural repression of Basqueness, and frustrated by the conservative tactics of the Partido Nacionalisto Vasco (PNV) as a government-in-exile, a group of young Basques in 1959 formed a group called *Euskadi ta Askatasuna* (ETA) [Basque Fatherland and Liberty]. Its original goal was to promote the forbidden Basque language, and in its early years it emphasized social causes and gained support among labor unions. Angered by Franco's continued suppression of language and culture, ETA's political goals solidified behind a platform advocating outright independence of Basque Country from Spain. In 1968, ETA adopted a policy of armed struggle against the Spanish state and in that year killed its first target, a Guardia Civil (Civil Guard) police officer.

Since 1968, ETA killed over 800 persons in pursuit of its political goals of independence; almost 500 of these individuals have been police or military personnel while more than 300 of those killed have been civilians (Guardia Civil website, accessed August 4, 2005). The group targeted mostly national and regional officials and government buildings in Spain, and its killings have had deep psychological and symbolic impact in the country. ETA began by carrying out attacks against Spanish state officials and personnel in the Basque region itself, and most of its activities took place there (Council on Foreign Relations 2002). Its attacks have occurred, however, throughout Spain, particularly in Madrid and in popular tourist destinations. In 1973, ETA pulled off its most important assassination when it killed Franco's apparent successor in Madrid. The Spanish state under Franco met ETA violence with state violence, and terror suspects were at times tortured (Council on Foreign Relations 2002).

From Franco's death in late 1975 to the adoption of the Spanish Constitution in 1978 and the granting of regional autonomy to the Basque Country in 1979, radical nationalists' aspirations for a fuller transformation of society and a move to Basque independence were countered instead by a gradual and negotiated evolution toward a Spanish political system that contained the aspirations for independence. Spain's Constitution stresses that it is "founded on the indissoluble unity of the Spanish nation;" it refers to Basques (and Catalans) as a "nationality" but only Spain is referred to as a nation. Accordingly, even the more moderate PNV called on Basque voters to abstain from voting when the Constitution came to popular vote for ratification. In December 1978, more than 40 percent of Basque voters abstained and in none of the three provinces did the Constitution gain more than 50 percent of the votes.[10] Subsequent to national adoption of the Constitution, the system of regional autonomy was developed and the Basque Country was given substantial self-government and financial competencies. This time, the PNV recommended approval when the Basque Statute of Autonomy (Estatuto de Autonomia) went to Basque voters for approval. The statute passed with 54 percent of eligible voters voting yes.[11]

The statute created the Basque system of regional and local government that exists to this day – a regional parliament and government (Gobierno Vasco) located in Vitoria, three provincial governments ("diputaciónes" [*Foru Aldundia*]), and substantial authority on education, language, and other cultural issues. Most significantly, the statute established in its "economic agreement" (Concierto Económico) a unique financial relationship between the Spanish state and the Basque Country. For 15 of the 17 autonomous regions in Spain, regional governments do not directly collect the bulk of their revenues and must rely on Spanish state budgetary decisions.[12] In Basque Country (and Navarra), in contrast, the region collects all nonsocial security income taxes and other taxes and, after negotiations with Madrid, remits to the Spanish state central treasury a previously agreed amount of money ("quota") to pay for national functions that benefit Basque Country (Agranoff and Gallarin 1997). All other revenue remains in the region. In 2003, less than 10 percent of all revenue collected in the region went to the central state (Kamelo Sainz, director,

Basque Association of Municipalities, interview). This retention of monies in Basque Country means that the per capita level of public expenditures in Pais Vasco is much higher than the Spanish average.

Regional autonomy for the Basque Country in a democratic Spain, however, did not appease strong nationalists. Blaming the PNV for acquiescence in the building of the autonomous state, a Basque political leader broke away in 1978 and formed a political party, *Herri Batasuna* (Popular Unity), that openly supported ETA violence. Three of the deadliest years of ETA killings were during the transition to democracy – 1978 (66), 1979 (76), and 1980 (92). And, over 600 persons have been killed since democracy and the granting of Basque regional autonomy; from 1978 to 1992, there were 2,459 ETA attacks resulting in 653 deaths and 1605 injuries (Urrutia 2004). The provision of considerable financial and tax powers provided to Pais Vasco appears at first glance to have had a negligible or no effect on Basque violence. P. Vilanova (professor, University of Barcelona, interview) observes, "fighting poverty and investing economically in the region has not produced the logical result of removing violence and the Basque political problem from the table." The Conservative Spanish government under Prime Minister Aznar (1996–2004) stringently refused to negotiate with the ETA or its political representatives. In May 2005, the new socialist government of Prime Minister Zapatero offered to hold talks with ETA once it renounced violence.

The Basque political party system has been described as "polarized pluralism" because of its marked fragmentation in support among the public, its significant ideological tensions and incompatibilities, and until recently, the presence of one party openly advocating violence and rejection of the existing system (Llera 1999a). Numerous interviews reinforced this view of a severe "democratic disability" in the Basque political world. Political parties in Basque Country are distinguishable along three axes – nationalism vs. constitutionalism, left-right, and the attitudes toward ETA violence (Mata 2004). The nationalist camp consists of parties which stress the advancement of Basque self-rule as a primary strategy and is composed of the moderate nationalist PNV, the now-banned radical nationalist Herri Batasuna (HB) [linked to ETA] and the Euzko Alkartasuna (EA) party, which developed after a split with the PNV in the 1980s. This group of parties garnered about 65 percent of the popular vote in the Basque parliament and Spanish parliament elections through the 1980s. Since 1994, its share of the popular vote has been more in the 55 percent range (Llera 1999a; Moreno 2001). In the 2004 general elections for the Spanish parliament, support for the nationalist camp fell below 50 percent (Jimenez 2004).[13] The PNV has been able to exert the most political control in the region since Spanish democracy. They have always had plurality support among the electorate, although they have never held an absolute majority of seats in the Basque parliament and thus have needed to form coalitions with other nationalist parties such as the EA (Llera 1999a).[14] The non-nationalist parties are those regional manifestations of Spainwide parties and they favor a constitutional solution to the Basque problem that would maintain or enhance regional autonomy in an otherwise united or federated Spain. The Partido Socialista de Euskadi (PSE), Partido Popular (PP), and Izquierda Unida (IU) have traditionally had a more difficult time finding support in Basque Country than in Spain overall.

This cleavage between nationalist and non-nationalist political parties breaks down when two other factors are introduced – the parties' location on the left-right continuum and their attitudes toward ETA violence. Three parties are on the left – the non-nationalist PSE and IU parties and the radical nationalist HB – and two parties are on the right, the non-nationalist PP and the nationalist PNV. And, in terms of violence, all parties except HB condemned it. Because of its opposition to violence, the PNV has at times conditioned any cooperation with HB on the full cessation of violence. In August 2002, the Spanish government banned Herri Batasuna as a political party due to its links to ETA terrorism. At the time of its termination, HB held seven seats in the 75-member Basque Parliament and hundreds of city council seats in the small towns and villages of Pais Vasco. Since Spanish democracy, electoral support for HB was usually around 14 to 18 percent of the total Basque vote. This electoral support commonly put it second among nationalist parties, behind the PNV that garnered between 25 and 45 percent of the total vote. Since the banning of HB, a new political party, Aralar, has formed which is leftist and pro-independence like the HB, but it rejects ETA terrorism. In the 2004 Spanish state elections, it received only 3.1 percent of the Basque vote.

Public opinion regarding Basque government performance, violence, and the region's preferred political configuration provides insights into whether or not the region will move forward in the future in politically constructive ways. Despite the polarization of Basque society and politics, residents are strongly positive about the effectiveness of their local and regional governments (EuskoBarometro 2005). Their level of satisfaction with government performance in Basque Country is one of the highest of any of the seventeen autonomous regions in Spain (Llera interview). The percentage of respondents who fully reject violence has increased through the past two decades, going from 23 percent in 1981 to 64 percent in 2003 and then to 56 percent in 2005 (EuskoBarometro 2005). During this time, there has been greater mobilization against violence and growing opposition to the ETA (EuskoBarometro 2005).

Finally, the Basque populous shows a marked ambivalence about the preferred future political configuration of the region. In the 2002–2005 period, there has been an approximately three-way split between those who advocate the current condition of regional autonomy, those who desire a federalism solution within the Spanish state, and those who desire Basque independence (EuskoBarometro 2005).[15] At no time in the 28 years of this poll has public support for independence gained more than 37 percent support. Satisfaction with the current Basque statute of autonomy has remained strong since the 1990s.

URBAN DYNAMISM, POLITICAL DISABILITY

One is struck in studying contemporary Basque Country by the existence of two seemingly parallel worlds – one of urban dynamism and capacity, the other of political gridlock and paralysis. In urban affairs, the major Basque cities of Bilbao, San Sebastian, and Vitoria have public bodies that are initiators of change, builders of coalitions, increasingly connected internationally, and able financially to affect change on the ground. At the same time, political violence (both its actual exercise and its potential threat) has distorted political debate and dynamics and suffocated larger political possibilities.

One would expect in a contentious region such as Pais Vasco that urban potentialities would be circumscribed as nationalistic conflicts and tensions disrupt cooperative efforts to achieve urban revitalization.[16] Yet, midst nationalistic conflict and the persistence of ETA activity, urban revitalization has occurred on a grand scale, particularly in the Bilbao area. Planning and urbanism in Basque Country have been able to provide at times during the past 20 years a space of rationality and even consensus in a society where political debate is dominated by nationalism and distorted by violence. In these cases, urban programs and policies have spawned cooperation between public agencies that have transcended differences on larger nationalistic issues. It thus becomes exceedingly relevant to a study of urbanism, conflict, and transitions to consider whether such urban partnerships and agreements constitute testing grounds or models for larger compromises that could move forward not just Basque cities but Basque society and polity generally.

Urbanism: an island in a troubled sea

Given the political instability and unease that we are suffering here, these urban transformation efforts are like an island.

Ibon Areso
Deputy Mayor, City of Bilbao
Interview

The city of Bilbao and its metropolitan area of thirty municipalities make up almost 50 percent of the population of Basque Country and have been its economic powerhouse and functional center. It is where urbanism is most active in the region and most potentially configurative of a different political future for Pais Vasco. Steel, shipbuilding, and textiles manufacturing drove Bilbao's economy for decades after the industrial revolution at the end of the nineteenth century. During the Franco years, Bilbao was artificially supported through import protection, strict anti-union laws, and the allocation of resources to Basque industrialists, who despite the political conflict, maintained strong ties with the Spanish central state and its largesse. When Franco and his regime died, these artificial supports were terminated and a severe crisis resulted during and after the transition to democracy. The industrial model of growth upon which Bilbao was based became obsolete, resulting in high unemployment (as late as 1995, the rate stood at 27 percent in metropolitan Bilbao), urban degradation, unaddressed environmental pollution, out-migration, and social marginalization. The industrial sector had brought good jobs to many persons during its productive years, but at the expense of quality of life and environmental and public health.

Revitalization of Bilbao needed to occur in a "socially traumatic" context that combined the hardships of industrial crisis with the disabilities associated with political violence (V. Urrutia, interview). Bilbao policymakers and planners had to face the urban area's deep post-Franco economic and social crisis midst a nationalistic politics that was unsettled and likely obstructive of the cooperation across government levels and between Basque political interests that would be needed to revitalize the urban region. To restructure Bilbao, there would likely need to be relocation of port and railroad facilities in order to connect the city to the river and to encourage the redevelopment of derelict industrial sites. Yet, many of these facilities were owned or run by Spanish state controlled entities. In a circumstance where radical nationalists were assassinating Spanish officials and attacking symbols of the state, the intergovernmental cooperation needed to resurrect Bilbao was a difficult proposition. Without agreement from Spanish state authorities, Bilbao would remain a city whose best days were rooted in its industrial past.

Bilbao policymakers realized in the first decade of democracy and regional autonomy that they could turn Bilbao's obsolescence into an opportunity to redefine the city. "What might be a handicap in other situations," says Ibon Areso (Vice-Mayor, Bilbao, interview), "became a plus and advantage here because it allowed us to pursue policies that were risky and at times harsh. We didn't have time to waste talking about political issues." Thus, a deep sense of despair, need, and social trauma laid the foundation for future public action. Yet, much more was required than objective need in such a complex political setting, in particular, collective ideas about how to move Bilbao forward, the organizational means and public leadership to implement revitalization goals, and financial capacity and self-sufficiency.

In 1987, the City Council of Bilbao drew up its first general land use scheme, which sought to restructure the physical city in order to facilitate new post-industrial employment growth. Its goals included urban regeneration that would integrate the river with commercial and residential areas, increased accessibility and mobility, and the development of cultural activities as the building blocks to collective vitality. Many of these land-use initiatives involved transforming the city from an industrial one that had its back to the river to a post-industrial city that used the river as a springboard for culture and public interaction. This plan identified two areas – the Abandoibarra waterfront area and the Ametzola central city district – as priorities for redevelopment; both of these areas had significant land holdings owned by public companies operated by the Spanish central state. The General Urban Plan of Bilbao (GUPB), approved provisionally by the city council in 1992 and accepted by the provincial government in 1994, maintained emphases on cultural identity as a means toward city revival, a focus on priority areas where "emblematic spaces" could be constructed, and physical restructuring as the key to socio-economic recovery (Rodriguez, Martinez, and Guenaga 2001: 168). City, provincial, and regional governments also sponsored strategic planning for the city and metropolitan area, a type of planning that encompassed more explicitly social and economic factors and how they relate to land use (Anderson Consulting 1990).

What happened next was a crucial step toward the actual implementation and revitalization of Bilbao – the creation in late 1992 of Bilbao Ria 2000, a publicly funded limited company comprised of 50 percent Spanish state organizations and 50 percent Basque public institutions.[17] It is a private firm whose shareholders are purely public sector bodies, and its mission has been to redevelop the city's obsolete waterfront. This is an atypical organizational arrangement and is possible because each of the shareholders owns or has control over major landholdings in the declared redevelopment area. Shareholders relinquish the land they own in the redevelopment zone, while the city modifies the uses proposed for these areas. Bilbao Ria 2000 then invests in the improvement of these sites and sells the properties to private developers through public tenders. Because of the attractive, central location of the redevelopment area and new public-provided infrastructure support, these sales generate profits for Ria 2000 that are then reinvested in new urban operations and improvement of rail infrastructure. Although the organization is officially a private firm, Ria 2000 operates more as a quasi-public agency because of its public sector shareholders and the public interest nature of its urban redevelopment activities (Rodriguez, Martinez, and Guenaga 2001).

From Bilbao city's perspective, landholdings were the key element to the process. Because much of the land was in the hands of the Spanish government (specifically the rail and port authorities), "without agreement from them, we would not have been able to move forward" (Ibon Areso, interview). There were times when a Spanish state entity was at first resistant to the urban plans; for example, the state rail agency *Red Nacional de los Ferrocarriles Españoles* (or RENFE) was unhappy with relocating their lines in order to open up waterfront properties for other uses. Yet, when such discordance happened, these state actors tended to abstain, rather than vote no, in final decision-making and often concur with an emerging consensus (Ibon Areso, interview). Thus, in an atmosphere of lingering nationalistic tension between the state and region, and among special interests within the region, there has been functional cooperation and consensus around the goal of physically resurrecting an aging industrial metropolis. Midst the perception that, among the overwhelming obstacles facing Bilbao in its redevelopment plan was the potentially obstructive relations between political parties at central, regional and local levels, Ria 2000 was created "to overcome in a cooperative manner the impending difficulties confronted by the partners involved" (Rodriguez, Martinez, and Guenaga 1999: 18).

The physical outcomes of these public initiatives have been extensive and are creating a new Bilbao. The port has been made larger and moved outside Bilbao city center in order to create more connections in the urban fabric, improvements along the river have included pedestrian walkways and bridges, rail lines have been depressed and relocated with new avenues and parks in their place, and a new urban subway system has been built. Most well known and symbolically important among the new urban projects that have been built or planned is the

Guggenheim Museum. However, the Guggenheim is actually just one piece of a larger process of cooperative planning that has taken place in this region more known for political disagreement and conflict.

Reimaging a city

In the mid-1980s, midst environmental degradation and industrial decline, the General Plan for Bilbao spoke a new language – one of cultural rebirth and centrality, mixed and integrated land uses along the waterfront, and the transformation of the city from a labor-intensive industrial city to a new post-industrial city that would be a cultural and tourist leader among the urban areas of the Atlantic Coast region of southern Europe. Among the significant projects carried out in Bilbao, the redevelopment of the Abandoibarra district along the Nervion River is the most recent manifestation of the city's renewal aspirations and its creative implementation and organizational tools. The area of about 86 areas is in a prime location identified as the new future center of the city; in former days, it was a zone dominated by shipyards, a container port facility and regional rail line. When this district was first considered for renewal, 95% of the land there was owned by public institutions, much by the central state. Sites for two "emblematic" structures were identified at the extreme edges of the waterfront zone – one would host a music auditorium and conference center, the other a museum. Criticism of such a reimaging effort was strong, with many feeling that public efforts during a deep industrial crisis should be going toward the rescue of productive activities and employment, not toward culture, commercial and tourism facilities, and tertiary employment (Areso, interview; Garcia 2003).

The planning and implementation of this ambitious redevelopment took form with the establishment in 1992 of Bilbao Ria 2000 to coordinate, finance, and implement specific physical projects. In 1993, an international planning competition sponsored by the city council and the Bilbao Metropoli 30 strategic planning organization was held. The winning entry built upon the cultural foundations of the earlier General Plan.[18] Formally begun in 1998, the Abandoibarra project will contain when it is completed about 200,000 square meters of mixed uses, including office, hotel, institutional, housing, and commercial along with almost 200,000 square meters of green and open areas (see Figure 5.5). The total cost of urbanization in the district is estimated at about 60 million Euros; about 40 percent will come from the public sector and 60 percent from the private sector (Bilbao Ria 1998).

Bilbao's revitalization shows how an early creative planning framework can catalyze and integrate multiple strands of urban regeneration and can provide unforeseen opportunities. The Abandoibarra plan uses as anchors the two symbolic structures envisioned in the earlier 1987 plan and which have been built at either end of the redevelopment district. During an earlier period when the larger Abandoibarra redevelopment plan was being drafted, an unexpected opportunity presented itself that would catalyze the larger renewal effort beyond what most public officials would have imagined. In 1991, high-ranking Basque

Figure 5.5   Conceptual plan for Abandoibarra district, waterfront Bilbao

public officials approached the Guggenheim Foundation, which had been looking for a site to expand its holdings in Europe. The Basque officials showed them their identification of a waterfront site for a major art museum, one of the two "opportunity sites" of the 1987 city general plan. A restricted competition had earlier produced an architect, Frank Gehry, and a building design for the possible facility. Yet, before this time the Foundation had no interest in locating their museum in Bilbao and had been negotiating with Tokyo, Madrid, and Barcelona. An added difficulty was the existence of strong public opposition in Bilbao to undertaking such an enterprise. Bilbao councilman Areso (interview) estimates that about 95 percent of Bilbao residents surveyed would have opposed having the Guggenheim in Bilbao, viewing it as an unnecessary diversion from needed industrial recuperation.

Despite such criticism, Basque leaders exhibited strong drive and, against all original odds, secured an agreement with the Foundation in 1992 to build this major art museum in Bilbao. The officials made the case to the local community that this investment was not a waste of money, but an investment in the future reurbanization and rejuvenation of distressed Bilbao. Public funding for the enterprise has been substantial – about 150 million Euros in construction costs, born largely by the Vizcaya provincial government and the Basque regional government, with about 30 percent paid for by 150 private companies (Rodriquez, Guenaga, and Martinez 2001). In addition, the regional government agreed to pay the yearly operating costs for the museum from its general budget. Backed by such strong public actions, the Guggenheim began construction in 1994 and opened October 1997.[19] To justify the large public outlay, a feasibility study estimated that 400,000 annual visitors would need to come to the Guggenheim. This visitation level has been far surpassed in the first seven years, averaging

900,000 to 1 million annual visitors (in its first year, it drew almost 1.4 million visitors) [I. Areso, interview]. The estimated gross domestic product increase owing to the first year of Guggenheim's operation alone exceeded the total cost of building construction and museum setup. In addition, the tax revenue attributable to the museum (in the form of increased spending by its visitors) paid back in five years total building construction and start-up costs. Initial investment in the Guggenheim has been fully recovered. Also, the number of jobs created has now surpassed the number of jobs that existed in the old shipyards at the site during that industry's years of strongest production.

The spectacular success of the Guggenheim "bet" has led to the coining of the term "Guggenheim effect." There is little doubt that the Museum helped turn the corner for the city and started a momentum that continues to this day in the various urban rejuvenation projects led by Bilbao Ria 2000. It has enhanced the capacity of the city to compete internationally and to diversify economically. In addition to the economic benefits, the Guggenheim has had iconic importance and has been instrumental in helping Bilbao reconstruct its image. These symbolic and psychological aspects of the "Guggenheim effect" are central planks in how urbanism may positively influence nationalistic group conflict. The television, tourist, and media images of the Guggenheim are priceless in directing attention away from the violence of the ETA. If there was no Guggenheim, the cost of positive advertising to repair the Basque Country reputation across Europe and the world would have exceeded the museum price tag itself, and likely would not have been as effectual in restoring a sense of city and regional pride after the social trauma of decades of industrial and social crises.

In its pursuit of the Guggenheim, its use of an international planning competition for the Abandoibarra district, and its use of culture as a central planning objective, the aim by Bilbao has been to be part of a network of cities on the cutting edge of culture that would thrive in the globalizing world. Culture has been at the heart of efforts to revive the city and transform its image (Gomez 1998). The logic behind this effort is based on the fact that culture, music, and the arts are direct and explicit tools by which to reimage a Basque Country of notorious political violence and industrial decline.[20]

Despite significant progress in revitalizing and reimaging Bilbao, the strategy carries with it limitations. Many of the jobs created in the culture sector are ill matched to the skills of laid off industrial workers, resulting in a large pool of displaced workers who face underemployment in their remaining work life. In addition, the institutional mechanisms used to pursue the revitalization strategy have intentionally decreased the public nature of public authority as a way to bypass deep political and intergovernmental conflict. This lessening of account-ability and democracy may harm in the longer term the development of Basque Country's civil and political society. These limitations notwithstanding, the Guggenheim and other cultural improvements have surely moved the region toward a more cosmopolitan character with a closer relationship and openness to contemporary culture, a quality that appears conducive to advancing a moderate and non-violent Basque identity agenda.

The planning of Bilbao's, and by extension Basque Country's, transformation displays the intersecting of urbanism, nationalism, and political conflict. Urbanism has responded to both challenges and opportunities provided by Basque nationalism. The challenge of nationalistic conflict and violence created the need to transform the image of this Basque metropolis. To accomplish this goal, policymakers and planners selected a controversial means, based on culture and identity, that met initially with considerable local skepticism midst unemployment and economic displacement. At the same time, nationalistic politics created space for innovative urban intervention. Such a major urban effort to refocus the region away from nationalist conflict in Bilbao and Basque Country likely would not have been possible without the achievements of Basque nationalists themselves in the late 1970s to carve out a region of significant political and financial autonomy. Nationalist aspirations were accommodated in the creation of post-Franco Spain, creating a Basque Country with substantial regional autonomy and a financial self-sufficiency and potency. This level of self-government and financial capacity has led to a flowering of urbanism. The importance of financial self-autonomy in the form of the *Concierto Económico* was highlighted in many interviews. Ibon Areso (interview) asserted emphatically that urban improvements in Bilbao would "quite certainly not have been possible without the Spanish-Basque economic agreement." University of Pais Vasco professor Francisco Llera (interview) supports this view, stating provocatively, "Our regional success over the past 20 years is a result of support by the Spanish state because Basque redevelopment has been paid for by Spanish money that stays here due to the economic agreement."

In the larger political forum and in day-to-day rhetoric, the Basque Country and the Spanish state remained at loggerheads through the post-Franco decades about the future status of Pais Vasco. Within Basque Country politics, the nationalism issue created an internal faultline to contemporary life. Yet, ironically, the farsighted and innovative urban transformation initiatives over the past twenty years would likely not have occurred here without the acquiescence and cooperation by the Spanish state. Urbanism in the region has advanced due to agreements and partnerships between parties often antagonistic to one another on other fronts – first, during negotiations over the post-Franco state in the 1970s, and second in the organizational mechanisms used to implement Bilbao urban improvements since the early 1990s. Regional autonomy and financial strength negotiated during the 1970s transition created a foundation that has enabled and catalyzed impressive urban revitalization in Pais Vasco. Then, over ten years later, organizational structures and tools established to implement Bilbao revitalization (most specifically, Bilbao Ria 2000) include a level of Basque and Spanish shared interests and goals that is exceptional in a region known more for disagreement and strife than consensus. Such urban-based cooperation opens up a channel with the potential over the longer term to broaden the ground of public condemnation of overt violence as a necessary tactic in Basque society. By proving its value at the local level, effective Basque self-governance embodies a viable nonviolent path for the public at-large and may narrow the ground of acceptance for extreme nationalists.

URBANISM AND THE BASQUE CONFLICT

Urbanism has engaged with several important societal challenges both during the Franco years and since the start of democracy. In this and the next section, I examine how urbanism and local policy influence nationalistic conflict in the Basque Country. I investigate the impact of urbanism vis-à-vis two types of such conflict, a violent strategy associated with ETA ("militant nationalism"), and the democratic struggle for self-determination advocated by the PNV and mainstream Basque interests ("democratic nationalism").

The Franco years and transition to democracy

The years of repression and authoritarianism in Spain tore particularly deep wounds in the Basque Country. In comparison to Catalonia, where the indigenous industrial elite and middle class maintained a certain degree of regional autonomy vis-à-vis the central state and its economy, the industrial elite in Basque Country was more right-wing politically and maintained many economic links with Madrid (Joan Subirats, professor of political science, Universitat Autonoma de Barcelona, interview). Without an "umbrella of protection" provided by a native bourgeoisie in Pais Vasco, the working class was left on its own and the conflict with the Franco regime was rawer and more violent. Between 1956 and 1975, the regime declared ten states of emergency in the Basque Country (Kurlansky 1999). Further, beginning in 1968, an additional tension not present in Catalonia was added – the start of political violence by ETA, a phenomenon that would generate waves of state repression and counter-violence for over three decades.

The coherency and size of urbanist opposition in Pais Vasco was not as developed as in Catalonia. Urbanist critique was not incorporated into larger political opposition movements in the region to the same extent as in Catalonia and thus progressive urbanism took hold in Basque Country later than in Catalonia. Nonetheless, Basque urbanists who favored the nationalist cause constituted pockets of opposition to the "official" and expansionary planning of the Franco regime.

In Spain during the Franco years, planners for the regime developed plans for the cities and towns of the Basque Country that were expansionist and built upon speculation. The regime overrode, partially applied, and distorted through corruption the legal-technical planning system established in 1956 (Madariaga 2004). Plans and projects were done that conformed to Francoist centralist visions and industrial and economic needs, not to the particular cultural and physical heritage of the region. In urban areas, plans slated historic centers for clearance and proposed high-rise residential towers. Franco regime planning practices resulted in high building intensities with little infrastructure, sprawl of growth along corridors with neglect of natural attributes, and limited cultural amenities and consideration of public goals. In rural areas, the regime whitewashed and transformed vernacular architecture in the 1960s in its endeavor to transform rural picturesque zones into tourist attractions and to cosmetically hide rural poverty for international observers (Alonso, Arzoz, and Ursua 1996).

In response to Franquist expansionist goals, counter-planning efforts emerged to provide alternative visions regarding growth. "We countered, and improved upon, the Franquist ideas," says Xabier Unzurrunzaga (professor of architecture, University of Pais Vasco, interview). One focus of these efforts was to protect old towns in Basque Country from the speculative assault by Franco planning. In the town of Mondragon, for example, a group of city councilors not part of the Franco establishment hired Unzurrunzaga in the late 1960s to do a "Mondragon Futura" plan. This project pictured a different future for the town, one more connected with the cultural values of Basque nationalism than the expansionist, speculative orientation of Franco planners. "This was a small crack in Francoism," says X. Unzurrunzaga (interview). These alternative plans were at times persuasive even to local officials aligned with the Franco establishment, especially during the latter years of the regime when technicians were in power more than extreme ideologues.

In addition to the endeavors of individual professionals, associations of experts have the ability to progressively shape public opinion due to their central placement in the civil society of a country.[21] Franco establishment professionals up until the early 1970s dominated the professional organization of architects in Basque Country – the Collegio Architectura Pais Vasco – and this limited its ability to be a change agent or source of professional criticism of Franco cities. Nevertheless, networking opportunities existed within the organization to connect members to other urban professional associations, such as those in Catalonia that were pushing their own nationalist agenda, and in other European countries where planning for historic and cultural preservation was gaining increased attention. Such organizational linkages provided nationalist Basque urbanists with ideas about how regional identity and cultural heritage could be preserved midst the onslaught of Franco regime strategies. These professional dialogues with the outside world provided "small theaters" within which innovative professional ideas and strategies could be discussed (X. Unzurrunzaga, interview).

Although pockets of opposition existed, the fact that Basque urbanist critique was not as developed during the Franco years as in Catalonia means that, once the transition to democracy commenced, urbanists were not as prepared to engage in the preparation of alternative visions of urban development. No major revisioning plans were launched during the transition years in Basque Country as we saw in Barcelona; such retooling of collective development objectives would have to wait for more than ten years after the death of Franco.

Since the political transition, urbanism and planning have nonetheless played important roles, midst hurtful industrial restructuring and political tensions, in recovering regional pride and collective memory and identity. The decades of Francoist repression and planning done without respect to the particularities and history of the region was an assault on a people's collective memory – in the forms of their cultural identity, traditions, and language. Planning, in its understanding and treatment of land, territory, and culturally historic places, has been able to help recover Basque ethnic memory. Because of this link between planning and collective identity, it is likely no coincidence that the two regions in Spain that experienced the greatest amount of suppression during the Franco years have developed, after the granting of regional self-government, the most thorough territorial planning frameworks in Spain. The 1990 Basque law, like the Catalonian one, articulates significant standards regarding future land stewardship (interviews: Martín Arregi, director of territorial organization, Basque regional government; Sabin Intxaurraga, minister of planning, Basque regional government). This emphasis on Basque planning and stewardship, after the many years of Franco domination and repression, may be likened to a storeowner who, after being robbed or subject to a fire, starts his life anew again by taking a full inventory of his goods and re-examining his business plans. Planning plays an instrumental role in the collective recovery process after political repression; its importance lies in its ability to fully document the physical and cultural assets of an independence-minded, nationalistic region.

Contributions of urbanism to peace building

I now consider the extent to which Basque urbanism over the past 25 years has contributed to societal progress on the nationalistic question and the means by which it may be moderating political dynamics in ways that lessen political violence. Theories of ethnic violence and conflict attribute the phenomenon to both material and nonmaterial causes (Toft 2003). In the first case, uneven or limited economic development and material conditions lead to inter-group conflict. In the second case, violence and conflict come about as the cultural identity and historic foundations of an ethnic group are threatened or disrupted. At first glance, one might surmise that urbanism amidst continuing Basque violence and a polarized regional politics has not had the positive effects on inter-group conflict evidenced in Catalonia. Yet, it is worth looking deeper beneath this superficial conclusion.

While political debate is often absorbed by the Basque nationalism question and distorted by the use and threat of violence, planning and urbanism appear to provide spaces for rationality and consensus around shared goals. As illuminated in the Bilbao case, the urban dynamism and partnering in Pais Vasco is based on functional and tangible city issues where cooperation can take place between parties who otherwise would not concur on the larger nationalistic question. These agreements and partnerships create joint shareholders of interest and can connote an openness to innovation and social learning that is anathema to the hardened and rejectionist politics of extremist sectors. In this way, urbanism and its required give-and-take between different political interests seeking mutual gains provide testing grounds for compromises that may over time move a society forward on other, non-urban issues. A substantial asset of urban planning and policy operating in a context of differing nationalistic aspirations is its reputation as a functional and technical enterprise distanced from larger politics.[22] Agreement on functional aspects of city building and urban revitalization can lead to shared understandings of coexistence and to trust and belief in political means toward addressing inter-group conflict.

Another illustration of local cooperation midst larger political conflict comes from the Basque Association of Municipalities (EUDEL). In May 2002, two nationalist and two non-nationalist political parties signed a "Civic Declaration in Defense of Democracy and Liberty, and with Respect for Plurality in the Basque Country" (EUDEL 2002). This is a declaration of civic norms that speaks of cities as the basic building blocks of a democratic system, defends plurality in ideas and identities and respect for differences, and holds as unjustifiable the use of violence to address political differences (Karmelo Sainz, interview). The Civic Declaration constitutes an important foundation and model, constructed and put forth at the municipal level, for a tolerant Basque society accommodating of both Basque nationalist and non-nationalist views.

Local and regional government actions and institutional relationships may positively influence the playing out of nationalism and its larger politics. Local governments were "the most poorly defined part of the post-Franco puzzle of reform" (Carrillo 1997: 39) and yet are key actors in translating to residents the tangible benefits of democracy and inter-group tolerance. In this translation function, Basque local governments appear exceedingly successful. Residents in the Basque Country, despite fear and violence, are strongly favorable about their personal situation (in 2005, 76% described it as "good" or "very good"). And this assessment improved significantly in the 1990s, a period of considerable governmental actions and programs in the Basque Country (Euskobarometro 2005). Further, Basques feel that their regional economic situation is better than the rest of Spain, a pattern of public opinion that started in 1999 and has persisted (Euskobarometro 2005). Compared to the other 18 regions of Spain, Basques are highly favorable in their satisfaction "with the way their region works" (Mota and Subirats 2000; Subirats and Gallego 2002; F. Llera, interview). Certainly this is in part due to regional pride in the face of Spanish centralism; yet, it also likely is a realistic assessment of their quality of life and of the effects of higher public expenditure levels and better public services. To the extent that urban projects and programs are successful in providing social and economic benefits, there are limited opportunities for the emergence of material-based grievances that lead to or exacerbate inter-group conflict (Gurr 1993; Burton 1990).

Significantly, effective local and regional governance would also over time change the calculus used by individuals in assessing future paths for Basque society. While it is unclear that effective democratic governance would directly moderate actions by extremist groups (indeed, democratic success may be seen as a threat to militant groups and increase violence), it seems more likely to increase the public's allegiance and trust in local and regional government and thus the general public's buy-in to political, rather than violent, means toward resolving conflict. As Basque residents gain trust in their local and regional governments, they would have less tolerance for extremist groups and actions that seek to disrupt these productive governmental channels. Strikingly, for a region known for political violence, 89% of those polled strongly support the notion that the region can move forward positively without the necessity of violence (Euskobarometro 2005). This indicates a strong optimism about political mechanisms for addressing society's contentious issues.

Another positive feature of the Basque Country in terms of its ability to build a peace culture is the high degree of citizen involvement. Indeed, studies have found the region highest among all Spanish regions in the degree of social and political capital (Ajangiz 2001; Mota and Subirats 2000). Among the pro-peace, anti-violence organizations is the Association of Peace in the Basque Country, a civil, pacifist, plural movement that is independent from any political party and institution. Since 1986, this organization has orchestrated numerous "silent gatherings" (or "Gestos por la Paz") that bring people together in silence for 15 minutes on the day following murders or deaths caused by Basque political violence. These public demonstrations against violence take place simultaneously in many locations, including cities, villages, university campuses, and schools. Major municipalities are used as primary locations for these public demonstrations against violence, with 27 staging locations used in the city of Bilbao and another 12 in the city of Vitoria.

The organization of local government in the Basque Country – its multiple layering of municipal, provincial, and regional authorities – is a further regional attribute that may moderate the power of nationalism. Assuredly, such a structure increases competition between administrations for financial resources and political power and can add to political paralysis caused by violence. However, multiple sites of political power also disperse conflict and can moderate it as political powers seek out governing coalition partners that may differ across cities and provinces, depending upon local circumstances.[23] These dispersed relationships of differential governing coalitions lead to relationships and interests that cut across the common fault lines of polarization. Over time, Basque Country has witnessed shifts in governing coalitions across political geographies. The traditional pattern was one of coalitions between the PNV and the socialist party controlling each of the three capitals and the regional government (Victor Urrutia, professor of sociology, University of Pais Vasco, interview). It has since moved away from this equilibrium and become more complicated. As of 2004, a coalition of PNV, the nationalist EA, and the leftist Izquierda Unida governed the city of Bilbao. The centralist PP party and the leftist Socialist party controlled the city of Vitoria, and the Socialist party and either PP or PNV governed the city of San Sebastian. At the provincial level, meanwhile, the governing regimes have distinct characters and different political alignments. While Vizcaya province has traditionally been the shared domain of PNV and Socialist control, Gipuzcoa province has had greater influence by radical nationalism and Alava province has been the place of greatest electoral success by the centralist PP. This pattern of local and provincial interests constitutes a mosaic of different governing alignments that increase the opportunities to blur nationalist-centralist and left-right fault lines in the pursuit of common objectives. This blending can add fresh water to a stagnant political pool and move Basque society forward out of gridlock.

A further contribution of localism to building peace in Basque Country is that urban deliberations about the long-term future – often called strategic planning or visioning – can provide discursive spaces of reflection pertaining to the challenge of Basque political violence. The advantage of strategic planning and visioning is that these projects are able to include the ornery political challenges within a safer discursive framework that examines pragmatic and functional issues related to the region's long-term development. The Diputacion of Guipuzcoa has engaged in a continuous process of reflection about its future, a project called Guipuzcoa 2020. Four levels of societal and urban advancement are predicted, dependent upon progress achieved on the attenuation of political violence and radicalism, and the expansion of Basque political autonomy (Diputacion Foral de Gipuzkoa 2002). This is a provocative project in how it links explicitly political issues – of violence and political autonomy – to economic and social quality of life issues that are the typical domain of planners in visioning exercises. Guipuzcoa 2020 keeps a safe distance from the contentiousness of political negotiations (indeed, radical nationalists have largely chosen not to participate) and it is not a political process, says Minister Beloki (minister of territorial management, Diputacion Guipuzcoa, interview). Nevertheless, its reflection on the roles of violence and political nationalism in Basque society is an important and constructive discussion not found in the formal chambers of Basque policymaking. Another strategic planning project, this one specifically within the City of San Sebastian, has proposed making the city "un espacio para cultura de la paz" (a space for the culture of peace), stressing the importance of a socially open and cohesive city, with respect for life and human rights, and the integration of all residents. The director of the project, Kepa Korta, attempts to maintain communication with political radicals. He sees strategic planning as potentially able to carve out spaces for dialogue and peace and thinks that these interactions can counter long-entrenched dynamics in the region. In a circumstance where political parties are hamstrung by these dynamics, consensus-based projects such as the city's strategic plan, which operate outside formal political channels, and efforts by nongovernmental groups provide places for constructive dialogue and actions.

In a region politically polarized and paralyzed by the question of Basque nationalism and violence, the urban level offers one of the few laboratory spaces for working through differences and creating new relationships not contained by the larger nationalist-centralist divide. Experimentation and innovation can occur through new organizational structures that pursue urban goals of common benefit and through the articulation of civic norms of tolerance and non-violence. In addition, effective local government, through its betterment of quality of life, enhances the general public's view of political mechanisms for working through societal problems and thus lessens tolerance for extremist violence. At the institutional level, different combinations of governing coalitions across the multiple political geographies of the Basque Country open up opportunities for moving the region out of political gridlock. Finally, a specific type of planning – strategic and long term in nature – provides deliberative forums where larger political issues can be linked to social and economic considerations in contemplating alternative futures for the region. Urbanism, because it focuses on concrete issues and challenges, provides opportunities for relationships and deliberations not possible in the more abstract and rigid world of political ideology.

Urbanism and political violence

I focus further now on the specific relationship between urbanism and violence. Basque violence has taken place in a region that has been, along with Catalonia and Madrid, one of the traditional economic powerhouses of Spain. In 2002, the average annual income was 18,775 euros, 20 percent above the Spanish average and above the European average (Barberia 2002). Modernization and development that are thought to diminish ethnic group loyalties and transfer them to the state have been present in most of the twentieth century, yet this was a period when Basque extremist nationalism and violence emerged and took root. Since the 1970s, the Basque economy has gone through a wrenching restructuring of its economy that has undoubtedly produced a good share of alienation on the part of industrial workers who now face uncertain futures. Part of this pool of dispirited workers may provide some support for more militant approaches to Basque nationalism. However, analyses do not uncover associations between the local intensity of street violence and rates of unemployment (Mees 2003) or between the radical nationalist vote and the microgeography of unemployment (Beck 1999).

As violence has sustained itself in an economically advantaged region and midst the economic recovery of the 1990s, non-materialist explanations of Basque violence become apropos for study. Identity, history, and culture are important variables in this tradition, as well as the motivations and tactics of group leaders (Toft 2003). At first look, this explanatory model suffers because violence has coexisted with wide-ranging state accommodation of regional identity and power. As Vilanova (interview) points out, the provision of substantial political and financial autonomy to the Basque Country seemed to have had little effect on Basque violence. The 1979 statute included amnesty for former ETA members who demilitarized, established separate Basque political institutions, an independent Basque police force, tax autonomy from the central state, official recognition of the Basque flag and anthem, and regional control over language, education, the media, and culture. Further, the period leading up to, and immediately after, approval by the Basque electorate of this substantial grant of regional autonomy in October 1979 turned out to be ETA's most violent period. In the 1978–1980 period, ETA killed over 200 people (compared to the period 1968–1977, when 19 was the greatest number of killings in one year). In 1980, the year after acceptance of the regional autonomy statute, ETA killed 92 persons, its deadliest year to this day. And, more than 700 of the over 800 ETA killings have occurred during the years of democracy.

Part of the explanation for violence after the gaining of Basque political goals is that even a substantial grant of autonomy fell short of the objective of outright independence held by the most vigorous nationalists. Yet, there appears to be another dynamic of extremism that has to do with how Basque separatists viewed their range of opportunities. Consistent with what Toft (2003) labels as the "security dilemma" catalyst of violence, Basque separatists may have feared losing out in the composition of the new democratic Basque governing regime. In this interpretation, increases in terrorism during the 1978–1980 period were due to "the threat democratic cooperation and conciliation posed to terrorists" rather than a "fundamental failure of the expansion of the democratization process to undermine the base of political violence (Newman 1996: 205)."[24] In other words, ETA feared that the very success of the new democratic state would be obstructive of their separatist goals. At the same time, while threatening to extremists, democracy can also provide opportunities to them because, as Mees (2003: 177) astutely notes, the freedom of press and expression in a democracy can provide a "considerable public echo" for violent activities. This implies that transitional processes toward democracy and its consolidation can be periods of greater violence, not less, as militants sense both threat and opportunity.

An additional challenge in relating material and political conditions to the frequency and pattern of political violence is that after some duration the organizational dynamic of nationalist militarism may be self-generating, and thus semi-autonomous from the realities of political conflict (Mees 2003; Crenshaw 1993). Violence is used by armed underground groups according to an inner organizational rationality, a logic that can defy understanding by those on the outside. The means of violence become as important as the ends as extremists use violence as recruitment and advertising tools to reproduce their own group structure.[25] Many of those I interviewed emphasized the importance of a self-generating and difficult-to-penetrate organizational logic to explain the sustainability of Basque violence midst economic wellbeing and political autonomy (interviews: V. Urrutia, P. Arias, and P. Ibarra). This self-perpetuating logic of the extremist agenda sustains it over the long haul but also provides certain entry points for those wishing to negotiate an end to such violence.

One entry point may be through erosion in overall public support for the legitimacy of political violence in Basque society. While violence has had a remarkable ability to sustain itself over the past 35 years, public support for it has lessened considerably through the years. I suggest that it is within this domain of public opinion that urbanism and Basque governance can have their most meaningful effects. Basque citizen rejection of ETA violence has been a fundamental attribute since the mid-1990s, and during this time there has been a significant increase in the public's image of ETA as "crazy/terrorists" and a substantial decrease in viewing them as "patriots" or "idealists" (Euskobarometro 2005). At the same time, the public substantially views the Basque Country favorably as a place to live and as a region that works effectively as a collective (Mota and Subirats 2000). It is important to note that this sentiment antagonistic to political violence and positive about the region has increased or been cemented during a time of significant urban and economic development of Basque cities and towns. The impetus behind many of the urban projects has been to project another image to both external audiences (Spain, Europe) and internal ones (the different political sectors in Basque Country). Urbanism, states V. Urrutia (interview), "is a very good protocol to use to change public image and, hopefully, political debate." Thus, 15 years of urban regeneration may have changed the public image of the region from one of terrorism and decline to one of rebirth based on culture and connectivity. Yet, political violence and intimidation remained a part of the Basque landscape in the 1990s and halfway through the first decade of the 21st century. If urbanism had an effect on lessening political violence through its effect on public opinion, this is not what we would expect.

I suggest the possibility that there may exist a time lag between improvements in objective conditions and the diminution of violence and intimidation in a society, and this time lag may be attributable to the medium- and long-term influences of urbanism to shape public opinion and narrow the ground of acceptance for militant radicalism. I base this assessment on my understanding of the conflict in Northern Ireland and the evolution of the Irish Republican Army (IRA) paramilitary.[26] One interpretation of the IRA's decisions to cease hostilities emphasizes the ability of social and economic improvements over a thirty-year time period to change the views of the IRA's constituencies (Bollens 2000). Urban policies by the British direct rule regime of Northern Ireland had substantial positive effects on the physical and socioeconomic landscapes of hard-line Catholic neighborhoods in Belfast that are the core constituency for the IRA and its political party, Sinn Fein. Over time, says Sinn Fein member Joe Austin (interview), these improvements broadened their followers' view of the Northern Ireland problem to include social and economic considerations and the perception that these benefits could be lost with continued hostilities. A time lag was involved in this relationship between urban betterment and the attenuation of political violence. Many of the rehabilitation projects in Belfast predated the IRA ceasefire by ten or more years. In the Basque Country, the March 2006 ceasefire by ETA comes several years after significant and visible urban improvements in many cities and brings hope of a fully nonviolent Basque nationalism.

BASQUE CITIES IN A NATIONALIST REGION

I investigate in this section three links between urbanism and the political dynamics of mainstream, nonviolent Basque nationalism. First, urbanism can be used to constructively reposition and modernize a nationalistic project for both external and internal audiences. Second, functional arguments of urbanism and spatial planning can be utilized to pursue cultural and political goals associated with Basque nationalism. And, third, internationalization and Europeanization are providing new footholds for Basque nationalists in efforts to distance their identity from Spain and in their ability to redefine Basque sovereignty in an interdependent world. There are both constructive connections and conflictive contradictions between efforts to modernize and urbanize the Basque Country and the advancement of the Basque nationalist project.

Urbanism and the nationalist project

> *The image of Basque Country is its rurality, its folk music and culture, language and country family house. But this is not contemporary reality.*
>
> Francisco Llera
> Professor, University of Pais Vasco
> Interview

The origins of Basque political nationalism and the moderate *Partido Nacionalistic Vasco* (PNV) political party lie in middle-class struggle against the oligarchy and the immigrant working class during the late nineteenth century. Nevertheless, the social movement took as its cornerstone the rural and small town village life of the Basque (interviews: F. Llera, P. Ibarra). There was, as described by F. Llera (interview), the "symbolic appropriation of rurality and rural images" by Basque nationalists. Their symbol and mother lode was the *Etxea* house of Basque rurality and strength and the movement's core constituencies were those from small villages and medium-sized towns in the region.

Electoral support for the PNV has varied from 24 to 44 percent in regional and municipal elections and from 25 to 35 percent in national elections.[27] Before its illegalization, Herri Batasuna (HB) had been supported by between 14 and 18 percent of the electorate. However, because there exists a connection between support for nationalism and smaller-scale settlements, it is important when examining the political composition of mainstream and extreme nationalism to go beyond regionwide data and examine the distribution of regional population across different size settlements. Urban structure influences political alignments and power dynamics; the smaller the settlement population, the more likely it is that nationalistic supporters will predominate (F. Llera, interview; Llera 1999b). Small villages (less than 9,000 population) across all three provinces are monopoly nationalistic areas; political representatives overwhelmingly are either moderate or extreme nationalists. Medium-sized towns (9,000–45,000) remain mostly nationalistic, although in certain towns there is intrusion by non-nationalistic interests (particularly socialists). In cities greater than 45,000 population (the three capitals of Bilbao, San Sebastian, and Vitoria plus six other cities mostly in Vizcaya), moderate nationalist constituencies for the PNV share space with constituencies for the non-nationalist parties – particularly the Socialist party in Vizcaya and Guipuzcoa provinces and the Popular Party in Alava province.

In terms of support for extreme nationalism specifically, Beck (1999) analyzes the spatial characteristics of its constituencies. Using data of voting support for the ETA-aligned and now banned HB party during the 1990s, he documents persistent support for ETA in the Basque-speaking smaller towns and villages of Pais Vasco, particularly in the interior of Guipuzcoa and Vizcaya provinces and in the San Sebastian area.[28] Radical and left nationalism came to Guipuzcoa in the late 1950s and early 1960s when ETA founders reached the limit of their frustration with what they viewed as the conservatism of the PNV during the Franco years. The small settlements at the core of HB/ETA support offered a context of social control, states Beck (1999), which sustains support for radicalism and political violence. In smaller villages, a socially and politically closed interaction system can more easily be perpetuated by small informal networks that keep the memories of Franco repression and support for ETA alive. In a small village context, radicals maintain control more easily than in urbanized settings through the political cleansing of local councils, the imposition of a "revolutionary tax" that provides a way to harass those holding opposing views, and the killing and isolation of those individuals seeking links to outside groups (Beck 1999). In addition, local Batasuna bosses could more readily assert political intimidation during elections amidst the greater visibility of voting in small villages (P. Arias, interview). In 2000, when the party asked for abstention from voting in the general election, high rates of abstention occurred in these small villages and towns.[29] In cities and larger urban areas, in contrast, such means of control run into counter-forces and opposing interests of too large a magnitude to be as effective in maintaining local control. With greater rates of Spanish immigration into cities and larger urban areas, the radical Basque nationalist message loses its appeal.

The foundation of small village support for radical nationalism has been deepened by the urban and rural structure of Basque Country local government. Only about 16 percent of Basque residents live in municipalities of less than 9,000 population. However, these small towns make up 82 percent of the 250 municipalities in the region and the number of town councilors elected in these towns constitute fully two-thirds of all elected councilors in the region. These local councilors, operating at the grassroots level in small towns, constituted the core of support for Batasuna and radical nationalism.[30] At the regional assembly level, voting procedures that weigh rural votes more than urban ones also constituted a built-in advantage for such grassroots-based electoral support.

While the PNV and the HB strands of Basque nationalism pull significant support from small towns and villages, PNV pulls from a wider segment of the population and their support is more dispersed across all types of settlements – the three capitals and large cities as well as smaller scale geographies. This architecture of electoral support has important implications for the future evolution of the nationalist project in Basque Country.

The PNV faces a challenge midst changing times. While its rhetoric and electoral base is based in small town Basque Country, it is increasingly perceiving its future as more aligned with issues of functional, economic, and international connectivity and openness, aspects that are best promoted through attention to urbanization and the utilization of assets found in the bigger cities of the region. "PNV nationalists have had to change their mind," says V. Urrutia (interview), "as they realized they could not live in this grand world with a rural mentality." Sixty-seven percent of the local city councilors in the region come from smaller villages, but F. Llera (interview) points out, "This is not the country; the country instead is in larger cities." Indeed, 54 percent of Basque population resides in the nine cities with more than 45,000 residents, and fully 36 percent of the population is in the three capitals alone.

These urban areas are the battlegrounds of future electoral competition in Basque Country because it is here where the presence of non-nationalist parties is more able to contend with Basque nationalist politics. In Bilbao, for instance, industrialization in the Basque Country came first to this city and with it trade unionism, Spanish immigrants, and a greater proclivity toward socialism. Today, the left bank of the urban area contains areas of strong support for the Socialist Party or for Izquierda Unida (two leftist non-nationalist parties). In addition, inroads have been made by the non-nationalist Popular Party in higher income areas such as Getxo in the northern part of the urban area. This greater electoral competition in urban areas is forcing PNV to compete for electoral support by generating projects and ideas that appeal to urban Basques more than their core base of small town or rural constituencies.

A clear indication of PNV's more modern, and urban, message lies in the set of urban and regional strategies of Euskal Hiria (Basque City) being developed under its political leadership. The plan's primary author is the Department of Territorial Management within the Basque Country regional administration. The proposed regional planning strategy is urban-centric and envisions the develop-ment of a city-region with the three capital cities as anchors and catalysts (see Figure 5.6). A detailed 245-page document (Gobierno Vasco 2002) lays out the rationale and spatial forms of this regional vision, as well as articulating the relative strengths of each of the three capitals. In order to increase the economic innovativeness and competitiveness of the Basque region vis-à-vis Spain and Europe, this strategy aims to increase connections and complementary activities among the Basque capitals. Planners view such economic and functional integration within the Basque Country as key to the region's exploiting more fully its strategic "hinge point" location relative to two axes – the north-south Paris to Madrid corridor and a lateral corridor that runs along the northern Spanish coast of Cantabria to the west and along the Ebro River towards Catalonia on the east. In a future where there is "the progressive dissolution of frontiers between countries," opportunities will expand for intra-state regions to be active and independent agents in determining their economic futures (Gobierno Vasco 2002: 49).

To bring this polycentric spatial model into being, the regional government anticipates a set of public and private investment actions – including train, road, airport, port and telecommunication improvements – to both increase internal connectivity within the three-pole city-region and external connectivity to areas in Spain and Europe. A key part in creating the city-region is the proposed location of high-speed (*alta velocidad*) railway infrastructure in a "Basque Y" spatial pattern that would connect each of the three Basque capital cities to each other,

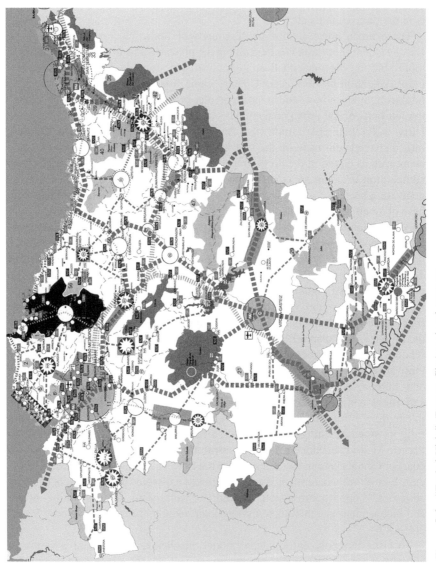

Figure 5.6   "Euskal Hiria" (Basque City) regional strategy

and to Madrid and the French high-speed system. The "Basque Y" would be part of the Spanish national high-speed train network, which has an intended goal of 4,500 miles of track along five major corridors in the country. These trains cut travel time in half and promise to stimulate regional travel activity and economic development around train stations.

The regional strategy illuminates how the PNV sees the region as best able to move forward economically and socially, and it underscores the vital influence of urbanization, connectivity, and openness to the party's modernizing nationalist project of the early twenty-first century. The strategy notes that medium-sized towns and rural settlements would benefit from the overall increase in functional integration in the Basque Country. And, planners describe the network of small towns and rural settlements as "essential for the survival of our traditions, customs, and idiosyncrasies" (Gobierno Vasco 2002: 57). Despite such language, however, this is unabashedly an urban strategy that positions the three main cities as primary beneficiaries and engines for change. It shows that after over 20 years of near monopolistic political control in Basque Country, the PNV is moving its ideology closer to cosmopolitan urbanity and further from its traditional cultural foundations. The vision's appeal will be strongest in those urban areas where PNV competes most directly with non-nationalist parties, and in this way the strategy is logical from a political view. At the same time, this strategy has electoral risk for the PNV because it does not offer substantial direct benefits to its traditional small town and rural constituencies. Indeed, urbanism may run counter to a project of nationalism, whose maintenance is often dependent upon a degree of exclusivity and parochialism. As stated by V. Urrutia (interview), "Urbanists seek to connect people and regions; nationalists seek to split people, civil society, and cities. Urbanism and its openness is the permanent contradiction of nationalism."

Despite these risks, the PNV sees its political future tied to the inevitable growth in interdependent links within the region and between the region and Europe. Based on political calculus, this strategy of connectivity and openness may provide an additional benefit to the PNV – an ability over time to squeeze out radical nationalists and the ETA (V. Urrutia, interview). With greater openness comes a greater web of economic and social opportunity that may help take the power out of ETA's message of radicalism and violence. Yet, the effect of increased urbanization and societal connectivity in diluting the influence of radical, militant nationalism should not be assumed. In the larger theoretical literature on the relationship between globalization, modernization, and inter-group conflict, some argue that increased global connectiveness will over time erase the imperatives for radical group-based actions, while others claim that it may intensify and stimulate greater ethnic self-consciousness and more counter-responses by increasingly threatened radical groups (Huntington 1997). Appadurai (1996: 10) observes that at the same time we experience the "mega-narratives" of modernization (involving economic growth, high technology, and education), we bear witness to subversive "micro-narratives" that fuel oppositional movements.

This "Basque City" project spearheaded by the PNV points to how urbanism can be utilized within a nationalist context. The PNV is urbanizing and redefining its nationalist project within an increasingly interdependent region and world as a way to strengthen itself electorally in those areas of greatest electoral competition with non-nationalists. Yet, it is a difficult challenge for a regional nationalism, rooted in the imagery and traditions of the small town, to evolve toward a more pragmatic and modern nationalism. In the end, the role of urbanism in the Basque Country as a platform for revising and modernizing political nationalism exposes the bidirectionality of the relationship between urbanism and nationalism. Earlier, I concluded that political nationalism in the negotiations over a new Spanish state created an opportunity space of self-government and financial resources that facilitated a flowering of urbanism over the past 20 years. Nationalism facilitated and promoted urban innovation and agency. Now, in the early years of the new century, such urban creativity is being used to move Basque mainstream nationalism along an evolutionary path.

function and culture

> *When Basque nationalists were not in power, the map finished at the border. This is ideological, not logical. Geographically, there are no borders. They are in the mind; we do not believe in that state border.*
>
> Jose Aranburu
> (referring to Spain-France border)
> Planning Analyst, Diputacion of Guipuzcoa
> Interview

Functional and economic rationales for planning policies intermingle with motivations linked to the protection and advancement of Basque culture and nationalism in certain ways: (1) a public emphasis on functional criteria that mask cultural and nationalistic motivations; (2) the use of spatial imagery to catalyze new ways of conceptualizing an autonomous region vis-à-vis its national state; (3) an emphasis on cultural and territorial integrity and protection as guideposts for functional planning; and (4) an awareness that increased European functional integration can support Basque cultural nationalism. In each case, there is the conflation of functional and cultural/nationalistic rationales for government intervention.

In 1993, key public sector actors in the urban corridor that runs along both sides of the French-Spanish border on the Atlantic coast signed a cooperation protocol. Thus started a cross-border cooperative process – "Basque Eurocity" – with the aim of structuring and uniting the areas that are located in a linear pattern between San Sebastian and the French city of Bayonne (see Figure 5.3, page 145). Such a spine would be a key bridge linking parts of the larger historic Basque region. This region, although divided by national borders, includes the three Spanish Basque provinces as well as three provinces in southwestern France – Basse Navarre, Labourd, and Soule. The Eurocity corridor is over 30 miles in length and contains between 500,000 and 600,000 residents. In 1995, a treaty was signed between the countries to allow for cross-border cooperation, and in 1997 a cross-border agency was created. Partial funding for this enterprise has come from a European Commission program that targets regional development. The Eurocity project, states its literature, "proposes a future in which we should overcome the scars that the border has represented throughout history..." (Cross-Border Agency, undated, 5). To overcome this historic cultural cleavage, Eurocity seeks to transform what now is an uncoordinated set of conurbations along the coastal corridor through a series of government interventions, including the development of a new road network to increase economic connectivity, the building of an inter-modal (highway, sea, rail) logistics platform at the Bildasoa River that is the border between Spain and France, and a set of cooperative cross-border technological, cultural, education, and cultural and physical heritage preservation policies.[31]

These actions aim at producing a new type of city or "linear polycentric metropolis," states Agustin Arostegi (co-director, interview) that would be "European in its openness and in its competitiveness." An interesting intermingling of functional and cultural/nationalist rationales of cross-border planning is evident in my discussion with Mr Arostegi. He describes the main focus of this work as functional, but he is cognizant that he is using these objective considerations as a basis upon which to develop what is in reality a cultural project. "Most of my day I speak the language of function," he explains, "but for me it is more of a cultural than a functional project." He details further the cultural importance of Eurocity to Basques on the Spanish side:

> *We always talk about institutions and it being an urban project with cross-border benefits. But for us it means we will be in contact with people whom we have been back-to-back with for hundreds of years. We have to put them face-to-face. With this contact, there will be an increase in the Basque language and those on the other side will understand better our feelings. We will be more a region than we are now. This is why I say Eurocity will support the Basque nationalist project.*

Mr Arostegi is comfortable working daily at the point of nexus between functional and more political cultural factors, using economic arguments and terminology in public forums in order to build a foundation upon which cultural nationalism can blossom. An additional benefit of economic reasoning is that it is compatible with urban and regional policy objectives of the European Union. Such compatibility between the Basque Eurocity and EU regional policy puts the Spanish central government in a difficult position because, if it were to oppose Eurocity, it would be opposing long-held EU goals of openness and cross-border integration.[32]

Functional and cultural rationales also intermingle when spatial and planning imagery is used to reinforce the self-sufficient nature of a region relative to its national state. Jose Aranburu (geographic analyst, department of territorial management, Diputacion of Guipuzcoa) stresses that the axis, or corridor, between Madrid and the Basque Country has been over-emphasized and built up to reinforce the centralist idea of a unified Spain. What is needed, he suggests, is to change the spatial imaging and representation of the Basque region to recognize and strengthen its linkages within an economic space that does not emphasize Madrid. Rather than an axis, Aranburu recommends the image of a "net," with Pais Vasco embedded in a network of relationships with the Atlantic Arc to its west and east and the northern Mediterranean region to its east. With such an image to guide regional strategy, investments in services and infrastructure would reinforce this vision rather than the corridor link with Madrid. Primary among these investments in importance is the Basque Y plan for high-speed rail that would connect the Basque Country internally and to the French train system.[33] It is evident that nationalist planners like Aranburu view routing and investment decisions regarding rail and other infrastructure as key elements in creating connections and boosting their claims of a Basque economic space of sufficient independence from Madrid. There is cognizance that there exist strong cultural and psychological implications of spatial reimaging, and that public actions that follow are capable of giving spatial form to Basque nationalism.[34]

In analyzing the functional-cultural nexus, professor of architecture and self-proclaimed nationalist Xabier Unzurrunzaga (interview) asserts that the ability of planning to support and rationalize the nationalist project should not hide behind functional arguments. Instead, the explicit nurturance of the cultural heritage of the region should be the basis upon which to functionally build the country. Planning should document and preserve the cultural landscape at a scale that encompasses human communities of likeness irrespective of "artificial" borders. He emphasizes cultural and territorial integrity instead of functional links, self-sufficiency instead of dependence, and focuses on the humanistic and psychological elements of Basque group identity and history.[35] "We must know very deeply our territory, our land, and our physical conditions," Unzurrunzaga asserts. "This is how we preserve and strengthen our Basqueness." In 2004, he was involved in the Eurocity project and recalls, "it is very nice for a nationalist to see the disappearance of that frontier and to talk about our 'others' on the other side. This is a passionate question, both from a nationalist and a planning perspective."

These cases of planning approaches and of individual professional planners exhibit how functional and cultural issues intermingle and interpenetrate in a setting of heightened nationalist sensitivity and tension. They also show how the functional basis and reputation of town planning provides a cover for urbanists to strategize about how to address and pursue nationalist goals. Urbanism and its protocol of functional objectivity have provided an effective anchor and tool for those professionals empathetic to the nationalist perspective. There exists one more increasingly important element in these nationalist urbanists' arsenal – the opportunity provided by internationalization and Europeanization.

Urbanism in the new Europe

You can be a nationalist in Europe in the 21st century, but you must be different from being a nationalist in the 19th century. Nationalism must change and adapt for it to complete its way from the 19th century to the 21st century.

Pedro Arias
Gesto por la paz (Dialogue for peace)
Interview

The growing institutional and economic interdependence of today's world promises to create openings and new linkages for Pais Vasco. Many of those interviewed saw new international and European connections as supportive, not erosive, of the Basque nationalism project. They see the European platform as creating new networks of interaction and interdependency, and new territorial scales, which would compete with the region's traditional links with Madrid. With Pais Vasco increasingly interconnected with other regions throughout the continent, Basque Country would be able to redefine what sovereignty means in an interdependent world, entailing a region not bound by state dictates but free to interact across European space.[36]

The use of the new Europe as opportunity is evidenced in several ways. First, the effort to modernize Basque political nationalism by the PNV is premised, in part, on the desirability of growing cosmopolitanism and European integration. Nationalism today, says P. Ibarra (interview), means "being connected to the modern Europe, not to the old stories and the music and culture of small towns." PNV adherents are using Europeanization as a key foothold in changing their image from an old, ethnic, and traditional party to one more modern and open. Second, in the revitalization program of Bilbao, we saw how cultural centrality within a new sphere of cosmopolitan and European audiences is playing a key role in that city's redefinition and reimaging. Third, the Eurocity project is a tactical foray into this new European space. A cross-border urban system would have "Basque as a common identity, but would also be European in its openness" (A. Arostegi, interview). Eurocity is viewed as a lobbying agent midst expanding Europe, allowing Basques to be present in Europe as Basques, not as Spanish.

The strengthening of regional identity, together with an increased embeddedness of Basque Country in Europe's institutions, promises forward movement on the nationalist project. The preferred route to globalization for many Basques is through local capacity building. J. Aranburu (interview) states: "You must be strong locally to be oneself in the global world and this is very difficult under a state-based system that spawns dependence." Because of the importance of local and unique assets in a globalizing world, visioning exercises such as Guipuzcoa 2020 that provide for regional self-reflection are an excellent means by which to put oneself forward into European space. For Basque nationalists, J. Beloki observes, "This visioning is the best way to be internationalist."

Such cultural optimism must be tempered by the realities of the "new Europe." The ability of Basque nationalists to undermine Spanish state dimensions appears to be an uphill struggle because the construction and institutionalization of the EU has become more and more a state-centered enterprise, with regions left at the sidelines. Even if regions emerge as semi-autonomous actors within the EU, Basque aspirations may be stymied by the institutional design of regional participation in the EU. Pere Vilanova (interview) explains, "If you go to a separate regional level within the EU, it pushes you toward a functional approach because there are 160 regions in the 15 member states. It's no longer an issue of historic nationalities or Basque and Catalan linguistic rights because less than 10 percent of the EU regions are historically regions with strong identity. The EU instead would engage in policies and processes for all 160 regions." In this scenario, EU regionalization becomes in the end more a programmatic and functional issue rather than a dramatic and political one. In moving from a Spanish realm to a European one, it may then be not the central state of Spain but the regional institutional architecture of Europe that dampens historically based aspirations in the Basque Country.

FROM ETXEA TO EUSKAL HIRIA

I have examined the relationship between urbanism, political nationalism, and political violence in Pais Vasco. This is a region that has experienced two transitions, one in the 1970s and early 1980s involving the shift from authoritarianism to democracy, and one that is still ongoing and involving endeavors to normalize Basque society and polity after decades of political violence and intimidation. I have found that urbanism can operate effectively midst political gridlock and nationalistic tension. Indeed, a politically contentious context at times can inspire organizationally innovative responses in order to overcome gridlock. At other times, nationalism was effective in creating a foundation of fiscal and institutional autonomy that has through the years catalyzed urban activism. Within a region of historic and still unresolved conflict, the major cities in Basque Country are areas of general stability. The three capital cities have been able to counter obstacles in pursuing development policies (indeed, their relative fiscal wealth creates positive opportunities) and in carrying out urban services or pursuing redistributive policies.

I have found that, as in Barcelona, urbanism is an effective means by which to bring the benefits of a political transition and democracy to the people. While Basque urbanists were not as proactive in their transformational activities as in Barcelona, urban revitalization over the past 20 years in Pais Vasco has been substantial and promises significant social and psychological benefits in a region hampered by violence and intimidation. Urban development creates facts on the ground that produce a growing momentum toward social and political normalization. Urbanism may also be able, by providing concrete and visible benefits, to narrow the ground upon which radical nationalism and political violence is exercised. Urbanism does this not by integrating radicals into urban partnerships and projects but by widening the acceptance by the general public of regional and local government. With greater approval of Basque political institutions and policymaking, claims by radicals that they must respond with violence hold onto fewer and fewer followers.

Finally, we have seen how urbanism can be an anchor and pivot point in projects aimed at modernizing Basque political nationalism, and that urban and international connectivity is antagonistic to the small-scale context within which radical nationalism thrives. I believe the key to whether Pais Vasco politically normalizes will be how well cities are able to protect and incorporate their Basque identity – in their architecture, urban iconography, local politics, and civic society – in a future when the region will be increasingly connected to other areas in Spain and Europe. An urban-based, networked, and externally linked future for Pais Vasco holds promise for normalizing society and polity if those persons who hold Basqueness near and dear to their hearts and who are exhausted by violence feel they are included in this path from the *Etxea* home to the *Euskal Hiria* city.

Notes

1. During the writing of this book, Basque extremists announced a "permanent ceasefire" effective March 24, 2006. It remains to be seen what will come out of this pronouncement, given the breaking of ceasefires in the past (*New York Times*, 3/22/06). On 12/30/06, a bomb destroyed a parking structure at the Madrid Barajas Airport; the Spanish government has blamed ETA and suspended plans for negotiations.
2. Field research in Basque Country occurred February 2004 midst political flux and uncertainty. This was 1½ years after a Spanish judge and the Spanish Parliament outlawed Herri Batasuna, a political party linked to Basque violence. It was about three months after the Basque Parliament passed the Ibarretxe Plan of Basque free "sovereignty-association" with the Spanish state. This plan proposed a type of cos-overeignty with the Spanish state and was subsequently blocked when it reached the Spanish Parliament early 2005. Field research occurred one month before Spanish national elections, which brought the Socialist Party to power and dislodged the conservative Popular Party of Jose Maria Aznar.
3. I focus my attention on the Basque autonomous community in Spain consisting of three provinces. Strong Basque nationalists would point to a larger region inclusive of Navarra province in Spain (population of about 375,000) and Basse-Navarre, Labourd, and Soule provinces in France (about 250,000 population combined).
4. For the most part, I use the Spanish/Castellano version of Basque names in this chapter. In some cases, I put in italics after the Castellano name its equivalent in Basque (*Euskera*) language.
5. For comparison, 44 percent of the persons in the Barcelona urban region considered Catalan or both Catalan and Spanish as their native, mother language.
6. Basque identity in the region may be stronger compared to Catalan identity in Catalonia. For the 1990–1995 period, 27 percent of respondents felt they were "only Basque" compared to 13 percent feeling "only Catalan" (Moreno, Arriba and Serrano 1998).
7. I use for resources on Basque history two insightful books – Kurlansky (1999) and Clark (1979).
8. This strategy was undeniably effective in the short term, with the industrial product growing 10 percent annually throughout the 1960s decade (Clark 1979). Industrialists

were favored by a series of measures protecting them from competition (including the highest tariffs in Europe) and by repressive labor legislation that kept costs down.

9. By 1970, due to in-migration, the percent of provincial population born in that province had decreased to 61 in Vizcaya, 65 in Guipuzcoa, and 59 in Alava (Clark 1979).

10. The results were Vizcaya (31% yes, 9% no, 55% abstain), Guipuzcoa (28% yes, 13% no, 57% abstain), and Alava (42% yes, 11% no, 52% abstain) [Salaberria 1991].

11. Fifty-nine percent of eligible voters participated in the statute referendum, with about 90 percent of those who participated voting in support (Lamarca and Virgala 1983).

12. At the time of writing, Catalonia and Madrid were restructuring their financial relationship to provide greater authority to the region.

13. Public opinion polls also show that the populous is becoming less nationalistic over time. In 2005, 44 percent of the 1,800 survey respondents labeled themselves as "nationalistic" compared to 47 percent who characterized themselves as "non-nationalistic" (EuskoBarometro 2005).

14. At times, such as in 1986 and 1994, the PNV has formed governing coalitions with the non-nationalist Socialist Party.

15. When this question was asked using a different configuration of options, 34 percent stated "autonomy as current", 22 percent desired "more autonomy" and 19 percent stated "independence." In this survey, "federalism" was not listed as an option (Palleres *et al.* 1997).

16. Strained relations between the state and region have characterized periods of both Popular Party (1996–2004) and Socialist Party (1982–1996) control in Madrid.

17. State agencies and public companies are the Ministry of Development, Public Land Use Enterprise, the Port Authority, and the national and regional railway companies. Basque institutions are the regional, provincial and two city governments (Bilbao and Barakaldo).

18. The selected plan was by architect Cesar Pelli, who has designed such places as Battery Park City in New York City.

19. The emblematic building at the other end of the Abandoibarra district is the Euskalduna Palace Conference and Music Center, opened in 1999 with public financial support for construction of some 56 million Euros. Like the Guggenheim, it also has distinct architecture and was designed by a world-renowned architect.

20. Bilbao is not alone. Neill, Fitzsimons, and Murtagh (1995) show that other "pariah cities" of deep political or social conflict have emphasized reimaging strategies.

21. This impact of professional associations on public opinion in nationalistic regions can also be strongly negative, as witnessed by the key role of a report written in the mid-1980s by the Serbian Academy of Arts and Sciences in energizing and legitimizing radical Serbian nationalism (Covic 1993; Weine 1999).

22. I am not arguing that urbanism is such an enterprise in reality, but that its professional moorings and reputation are tied in many peoples' minds to standards of objectivity and rationality. Such a perception, I suggest, is sufficient to bring to negotiations over city issues different political interests which are otherwise polarized by larger politics.

23. Because of the multiplicity and fragmentation of political parties in the Basque Country, coalitions are the common requirement for obtaining governing majorities.

24. This thesis that democratization poses a threat to extremism is also supported by the fact that political killings by the ETA spiked again in 1987, a period of time when all political parties except for Batasuna signed the Pacto de Ajuria-Enea committing them to work for the end of terrorism (Newman 1996).

25. This conceptualization does not rule out the influence of external factors on an extremist group. Pertaining to the 2006 "permanent ceasefire" announced by ETA, it

6 Mostar: urbanism and the spoils of war

Figure 6.1 Šantića Street, 2002

Mostar is the only multicultural city in Bosnia Herzegovina today because nobody won here.

Neven Tomić
Former mayor
(April 6, 2003)

The problem with Mostar is that it is a 50–50 town. You can't be generous to anybody, only nasty to each other.

Murray McCullough
Head, Delegation of the European
Commission to Bosnia and Herzegovina
Interview

may be that the extreme public abhorrence of terrorism after Al-Qaeda's attack on Madrid train stations March 2004 led ETA to recalibrate whether it should continue its militant campaign.

26. Whereas terrorism is an increasing part of the world landscape, the Basque and Northern Irish cases stand out in their strong public support by a definable segment of the population in favor of political violence. The link between violence and public support increases the salience of examining public opinion as a potential entry point into the violence dynamic.

27. The overall nationalist vote has usually been over 50 percent (ranging from 53 percent in 2001 to 68 percent in 1990) for regional elections. This means that PNV has been part of most governing coalitions and usually as the lead partner. In the 2004 national elections, the non-nationalist vote for the Socialist Party and the Popular Party reached over the 50 percent mark, an occurrence that did not happen from 1979–1993 but now has occurred for the last three national elections.

28. Beck also found speaking of Euskera (Basque) language to be an important local correlate of ETA support.

29. Such abstentions can cumulatively be significant regionwide, such as in the regional elections of 2003 when 556,000 abstentions and 131,000 null ballots were filed alongside the 1.11 million candidate votes.

30. The base of nationalist support in small Basque towns constituted an important component of an ultimately unsuccessful agreement between the PNV and HB in 1998 (*Pacto de Lizarra*) based on the perceived ability of a pan-nationalist "assembly of local power" to push for Basque independence (F. Llera, interview).

31. For details, see i3 Consultants *et al.* (2000); Basque Study Society (2001).

32. Arostegi does note that although the Spanish government officially supports Eurocity, some of its actions have been slow, such as Madrid's approval to use Spanish government-owned land to build the intermodal logistics facility at the Spanish–French border.

33. Basque planners are also advocating a high-speed train connection between Pais Vasco and Barcelona in order to strengthen links in that direction.

34. Such reimaging is a significant part of the Eurocity project. Its stunning digital orthophoto map of the Spanish-French coastline (with no international border designated) was hanging in almost every office I visited at the offices of the Diputacion and is on the cover of many reports and documents produced by the project.

35. I borrow from the insights of Friedmann and Weaver (1979).

36. As a component of the Ibarretxe Plan of cos-overeignty or free association, put forth in 2004 by the Basque government, there would be the creation of a new, direct relationship between the Basque Country and Europe, unmediated by the Spanish state. Despite the gains envisioned through strengthened ties to Europe, the costs of separation from the Spanish state could be substantial. A 2004 study estimated that outright independence would cost the region between 10 and 20 percent of its gross domestic product due to its severing of economic links to Spain. The dependence of Basque Country on the Spanish economy was found to be 11 times greater than any other region in Spain (Buesa 2004).

THE CITY OF NO WINNERS

Mostar is a story of urbanism amidst, and as a contributor to, rupture. Although not physically partitioned, the city of Mostar in southeastern Bosnia Herzegovina was ten years after war a divided urban area – psychologically, economically, and governmentally. It is a city of greater de facto division than Sarajevo and more lasting and visible physical damage and destruction. It is a city where Bosniak Muslims and Bosnian Croats live in parallel universes. As an international community official explained, "Sarajevo was and is not divided, but Mostar most assuredly is" (Gerd Wochein, Office of High Representative [OHR], interview). In a unilateral decree reforming the city's institutions, the OHR (2004a: 2) described the "profoundly conflicting interests among its constituent peoples."

Mostar presents a dispiriting story not just of the devastation of war, but also of how post-war local governance and urbanism can become means by which war profiteers solidify their power and reinforce nationalist divisions. The collective interest of the city collapsed and dissolved, being usurped and exploited by nationalist political leaders who used the urban area to construct new demographic, social, and psychological realities. Planning and urbanism have been at the core of this fight over the post-war city and have disintegrated into absurd conditions of parallelism. The active form of war stopped in 1994, but the antagonisms that created the war continued and found other means by which to implement their hatred. Primary among the spoils of the war has been the city and its collective sphere.

Mostar is an instructive example of direct international management of a city for ten years. During that time, the international community (IC) and its lead entity, the European Union, have sought to counter the development of a "parallel" city, but in the end the IC has been used as a partner in the creation of just such a city. The IC proposed an urban strategy using a neutral "central zone" intended to be a seed for future normalization of Mostar and this approach was, in principle, sound. However, in practice, this "neutral" spatial planning approach fell victim to ethnic polarization and constitutes a lost opportunity to construct a buffer and bridge between the antagonistic urban groups. Mostar has revealed the limits of IC involvement in the complexities of urban management. Nonetheless, due to its geographic and political setting within the Muslim-Croat Federation (one of the two entities within the state of Bosnia Herzegovina), Mostar is today an important potential model for figuring out multinational governance in Bosnia and Herzegovina. Yet, the challenge facing the international community and domestic leaders to normalize Mostar and make it work in a unified way confronts policymakers with unsettling conclusions about the nature of ethnic space, displacement, and inter-group stability.

Bridges and cultural fault lines

Mostar is less populated than Sarajevo; the city's pre-war population was about 126,000 according to the 1991 Census. It is a provincial locale in the Herzegovina region of Bosnia that lacks the central geographic location and international reputation of Sarajevo. Such a feeling of being out of the way makes the violence and physical destruction that occurred here over a two-year period feel even more barbaric. Yet, Mostar is not actually remote; rather, it has for centuries had the fortune and misfortune of residing at the fault lines between cultures and peoples, historically between Turkish and European dimensions, and more recently between Croat, Serb, and Bosniak (Muslim) territorial and political aspirations. Four centuries of Ottoman rule over Bosnia commenced in the later half of the fifteenth century, and this transformed Mostar from a minor river crossing to a thriving colonial crossroads. In 1566, the Stari Most (Old Bridge) was constructed over the deep and fast-running River Neretva, shortly after the Ottoman conquest of the Balkans, to act as a key link in one of the Turkish Empire's main east-west trade roads. Stari Most substantially facilitated travel, trade, and the movement of military troops; it also was a symbol of power of the Ottoman Empire and ensured Mostar's primacy as the capital of the county of Herzegovina.[1] The city became one of the foremost outposts of the Muslim presence in the West and Islamic architecture took root in the building of mosques and houses and neighborhoods (*mahalas*) developed on both sides of the Neretva (Aga Khan Trust 1999).

In 1878, European powers decided after a long period of Ottoman decline to have Austria-Hungary assume the administration of Bosnia-Herzegovina. New, more western planning schemes were introduced that created broad avenues and an urban grid on the west side of the Neretva, improving circulation and facilitating city growth. An important de facto division within the city emerged during the development of the Austrian parts of the urban area (Plunz, Baratloo, and Conard 1998). Today, just to the west of and parallel to the Neretva River is the "Boulevar;" at the time of Austrian expansion of the city, this was a railroad line that separated the old city from the new Austrian extension of the city to the west (Plunz, Baratloo, and Conard 1998). At the close of World War II, a new regime took hold in Bosnia and Mostar, with Tito at the heart of a Socialist Yugoslavia. During the years of this regime, Mostar grew significantly, from a population of 18,000 in 1945 to about 100,000 in 1980. Because the eastern side of the river was crowded and hemmed in by a mountain range, the city expanded on the western bank with the construction of large residential blocks and commercial buildings.

One year before the outbreak of the Bosnian war, Mostar was an ethnically mixed city. Bosniak Muslims and Croats were each about 34 percent of the city population, Serbs were about 19 percent, and about 14 percent identified themselves as Yugoslavs or others (1991 Census). The city constituted a melting pot in that about one-third of marriages were ethnically mixed. With the exception of old town Mostar of mostly Muslim residents, the rest of the city was fairly mixed ethnically and one could not discern an east-west ethnic divide (Nigel Moore, former head, Reconstruction and Return Task Force, OHR Mostar, interview). Although there existed identifiable districts where either Croats or Bosniaks were the clear plurality, even in those districts the minority ethnicity was well represented (Ministry of Human Rights and Refugees 2003). For example, in the old town and central zone district, the Bosniaks clearly had the greatest presence (47 percent of the district population); however, 21 percent of the residents were Serbs and 16 percent were Croats. In Mostar west, Croats were in the plurality (42 percent), but Bosniaks (22 percent) and Serbs (20 percent) were well represented. It is also notable that significant ethnic mixing occurred at the smaller neighborhood scale. In two of the neighborhood areas of the city that would be near the frontline of fighting, pre-war populations were mixed. Bosniaks occupied 44 percent of the 177 dwelling units destroyed in the Šantića/Milosa neighborhood, Croats 35 percent, and Serbs 21 percent (Mostar Urban Planning Department 2004). And, in the Boulevar area, Croats occupied 38 percent of destroyed dwelling units, Bosniaks 35 percent, and Serbs 26 percent.

"Mostar was always going to be problematic"[2]

One look at a map of the ethnic geography of Bosnia Herzegovina displays the precarious location of multi-ethnic Mostar relative to the military campaigns of ethnic territorial consolidation and aggrandizement that constituted the 1992–1995 Bosnian War (see Figure 2.3, page 27). The city is proximate to the settlement zones of each of the three nationality groups, located near the intersection of western Herzegovina (majority Bosnian Croat and next door to Croatia 20 miles to its south), Central Bosnia (of a majority Muslim population), and eastern Herzegovina (majority Bosnian Serb and proximate to Serbia).

The city from April 1992 to February 1994 was ravaged by two wars. In the first one, Serbian elements of the Yugoslav National Army attacked the city with heavy artillery and multiple rocket launchers. Shelling killed an estimated 1,600 persons, industrial capacity was destroyed, and historic and sacred buildings were targeted (Aga Khan 1999). Croats and Muslims together fought in defense and the Serbs eventually withdrew for tactical reasons; at the same time, a majority of Serb residents of the city left. About one year after the Serb withdrawal, a "war within a war" or the "second battle of Mostar" began when the Bosnian Croatian Militia (HVO) occupied the west bank of the Neretva and began expelling Muslim families from their homes. Former allies, the Croats and Bosniaks turned on each other and a close-fought war – street-by-street and building-by-building – ensued for nine months. These hostilities killed about 2,000 individuals, radically changed the demographic profile of the city through forced displacement, and physically decimated large parts of the city. The confrontation line originally was down the Neretva River, dividing the city into two camps – one Croat and western, the other Muslim and eastern (with a significant Bosniak enclave west of the river). Subsequently, the frontline of fighting was established along the "Boulevar" in west Mostar, reinforcing Croat and Bosniak parts of the city.

The main aim of the Bosnian Croats was to preserve their national culture through the creation of a "Croatian Republic of Herceg Bosna," declared by Croatian President Franjo Tudjman. The political power base of the Bosnian Croats was in west Mostar and its leaders were concerned that they could lose the only city in a post-war Bosnia Herzegovina where they were a high demographic presence. Threatened by their future status as a minority population in a post-war Bosnia Herzegovina, they surmised, "The Muslims have Tuzla, Zenica and Sarajevo, why can we not have Mostar?" (Garrod 1998). Bosniaks' aspirations, in contrast, were more for a unified Bosnia Herzegovina such as existed in the republic before the war. Lacking a "mother country," the Bosniaks under this unified Bosnian scenario would constitute the largest population. While feeling under threat by Croat territorial motives, Mostar Bosniaks' anxieties increased further because they felt remote from, and marginalized by, the Sarajevo-centered power base of Bosniak leader Alija Izetbegovic (Nigel Moore, interview).

When Croat-Bosniak hostilities in Mostar ended in February 1994, the demographic and physical composition of the city had been severely reconfigured. The city had been ethnically sorted and cleansed, with the Boulevar as the line of division. An estimated 15,000 individuals, overwhelmingly Muslim, were expelled from west to east during the war, according to Javier Mier (OHR Mostar 1994–2001, interview). According to a local NGO, east Mostar was now 98 percent Bosniak and less than 1 percent each Croat and Serb; west Mostar was now 84 percent Croat, 11 percent Bosniak (primarily within an enclave just west of the river but east of the Boulevar) and 3 percent Serb [Repatriation Information Center 1998]. Another survey documented similar results, with Croat percentages in the three western districts ranging from 77 to 81 percent and Bosniak percentages in the four eastern districts ranging from 93 to 98 percent (International Crisis Group 2003).[3]

Due to the displacement of individuals during the larger Bosnian War and the Mostar war-within-a-war, it was estimated that post-war east Mostar contained over 30,000 displaced persons, coming from Eastern Herzegovina, Stolac, and the Capljina region in addition to west Mostar. In west Mostar, about 17,000 displaced persons resided there after the war, coming mainly from Central Bosnia, Sarajevo, Jablanica, and Konjic (Repatriation Information Center 1998). On the western side, a portion of the resettlement of displaced persons there appears to be the result of an intentional project by the Croatian government, which had subsidized resettlement in order to strengthen Croat demographic and political control over the city's destiny (Interviews: Wolfgang Herdt, Malteser Hilfsdienst NGO; Gerd Wochein, former OHR Mostar). As a result of the war, displacement, and post-war resettlement, one nongovernmental group estimated that for the city overall "the narrow Bosniak plurality of 1991 has become a substantial Croat majority" (International Crisis Group 2003: 7). Other sources are silent on this current demographic reality or feel that the proportions for the two groups remain roughly equivalent. A particularly alarming result of the city's trauma is that its total population has decreased by almost 20 percent since 1991, with estimates for 2003 standing at about 105,000 residents (Commission for Reforming the City of Mostar, 2003).

Mostar was the most heavily destroyed city in Bosnia Herzegovina (Garrod 1998; N. Moore, former OHR Mostar, interview). The area of greatest destruction was Muslim east Mostar and the Bosniak part of west Mostar, where between 60 and 75 percent of buildings were destroyed or severely damaged (Interviews: M. McCullough, EU Commission to BiH, Mostar office; N. Moore). Croat west Mostar, in contrast, sustained about 20 percent severe damage or destruction, with most destruction concentrated along the western side of the Boulevar line of hostilities. Estimates are that about 6,500 individual housing units (of a total of 17,500) were damaged or destroyed in the city, while significant numbers of larger collective housing complexes were damaged or destroyed, particularly those near the banks of the Neretva River (Vucina and Puljic 2001, Aga Khan Trust 1999):[4] The area of greatest and most concentrated destruction was Šantića Street, an area of mixed ethnicity along the river on the western bank (see Figure 6.1). The former EU Special Envoy in Mostar, Sir Martin Garrod, called this "the most sinister street in Europe." There was deliberate targeting throughout the city of historic monuments, cultural property, and religious buildings during both wars in Mostar, including the Bishop's Palace, cathedrals, mosques, orthodox churches, Austrian and Ottoman baths, orchestra buildings, museums, and numerous historic residential buildings. Bridges having economic, cultural, and military importance were also targeted. The Serbs in the first war dynamited nine bridges. Stari Most, the old bridge, survived the first war but, in November 1993, collapsed into the Neretva River after suffering sustained bombardment from Bosnian Croat militiamen. At that point, Mostar, a city defined by bridges and crossings, had not a single one left.

Stopping the bloodshed, governing the city

The "Washington Agreement" of March 1994 stopped the fighting in Mostar and created the Muslim-Croat Federation to jointly administer areas then under Muslim and Croat military control.[5] The Agreement recognized Mostar as the most seriously divided city in the Federation and stated that it needed to be directly administered for a two-year interim period by an international body. This body – European Union Administration of Mostar (EUAM) – would facilitate post-war transition, coordinate reconstruction, and establish essential structures of governance in the early years. A Memorandum of Understanding (MoU) signed in April committed the signatories to the development of a unified, multi-ethnic city of open return and freedom of movement and officially established the EUAM and its powers.[6]

The EUAM began operations in July 1994 amidst unstable and volatile conditions. With blockades enforced by militias in place, travel from east to west Mostar required a circuitous 30-mile trip (Javier Mier, former OHR Mostar). It was shocking to see groups who were allies in Sarajevo bitter enemies in Mostar (J. Saura, interview). Martin Garrod, former EU special envoy in Mostar, recounts the "climate of fear and intimidation" that existed in Croat west Mostar, created by criminals and gang leaders who had achieved their power and money during the war (Garrod 1998). These ethnic strongmen obstructed any effort to unify the city, continued a spate of evictions of Bosniaks from west Mostar apartments throughout most of the interim EU period, and were likely behind the anti-tank rocket that was fired at the apartment of EU administrator Hans Koschnick in September 1994. Explosions and shootings continued unabated through the end of 1996 (Garrod 1998).

International agreements such as the Washington Agreement and its successors focused on the ending of active conflict. "These negotiated agreements were a good option to continued fighting and they stopped the shooting war," says Murray McCullough (interview), "but they did not bring the peace." Thus, EUAM personnel tasked with building the peace and reconstruction felt a sense of isolation and lack of guidance. The director of reconstruction for EUAM recalls, "No one told us what we should be doing. There were no precedents and no prior experience on which to decide what action to take" (Yarwood 1999: 10). Operating out of a former home for the elderly that was spatially between the two ethnic halves of the urban area, the EUAM established seven departments – city administration, finance and taxes, reconstruction, economic and transport infrastructure, education and culture, health and social services, and public order – in order to run the city (Garrod 1998). An advisory council of local leaders was created to support the EU administrator, and principal counselors (one from each of the three ethnic groups) would advise the administrator on a daily basis. Each department head within the EUAM had two co-heads (one Croat and one Bosniak) as a way to cooperate on technical matters and to act as liaisons with their respective ethnic groups. Although the EUAM was based on a political mandate to organize city governance and run the city rather than a humanitarian one, it came with substantial funds to help rebuild the city, approximately 150 million Euros that it spent over two years (M. McCullough, interview).

The Dayton Agreement in November 1995 stopped the larger Bosnian war and divided the state into the two autonomous regions of today. Mostar was again on the table for international discussion. Dayton negotiators considered the city to be of such importance to Bosnia's future that a specific annex addressed the city and reaffirmed agreement on a set of principles for an interim statute and eventual city unification.[7] There was the sense that "if Mostar wouldn't work, the Federation wouldn't work," and attention turned in 1996 to substantial reconstruction of the city as a way to bring people back (N. Moore, interview). In February 1996, the EUAM administrator and the de facto mayors of Croat Mostar and Bosniak Mostar signed an Interim Statute for Mostar. This designated temporary arrangements of powers to guide Mostar's gradual recovery while a permanent legal structure for the city could be adopted. In that same month, the Rome Agreement promoted the goals of unifying the city and facilitating returns and demarcated a "central zone." This zone is to play a central role in this story of urbanism in post-war Mostar. It was demarcated to consist of a common strip of land around the former confrontation line, where joint Federation, Canton, and City institutions and administrations were to be located. It was envisioned by the IC as a buffer between the two sides and as way to not validate what the war did by preventing the establishment of ethnic municipalities right up to the confrontation line (J. Mier, interview). To be composed of approximately equal numbers of Muslim and Croat residents, the central zone was intended to be a seed of mixed and joint activities that might grow over time. It was therefore "the symbol and key to a unified and multi-ethnic Mostar" (International Crisis Group 2000: 43). A city council would govern the zone, and a city administration would manage it. This council was created to assure power sharing across the nationalistic divides and consisted of 48 members – 16 Bosniak representatives, 16 Croat members, and 16 "others."[8] The mayor and deputy mayor for the city would come from each of the two main ethnic groups and these positions would rotate each year.

Despite the goal of unification, the reality of a severely ethnically sorted city meant that Dayton and the Interim Statute had to accommodate war-imposed ethnic territoriality. These agreements created, in addition to the central zone, six municipal districts, or city municipalities (hereafter municipalities) – three in Croat-controlled areas (Mostar South, Mostar Southwest, and Mostar West) and three in Muslim-controlled areas (Mostar Southeast, Mostar North, and Mostar Old Town).[9] In order to assure that the voice of each group would be heard in political deliberations, the composition of the municipal councils was to reflect the proportions of each nationality group that resided within the municipal boundaries in 1991. Nevertheless, the geographic boundaries of these municipalities reflected the demarcation lines established through war and rein-forced the power of the dominant group in each area (Commission for Reforming the City of Mostar 2003).

The intent of the Interim Statute was that the city council, at the end of the EU mandate, would be in the dominant position vis-à-vis the six municipal councils (in terms of powers and responsibilities), and that the city-wide administration would control the municipality administrations, in addition to directly adminis-tering the central zone. During the interim phase, the six municipalities were expected to turn over to the city administration important powers, including finance and tax policy, infrastructure, urbanism/planning, economic policy, and public transport. The Interim Statute sought to increase the abilities and com-petencies of the central authorities through empowerment of the city council and development of the central zone (Commission on Reforming the City of Mostar 2003).

The war and the international agreements that had stopped the open conflict (the Washington Agreement, Dayton Accord, the Interim Statute, and the Rome Agreement) had transformed and reconfigured Mostar from one mixed pre-war municipality that encompassed all the urban area into a politically fragmented urban area containing six ethnically sorted municipalities and one city administration (with its central zone jurisdiction) that was to hold the urban area together politically. The Interim Statute outlined a clear and logical evolutionary path whereby Mostar would be unified under a single empowered administration of local leaders. The intent was to constructively undo the city's wartime partition and ethnic political rule during the interim period of direct international administration. The *city* of Mostar, governed by a council of mixed ethnicity, would progressively enlarge the scope of its responsibilities and the ethnically demarcated *municipalities* would gradually lessen in importance. Meanwhile, there was the hope that the development of the central zone would be an important land use model and catalyst for the normalization of the larger urban area.

The transition of Mostar from direct international rule to local rule occurred with the conducting of municipal elections in June 1996 and September 1997 and with the formal ending of EUAM tenure in the city January 1997.[10] The visions of the international community regarding Mostar's unification would now be put to a critical test.

URBANISM AND THE SPOILS OF WAR

On the political side, the war still goes on.

Murray McCullough
Head, Delegation of the European Commission
to Bosnia and Herzegovina
Mostar Office
Interview (May 6, 2004)

It is clear that the current situation is unsustainable and unacceptable.

Commission for Reforming the City of Mostar
Report of the Chairman (Norbert Winterstein)
December 2003, p. 15.

The international community concentrated its efforts to reintegrate the city on the "central zone." However, this effort was sabotaged and eclipsed because in the larger urban area the IC accommodated ethnic, wartime geographies in the form of the six "city-municipalities." Creation by Dayton of these six municipalities was the critical decision that determined the urban area's future. These ethnically delineated municipalities appealed not to a cross-ethnic collective interest of the larger urban area, but rather to specific ethnic group interests. In certain ways, this accommodation to ethnic rule at the local level followed a strategy first laid out in the Washington Agreement about how the larger Bosnian Federation would address ethnic rule. In that agreement, negotiators accepted that cantons would provide some degree of ethnic self-rule within an overarching Federation structure. For the city of Mostar, the same logic was applied on a smaller geographic scale – that the set of three Croat and three Bosniak municipalities would provide for a degree of ethnic self-rule required after warfare, while at the same time positioning these municipalities within an overarching city government structure. Both for the Cantonal system at the Bosnian Federation level and the municipality system at the Mostar city level, the question was the same – would the centrifugal tendencies of ethnic governments co-exist productively with the integration objectives of cross-national governance?

"The war still goes on"

> *Young boys and girls in west Mostar elementary schools know very little about the Neretva River. They tell me, 'the river is near the city'. It is strange and very sad to hear this — the river is not near the city, it is inside the city.*
>
> Muhamed-Hamica Nametak
> Director, Puppet Theater of Mostar
> Interview

Ten years after war, Mostar is a divided city. As Gerd Wochein (interview) describes, "The front line of the Bosniak-Croat war still exists today, ten years later, as a line of division." The ethnically fragmented local government structure has solidified and reinforced Bosniak-Croat differences and has created a divided Mostar possibly beyond repair and reconciliation. In accepting wartime ethnic geographies, Mostar is an illuminating microcosm of the problems that have been part of ethnically fractured Bosnia Herzegovina. And, in its belated efforts to transcend such wartime geographies (with a March 2004 unification decree), Mostar also exhibits the significant challenges facing BiH after ten years of ethnic entrenchment and obstruction. The city represents a significant missed opportunity to use an urban integration strategy as a local remedy and model for inter-group relations at broader political levels. What makes this failure more acutely felt is that the IC had an early awareness of the importance of Mostar to Bosnian peace-building efforts and of the measures needed to move the city away from embittered ethnic fragmentation.

The Dayton-created system of local governance in Mostar operated mostly as a shell and artifice. In reality, the ethnic municipalities retained all of the competencies intended for the city administration, thus hamstringing it as an integrative force. The city administration, configured as an overarching structure for the entire urban area, ended up instead as a spatially circumscribed seventh municipality whose borders were consistently exploited and ravaged by the six ethnic municipalities. Much of Mostar's governance and the operation of public power have been paralegal, connected to a war criminal elite that has shamelessly exploited public power to pursue private nationalistic ends. "While Dayton-arranged central city institutions operate for show," says J. Mier (interview), "the real power has been with the municipalities and even more, with the political parties that control these municipal leaders." The public, collective sphere that existed in pre-war Mostar has been fragmented, subordinated, and manipulated.

War profiteers adroitly used urban governance to solidify their power and reinforce nationalist divisions in and around Mostar. Traditionally, tensions in Mostar have been more between local Croats and Bosniaks, and less involving Serbs, and corruption and private abuse of political power have been elements of local governance. Indeed, one observer notes that the war, and the polarization since, may be as much about criminal and extralegal control of the area than about strict ethnic or nationalistic issues (N. Moore, interview). The 1996 local elections that were deemed of vital importance by the international community as a step toward the normalization of the city, instead brought into local governance the same interests that conducted the war. On the Bosnian Croat side, these included many ethnic leaders who were adamantly opposed to the European Union mission and mandate to unify the city. "The elections were absolutely too early; you don't break down psychological walls through political elections," criticizes J. Saura (interview). In a climate of fear and intimidation that exists after war, "you vote for who will defend you from the others, not for someone who will bring people together," states J. Mier (interview). Those elected into the ethnic municipal power structures after war were "strong hardliners whose power lay with protecting the status quo and who clearly were not the future," states Julien Berthoud (Head of Political Section, OHR Mostar, interview).

The ethnic municipalities have been virtual "fiefdoms" that have used communal resources and revenues to advance the good solely of "their own people" (Commission for Reforming the City of Mostar 2003: 13). The power base for Bosnian Croats has been in Mostar Southwest municipality; for the first six years after the war, this municipality spoke for and controlled all of west Mostar. The Croatian Democratic Union (HDZ) of Bosnia Herzegovina political party,[11] with its base in western Herzegovina and an instrumental protagonist in carving out a secessionist "Croat Republic of Herzegovina" during the war, controls Croat society and polity in both the western Mostar municipalities and in the Herzegov-Neretva Canton. Although local Croat leaders in Mostar from the HDZ signed the 1994 Memorandum of Understanding, committing them to the development of a unified, multi-ethnic city, the real ability to influence Mostar's future was more in Zagreb, Croatia, where war hero and President of Croatia, Franjo Tudjman, worked to create a Croat Herzegovina and obstruct Mostar's unification (M. McCullough, European Commission, BiH, interview). In the three Croat majority municipalities, the HDZ worked to exclude Bosniaks from municipal institutions and has resisted the imposition of minority safeguards outlined in the Interim Statute.[12] Throughout the years, the local Croat elite has obstructed the development of any central, integrative body, remaining separate through a system of illegal activities, split public companies, and privatization of public entities along ethnic lines (Commission for Reforming the City of Mostar 2003; International Crisis Group 2000). HDZ has further obstructed the return of Bosniak households to units in west Mostar in order to keep the city divided and their power base unimpeded.

Bosniak political power in the city, meanwhile, resides in Stari Grad municipality and within the city administration and is largely based in the Party of Democratic Action (SDA). The Bosniaks for most of the ten post-war years have been on record as supportive of city unification and the eradication of ethnic municipalities. However, the Bosniak elite has also contributed, through the use of ethnic patronage and policymaking, to the dual nature of the city and to the parallelism that exists in the exercise of public power. Interviewees from the international community point out, however, that the misuse and illegal use of public power for ethnic ends has been more a fact on the Croat side than on the Bosniak side (interviews: G. Wochein, N. Moore, M. McCullough; International Crisis Group 2000).

> *The six City-Municipalities function in reality as two separate blocks,*
> *a Bosniak and a Croat Mostar, whose interests are basically contrary to one*
> *another, and which have brought the City of Mostar to the edge of collapse.*
> Commission for Reforming the
> City of Mostar (2003, p. 13)

As of 2004, parallelism and division in Mostar were at absurd levels. There were almost 200 politicians elected in the seven local governments who run a city of a little over 100,000 people. There were over 700 public employees in the seven bureaucracies, many placed in their positions through patronage and ethnic vetting. Redundancy, inefficiencies, and waste accumulate in this ethnically fractured public sector. For the six municipalities, there is one public employee per 189 residents, a concentration far above the 1:500 ratio recommended by experts.[13] The City of Mostar, an entity that was to foster ethnic integration in the central zone, has been "a dead letter on paper" and itself been divided administratively (Commission for Reforming the City of Mostar 2003: 13). The city council for years "presented itself to the outside world as a multi-ethnic body, but in reality they didn't even sit together" (G. Wochein, interview). The Mayor and Deputy Mayor have worked in parallel to each other using their own ethnic administrations, which until 2002 were even physically and spatially separated. There existed two separate treasuries so that all spending by the central city administration was paid through ethnically separated accounts. Although a power-sharing arrangement was established for the City Council, in reality those with power in the council had their primary power base and allegiance elsewhere in one of the municipalities. Ethnic interest consistently trumped city-wide interest.

Because the six ethnic municipalities captured many normal functions of local government, public services that should be integrated and city-wide became ethnically fragmented. As of 2004, there were divided health care, educational and childcare institutions, dual urban planning and regulatory systems, and parallel public transportation, water supply, electricity, and sewage systems (Commission for Reforming the City of Mostar 2003). Urban planning institutions were split into two – the Institute for Urbanism serving Bosniak municipalities and Urbinig Mostar serving Croat municipalities – and they are under the significant influence of ethnic leaders (M. Raspudić, interview). These planning organizations are quasi-public, having been formed by the ethnic municipalities to provide technical support on reconstruction project planning. Raspudić estimates that there are about 200 urbanists and related professionals in these quasi-public institutes and the six municipal governments themselves, compared to about one-tenth that many in the city administration. She observes, "if all of these urban professionals were working competently, this would be a beautiful town."

Political connections, rather than sound planning, have driven development in both the Croat and Bosniak municipalities (interviews: Marica Raspudić, Zoran Bosnjak, Palma Palameta, Urban Planning Department, City of Mostar). This has resulted, particularly in the western Croat municipalities, in construction of bulky buildings incompatible with the adjacent urban fabric, carving up of green environmental space for development, and a lack of municipal tax revenue – in effect producing a "city of chaos" (M. Raspudić, interview).

Elementary and secondary education has been divided; students on the eastside use a "Bosniak" curriculum and those on the westside teach a Croatian one and use textbooks from Croatia.[14] This ethnic separation of public education is particularly dispiriting because, as one source notes, this likely means that "the effects of ethnic cleansing and the psychological fears and prejudices created through the war years will simply be passed on to the next generation" (International Crisis Group 2000: 50). With regard to electricity, although the three hydroelectric power plants in the urban area are in the designated "central zone" and thus were to be controlled by the city administration, in reality two separate state companies have controlled them. Two of the plants are administered by Bosniaks to supply electricity to east Mostar; the other plant is owned by a Croat concern that supplies power to west Mostar. The ethnic capture of electricity (and other public goods) means that the city administration lost substantial amounts of potential revenue that could be used to fund activities that promote a unified city. There has been a consistent pattern of private aggrandizement at the expense of the public sphere, both through the ethnic capture and delivery of public services and in the privatization of companies along ethnic lines that were formerly publicly controlled (International Crisis Group 2000).

The subordination of the collective city-wide interest to those of the two ethnic groups is also evident in the demographics numbers game being waged. Both sides have endeavored to increase their share of the city-wide population as a way to increase their claims to the city and to assure political control in a future when the city might be politically unified.[15] Croat settlement in west Mostar of displaced persons (DPs) from central Bosnia and Sarajevo, in particular, has been significant (Interviews: N. Moore and G. Wochein). In the early years after the war, the Croatian government created and carried out a deliberate, well-coordinated plan to settle Croat DPs into the area with the help of Croat municipal authorities, who provided land, building materials, and other logistical support (N. Moore, interview). Possibly up to 30 percent of all housing west of the Neretva River (an estimated 4,000–5,000 housing units) has been built through these politically inspired mechanisms (N. Moore, interview). Bosniaks have also engaged in these practices; however, the magnitude is thought to be much less, likely because there is not for Bosniaks the pipeline of money and support that Croat DPs enjoy from Croatia (N. Moore, interview).

The parallelism and division within the urban sphere also pervades Herzegov-Neretva Canton in western Herzegovina, one of the ten new cantons created by the Washington Agreement and formalized in the Dayton Agreement. Herzegov-Neretva Canton is one of only two cantons in the Federation that were designated as "ethnically mixed" and thus in need of special legislative procedures to protect each of the ethnic groups. Since the wars, it is difficult to document population in what today is the Canton; nevertheless, it is estimated that the Croat population has increased, Bosniaks have decreased somewhat, and the Serbs have declined significantly from pre-war levels (N. Moore, interview). One estimate puts the Cantonal population at about 55 percent Croatian, 40 percent Bosniak, and 5 percent Serbian (N. Moore, interview). post-war activities of the Cantonal government have been limited and ethnically circumscribed. Through 2003, the government of the Canton had not been formally composed; thus, there existed de facto two governments, one Croat, one Bosniak. A minister and a deputy minister headed each administrative department and each was the chief of his own ethnic group. Budgets were drawn up and allocated for each side of each department (Semin Borić, Minister of Finance, Herzegov/Neretva Canton, interview). The de facto division of Cantonal administration meant that Cantonal competencies (such as primary and secondary education) would be ethnically divided. It also fore-closed on any possibility to engage in joint activities, such as creating an urban development plan for the Canton (Jaroslav Vego, Ministry of Urban Planning, Herzegov/Neretva Canton, interview).

The hardening of hate

The life of ordinary people on this artificially
divided space has become absurd.

> Commission for Reforming the
> City of Mostar 2003, 14.

Rather than providing Bosniaks and Croats with a sense of security that might engender cross-community relations, the ten years of de facto division of Mostar have hardened antagonisms between the two sides. "The old front lines are not visible but they are still with us as our major boundary," asserts Wolfgang Herdt (Regional Director, Malteser Hilfsdienst [NGO], interview). Unlike in Sarajevo, where there is now a clear majority of one ethnic group, antagonistic sides in Mostar have remained in the urban arena, albeit displaced. The political division of Mostar after the war has galvanized greater inter-group economic, religious, and psychological differences through the ten subsequent years. "Before we all looked the same," says Palma Palameta (Urban Planning Department, City of Mostar, interview), "but now people engage in activities that symbolize their bigger group." For example, the hold of religion on people is now greater than before the war. There is some movement among Muslims in Mostar from a moderate Turkish and western-oriented Islam to a more Middle Eastern version. At the same time, the building of Catholic institutions and symbolic structures have been key elements of Croat rebuilding on the west side.

Economically, there has been a hardening and consolidation of separate workplaces (including public administration) and customer bases over the ten years (N. Moore, interview). The slight to moderate linguistic differences between the Bosniak, Croat, and Serb use of the Serbo-Croatian language have been highlighted and now three formal languages exist – Bosnian, Croat, and Serbian. With little in the way of reconciliation and minimal levels of civil society organizations that bridge the ethnic divide, the scars of war and division are open and painful (interviews: Jaroslav Vego, professor at University of Mostar; Mohamed-Hamica Nametak; Zlatan Buljko, head, United Methodist Committee on Relief).

If Mostar is ever to resemble a normal city and operate in a politically unified way, the distinctly different psychological worlds of the two sides will need to be bridged. Bosnian Croats, who lost population in Bosnia during the war and feel as a threatened minority, all have Croatian passports and thus have an "actionable option" to relocate out of the Bosnia state (M. McCullough, interview). International community observers are uneasy about Croats' genuine involvement in a multi-ethnic Bosnia and Mostar: "We have tried for ten years to convince the Bosnian Croats that they are part of Bosnia rather than the motherland and we have been unsuccessful" (M. McCullough, interview.) Bosniak Muslims, meanwhile, feel that, while they won in Sarajevo, they lost in Mostar (N. Moore, interview). M. McCullough (interview) describes this Bosniak perspective: "that Croats have engineered the situation through bestiality and violence and now want to be rewarded with majority control of the city. It is unacceptable to Bosniaks that Croats would achieve in the peace what they couldn't achieve in the war." Midst such hostility, the effort by both groups (particular the Croats) to change the pre-war demographic balance of the city had "become a tool to foster mutual fears and distrust among people" (Commission for Reforming the City of Mostar 2003: 15).

After numerous unsuccessful efforts by the IC to reform and unify the political and legal structure of the city, the High Representative of Bosnia established a Commission for Reforming the City of Mostar in September 2003. As a result of this Commission's recommendations, the High Representative decreed that, beginning March 15 2004, the City of Mostar would be unified within its 1991 boundaries under a single city administration.[16] The decree terminated the municipalities as separate local governments and reconstituted them as "city areas" that would be used as sites for branch offices of the unified city administration and as electoral districts. Elections to be held in October 2004 would elect 35 city councillors, three from each of the six city areas plus 17 city-wide councillors. To provide incentives for cross-group agreements, certain city council decisions – such as those concerning urban plans and the city budget – would require a two-thirds majority approval, while issues deemed of vital interest to an ethnic group would require majority approval within each constituent group block. This unification decree effective March 2004 is a bold imposition by the IC of its decade-long aspirations concerning the city. However, the restructuring may not be able to reverse the deeply ingrained pattern of deconstruction that ethnic leaders in Mostar have created throughout the decade.

THE FROZEN SEED: PLANNING, THE CENTRAL ZONE, AND RUPTURE

I scrutinize in this section the objectives and constraints of urbanism during the 1994–96 period of direct EUAM administration of the city. I explore the central zone strategy in detail to show how urban space was conceptualized as potential glue to hold the city together and catalyze future change. I also discuss the psychological and professional challenges that urbanists face when confronting development dynamics that are clearly driven more by nationalist politics than planning rationality.

Urbanism and the central zone

An architect and urban planner who was director of reconstruction for the EUAM recounts how the 1994 physical damage to the city was matched by the "shattering" of the institutional planning mechanisms of city governance (Yarwood 1999: 28). In such a circumstance, EUAM did not just need to repair damaged buildings, but also to repair the processes of urban management and planning. Optimistically, Yarwood opines that by catalyzing cross-ethnic cooperation in the strategic rebuilding of a hopefully unified city, planning could possibly constitute a foundation and forerunner for additional forms of integrated urban management. Located within the EUAM Reconstruction Department, a cross-ethnic strategic planning team aspired from 1994 to 1996 to create a "structure plan" for the entire urban area that would provide a common base of information and expectations regarding post-war urban growth.

Cooperation between Bosniaks and Croats on planning efforts was subject to purposeful obstacles by both sides (Marica Raspudić, interview). After many months of work sessions, east Mostar participants pulled out and developed their own planning document, "Platform for Reconstruction and Development of Mostar, 1995," rather than develop a joint cross-ethnic strategy. No city-wide structure plan was to be prepared during the EUAM period, nor for the remainder of the first post-war decade. There was also a less ambitious effort to create a sound analytic database that could at least provide both sides with common information. A partial product was developed,[17] but neither side wanted to engage in the process by sharing their "own" data.

Midst the severe fragmenting impulses of post-war Mostar, the IC put forth an excellent strategy – that of the central zone – for gradually integrating and normalizing a key core area of the city. The Rome Agreement and the Interim Statute created the zone in the traditional commercial and tourist center of the city and it was to be administered by an ethnically balanced city council and administration. About one mile long and one-half mile wide, the central zone was put forth as the "the symbol and key to a unified and multi-ethnic Mostar" (International Crisis Group 2000: 43). It consisted of a common strip of land along the former confrontation line where joint Federation, Canton, and City institutions and administrations would be located (see Figure 6.2).[18] The central zone was to act immediately as a spatial buffer between the two sides and to indicate to both

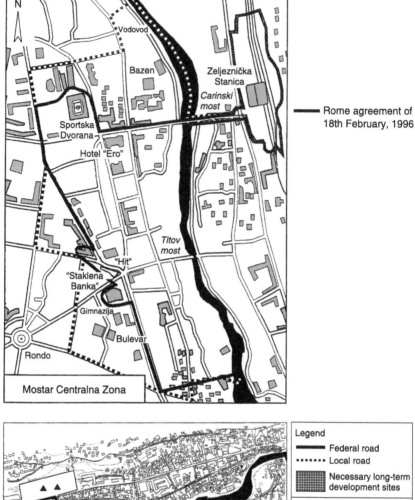

Rome agreement of
18th February, 1996

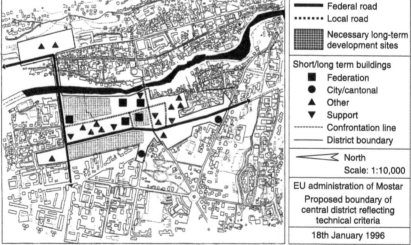

Figure 6.2   The Central Zone, Mostar

Croats and Bosniaks that no land would be allocated based on wartime positions. It would be a place of neutrality and ethnically balanced control and administration. Over time, through appropriate development, the central zone would grow like a seed and demonstrate that cross-ethnic activities could resume, first within the zone, and then hopefully in larger swatches of urban space within the "ethnic" municipalities. The reality of the central zone, however, was strikingly different than its intentions. Both ethnic sides acted repeatedly to freeze this seed and obstruct its ability to grow roots that would connect the two sides. In the end, the central zone strategy in post-war Mostar represents both the promise and lost opportunity of using spatial planning as a means of reconstructing a city of extreme division. The same forces that captured the six municipalities for ethnic gain also were able to warp and dismantle the integrative goals of the central zone.

The idea for the central zone came out of contentious negotiations over post-war local boundaries. In accepting the creation of six city-municipalities in the Mostar urban area in late 1995, the fundamental problem for the international community was where to draw the boundary between Croat and Bosniak municipalities. The Croats insisted it should follow the former confrontation line while Bosniaks felt that this delineation would confirm a division created out of war. Bosniaks also noted that a part of west Mostar was demographically Bosniak and that the boundary should include that population within the Bosniak municipalities. After weeks of negotiations, the EU administrator proposed a central zone that would break the confrontation line in those parts of the city where it did not follow the Neretva River. The size of such a neutral zone was then the point of contention, with Croats wanting a small zone and as much land as possible put within their own municipalities. Bosniaks wanted a larger zone that would include some of the Croat war territory and thus constitute a larger buffer against Croat territorial ambitions.

With progress by diplomats stalled on the size and boundaries of the central zone, urbanists moved the debate forward by providing an objective methodology and technical framework. Yarwood (1999: 32) recounts how urbanism and the "naïve logic of sensitive, neutral experts" were able to break the initial diplomatic gridlock about the central zone's size and shape. Several planning-related criteria – land availability, access to neutral highways, inclusion of cross-ethnic infrastructure, and enclosure of an equal amount of each side's territory – were used to determine the drawing of central zone boundaries in a more logical, less political, way (Yarwood 1999). Because the IC desired to have an ethnically balanced population in the central zone so that neither side would proclaim it as theirs, this led to the controversial incorporation into the central zone of a housing area within Croat territory.[19] Despite the support of EU administrator Koschnick of this central zone delineation, the EU meeting in Rome in February 1996 ("Rome Conference") supported the Croat view by overturning the proposed central zone boundaries and substantially reducing the size of the buffer zone.[20]

The central zone, intended to be neutral space immune from the ethnic compartmentalization occurring elsewhere in the city, became in the early days a target of ethnic territorial ambitions and remained that way for ten years. The six city-municipalities that withheld power and authority from the city administration ethnically carved up the central zone and turned it, as with the rest of the urban area, into a "political space" of contested territoriality (Zoran Bosnjak, Urban Planning Department, City of Mostar, interview). Concurrently, there has been obstruction by both sides, particularly by the Croats, of the establishment of shared institutions intended for the central zone, including offices of the courts, police, and those of the city, canton, and federal government.

Croats, more than Bosniaks, have strategically built ethnically exclusive institutions in the central zone, an outcome directly opposed to the intended use of the zone as an area of joint and mixed use (International Crisis Group 2000; J. Berthoud, interview). Much of the construction allowed in the zone has been illegal, in the sense that a municipality would issue the building permit, not the central city administration. G. Wochein (interview) calls this "policy-making through land occupation," explaining that much construction would be in place by the time the legal review by the city administration was completed. This strategy of ethnic land occupation was apparently worked out early during the negotiations over the central zone. Deals took place between political and business leaders that illegally allocated land and/or building rights in the zone to ethnically-based companies for the purposes of building specific projects (J. Berthoud, interview).[21] With such political deals on the front end, the city planning department was left hanging and was never able to rightfully take control of regulation and management in the zone. The EUAM, and subsequently the OHR, had full rights of review and denial of municipality actions, but in reality such oversight was absent because there existed only two staff members to review hundreds of local actions (G. Wochein, interview).

One glaring example of ethnic intrusion into the central zone is the effort to construct a massive Catholic cathedral. Construction commenced with the help of about 25 million euros, and before apparently running out of money, builders laid down huge cylindrical concrete foundations in the epicenter of the central zone. Today, with ownership suspect and construction incomplete, this ethnically exclusive building by its very presence is obstructive. Further south one finds another huge construction site, a proposed Croat National Theater with funding of at least 17 million euros. Begun in 1997, it too has run out of money, but not before putting a huge ethnic footprint upon the central zone. With millions of euros pouring in from Croat nationalist interests to build such landmarks, these "facts on the ground" have bypassed legal channels of city-wide decision-making and obscured the IC's visions of the central zone as a multicultural seed. Whereas Croat obstruction of central zone goals and their opposition to a unified city has been more explicit, Bosniak resistance to international community objectives has come in a different form. According to M. Raspudić (interview), the Bosniak mayor for Stari Most (and subsequently of the city administration), Safet Orucevic, saw value in having ruins remain in the central zone because they provided an incentive for potential international donors to invest. Once donated, however, much of this money went not into the central zone but to other areas and projects in east Mostar (Haris Kovačić, Head, Urban Planning Department, City of Mostar, interview).

Ethnic penetration of the central zone constituted a missed opportunity in establishing a foundation for the long-term normalization of Mostar. If the city administration supported by the international community had enforced the rules of governance laid out in the Interim Statute, then urbanism may have had a chance to change the ethnically distorted logic of Mostar's development. Such a "public interest" urbanism would surely have been a threat to entrenched powers; as explained by M. Raspudić (interview), "Urbanism considers the public interest, which means it becomes an obstacle to groups' own private interests and will be attacked." In the face of ethnically entrenched and war-hardened antagonists, G. Wochein (interview) asserts that "we as urbanists should have helped integrate the city quickly through the design of the central zone with public functions, coffee shops and meeting places, and mixed living areas. We had the chance, but I think we lost it." He also notes that the "international community is mostly politicians, legal officers and everyone else except architects and urbanists. We do not have the needed urban planning capacity here" (G. Wochein, interview). With an urban development plan for the central zone in place early in the process, the city administration and the IC could have more effectively pursued projects of joint city-wide application and directly targeted inconsistent and illegal projects. M. Raspudić (interview) laments that EUAM's biggest mistake in the early years was not reconstructing the front line areas of the central zone. She describes how "we did not have a plan for a central zone, but only a small document that asks for reconstruction money."[22] The pace of residential reconstruction in the central zone was dreadfully slow for the first eight post-war years and this left a yawning physical and psychological gap in the urban landscape. "With these holes in the urban area, Boulevar and Šantića, the city is divided," Raspudić contends.

International investment in erasing the war front lines within the central zone – namely Šantića/Milosa Street to the north and the Boulevar to the south – did not pick up until 2002. The author's visual inspection of these areas in 2002 and 2004 showed these areas emerging as strange zones where donor-driven reconstructed buildings reside next to war ruins (see Figure 6.3). Reconstruction had started to take hold, but many ruins remained ten years after the war and are daily reminders of what happened. The rebuilding effort has concentrated on damaged residential blocks and houses along those two streets. It had progressed by 2004 to the point that about one-half of the residential stock along Šantića and about 75 percent of residential buildings along the Boulevar had been rebuilt.[23] Interestingly, before the war these residential areas had a strong multi-ethnic dimension; thus, reconstruction of them as part of a return program for these former tenants would not only physically rebuild ruins, but also be a first step in potentially reconstituting the semblance of multi-ethnic mixing in the city.[24] Of the 177 dwelling units in the residential blocks along Šantića Street that were damaged during the war, 44 percent were inhabited by Bosniaks,

Figure 6.3   Šantića Street, 2004

35 percent by Croats, and 21 percent by Serbs (Mostar City 2004). This represents a high degree of mixing at this neighborhood micro-scale. There was a similar mixing in the residential blocks along the Boulevar that were damaged; of the 167 units, 38 percent were occupied by Croats, 35 percent by Bosniaks, and 26 percent by Serbs.[25]

The rebuilding of what were ethnically mixed buildings provides the opportunity to bring the sides together and prevent the hardening of ethnic segregation in the central zone. The international community was cognizant of Croat aspirations to harden separation along Šantica Street and acted, through reconstruction, to encourage ethnic mix as a way to eradicate the former front line (M. McCullough, interview). Nevertheless, it must be asked why such a reconstruction effort did not occur earlier and include all types of land uses, not just residential. Rebuilding of these two streets may be too late to make a real difference in undoing the damage caused by a decade of ethnic carving up of the central zone and the larger city. International observers are concerned that not enough has been done to prevent a redivision of the city should tensions rise in the future.

Much as ethnic entrepreneurs tried through war to kill old Mostar, their goal for ten years after the war was to kill the seed required for the urban area to move forward along a path of normalcy and ethnic tolerance. Because efforts by the international community and the city administration to counter these forces have been ineffectual and tardy, ethnic separatists have likely succeeded.

Urbanism and ethnic territoriality

We are not interested in politics. We don't think in that way. We do not understand politics and politicians well enough to put nationalist qualities into our work assessments. We make recommendations to the politicians.
Zoran Bosnjak
Architect, Urban Planning Department
City of Mostar
Interview

The urban area of Mostar has been torn apart and reconstituted based on ethnic territorial imperatives. There has been the ethnic fragmentation of the city into two parts, ethnic intrusion into central zone neutrality, and efforts to bolster both sides' populations as a way to gain demographic majority and a claim to being rightful controller of the urban region. The use of land and investment has been driven by nationalistic aspirations, and patterns of ethnic territoriality that have been created over the ten post-war years have significant implications for the viability of Mostar as a place of cross-ethnic interaction. Croat nationalists, to boost their numbers and their political claim on the city, have unilaterally constructed thousands of residential units for Croats displaced from other parts of Bosnia. New Bosniak housing has been built on the hillsides to the east and to the north. Based on planning criteria, much of the land available for future growth in the urban area is to the north near the area of a former military camp; yet, planning criteria bump up against ethnic territoriality because this land is "so-called ex-Bosniak war territory" (Z. Bosnjak, interview). Ethnic territorial consider-ations have dominated the spatial development of the urban area for ten years and threaten to continue in the future.

Urbanists in Mostar certainly recognize the potent ethnicity that has structured development, but they feel professionally constrained to incorporate ethnic factors into planning calculations in ways that might lead to a fairer and less stringently territorial urban area. "We know that people and their politicians act only on ethnic principles; planners respond to this reality and we draw the plans for them," states Z. Bosnjak (interview), indicating both an awareness of, and detachment from, nationalist motivations. Palma Palameta (civil engineer, Urban Planning Department, City of Mostar, interview) rationalizes this professional approach, stating that "no matter which nationality, the urban planning norms stay the same. Real urban planning is not interested in the nationality question. Nationality rightfully comes into play when consid-ering the design of buildings, but that is a question for architects and not planners." Here we have allowance for ethnicity in the design of buildings, but not in urbanists' considerations of how ethnic groups are distributed across space, how ethnic imperatives seek territorial control, and how ethnic territorial actions can harden inter-group intolerance. Such a hands-off approach provides, unin-tentionally, a professional legitimization of the ethnic hardening and fracturing of a city.

In cities such as Mostar, urbanists have an opportunity to use history as a means of bringing back the city to some semblance of normalcy. The historic core of a city, due to the solidity of its place identity as a district of ethnically heterogeneous users, may be less susceptible to ethnic capture than more modern parts of the urban area. This historic center of Mostar is the area on both sides of the Stari Most Bridge, an area of commercial and residential uses that dates back to the sixteenth century. An approach that utilizes historic reconstruction as a means toward inter-group reconciliation is favored and used by Amir Pašić, an architect and urban planner born in Mostar who studied at the University of Sarajevo and the University of Zagreb (Croatia). He is an expert in the Islamic architecture of Bosnia and manages a historic reclamation project jointly funded by the Aga Khan Trust for Culture (AKTC) and the World Monuments Fund (WMF).[26]

Pašić views historic reconstruction as playing an important role in societal and psychological reconciliation and normalization. He writes how the destruction of heritage "leaves the psyche rudderless in a disintegrating world" (Plunz, Baratloo, and Conard 1998: 73). In such circumstances, the "reconstruction of familiar architectural icons becomes essential to the national healing process, signaling the return to a more civilized environment where self and place can once again be reunited." The AKTC/WMF historic reclamation project has three objectives – rehabilitate historic neighborhoods on either side of the Stari Most Bridge, restore 15 priority historic buildings to act as anchor sites for larger reconstruction, and prepare a comprehensive conservation plan for the entire historic area (Aga Khan Trust / World Monuments Fund 2000). Pašić (interview) links reconstruction of the historic core to inter-group reconciliation because of its ability to stimulate the return of all three nationalities to this area for economic opportunities and tourism. "If we restore the historic core," Păsić says, "then people can see their future living and working in that area." However, because the historic core around Stari Most Bridge developed during the Ottoman times and neighborhoods on both sides of the bridge are inhabited primarily by Bosniak Muslims, a historic rehabilitation project focused only on those areas faces criticism by Bosnian Croats that the project is promoting a mono-cultural, Bosniak perspective on the city.[27]

Both through its definition of planning area and its identification of priority buildings for repair, the AKTC/WMF project is endeavoring to span the ethnic divides of the post-war urban area. The conservation plan applies to the historic part of the city included within its 1918 city boundary, an area that includes the Islamic historic core on either side of the Stari Most Bridge and large percentages of east Mostar. However, it also encompasses on the west side the central zone and several areas of revitalization in west Mostar municipalities (AKTC/WMF 2000). Meanwhile, in the component of the reconstruction project targeting "endangered priority" buildings, there is a similar attempt to span the Ottoman and Austro-Hungarian facets of Mostar culture and history; it is explicitly noted that Catholic, Orthodox, and Muslim sacred sites are among the identified buildings (AKTC/WMF 1999). Based on the author's appraisal of the 15 priority buildings described for potential international donors, at least one-half of them are of at least partial Croat or Serbian interest. In the end, however, although the expansive jurisdiction of the reconstruction program supports the assertion that both Muslims and Croats will benefit, the focus of reconstruction efforts is clearly where buildings are most at risk – in the Bosniak Muslim districts on either side of Stari Most Bridge.

This effort to use historic reconstruction as a path to societal reconciliation appears viable on the surface. However, buildings, new and old, can have multiple meanings that complicate efforts at physical reconstruction. In a potently nationalistic environment, seemingly benign efforts at preserving history can be contentious. I will illustrate this by describing a project to reconstruct the Stari Most Bridge.

This sixteenth century emblem and symbol of Mostar was meticulously reconstructed with the help of international funding and triumphantly reopened in July 2004. International community officials and Bosniak leaders promoted the interpretation that this event constituted both a physical and metaphoric bridging of the two divided communities. The rebuilding of the bridge symbolizes the reunification of Mostar and is viewed as a part of the healing process for this ethnically divided town. Paddy Ashdown, the then High Representative for BiH, said at the ceremonial reopening of the bridge that it was an important step towards reestablishing the "multiconfessional, multinational coexistence" in Bosnia-Herzegovina. The new Old Bridge has become a powerful symbol of the re-emergence of Mostar's multiculturalism (Makas 2005). This perceived association of the bridge with pre-war multiculturalism and its seeming neutrality as infrastructure meant that international organizations and donors viewed it as a surefire and non-contentious catalyst toward Mostar's normalization. The project was funded by a loan from the World Bank, donations came from more than five countries, including Turkey and Croatia, and UNESCO (United Nations Educational, Scientific, and Cultural Organization) worked on the technical aspects of the reconstruction. Hungarian, Italian, and American engineering firms worked on the preconstruction details, and two Turkish construction firms worked on the reconstruction of the foundation and of the bridge itself (Makas 2005).

Despite this wide-ranging support for rebuilding the bridge, many local Bosnian Croats felt that its multicultural symbolism is imposed and not genuine. The bridge does not link traditional Croat and Muslim areas, but rather two halves of the old town that are both predominantly Muslim. Since it was built in the Ottoman time, the bridge is also a symbol of the Muslim nation here. Indeed, the multicultural meaning of the bridge is not a historic one, but one more socially constructed since the war to advance a vision of a post-war Mostar. Makas (2005) describes how the bridge before the war assuredly played a significant role in anchoring the city's historic sense of place and identity. However, the bridge did not gain its multicultural symbolism until after it became a deliberate target by those pushing for mono-ethnic dominance. As part of a larger project by international organizations, governments, and media to advance post-war notions of a pluralistic Bosnia and Mostar, the bridge "was reconstructed as an *intentional* monument to an alleged Bosnian multicultural identity" [italics are author's] (Makas 2005: 67).

The local Bosnian Croat community felt antagonized by the celebratory exultations of the bridge's reopening. "The old bridge reconstruction is not a unifying symbol for Bosnian Croats," states N. Moore (interview). Makas (2005) criticizes the celebrated reconstruction as a "superficial symbol of unity which glosses over and simplifies the complexity of the still-divided city." The old bridge, by the nature of what it was, connotes connection between people. Yet, it also was a target in the war because it was more a symbol of the Muslim nation. The rebuilding and arising of the bridge represents a complicated mixture of both the re-emergence of the Muslim nation and the potential reawakening of cross-ethnic links. The bridge is not a mosque or a church and so is not an ethnically exclusive structure. However, neither is it neutral or benign in this brutalized city where reconstructing history, as well as planning for the future, is contentious.

INTERNATIONAL URBAN MANAGEMENT: PROMISES AND RISKS

We are both the solution and the problem here. Bosnians will need to be on their own at some time and see whether they can meet these challenges on their own.

Pablo Barrera
Political Officer
OHR, Mostar
Interview

Mostar is an extraordinary example of direct international management of a city. The early establishment of a European Union Administration of Mostar (EUAM) to directly administer the city for a two-year period was an unusual strategy. It has been highly uncommon in contemporary history for there to be direct administration by an international body of a non-state territory (J. Saura, member, Mostar elections monitoring team 1996, interview).[28] Since the end of EUAM and the taking over of international management by the OHR, the intent was for there to be more local administration of the urban area as it increasingly normalized. Yet, due to intransigence on the part of local ethnic leaders, the OHR has remained in a position of de facto direct control over the city, imposing its will when international goals become too distorted by ethnic motives. Ten years after the war's end, this legally superordinate position of the IC relative to local authorities was clearly illuminated in 2004 when OHR imposed a unification decree that created a single coherent city administration and terminated the six ethnic municipalities. This experience of direct international management and oversight means that Mostar provides an excellent lens through which to study the effectiveness of international intervention at the micro-scale.

International negotiators early in the peacemaking process were cognizant that Mostar was a special case. The Washington Agreement in 1994, the Dayton Accord in 1995, and the Interim Statute of 1996 all referred to the difficult conditions of this divided city. These agreements imposed direct international supervision, freedom of movement, and ethnic power sharing upon Mostar, and negotiations subsequent to the Interim Statute further led to the unique demarcation by the international community of a central zone district to be immune from ethnic prerogatives. It was felt that if this regional capital and the most populated city for Croats in Bosnia did not work, the Bosniak-Croat Federation (premised as it is on cooperation between Muslims and Croats) would not work either [Interviews: N. Moore, J. Berthoud]. Mostar was viewed with hope as an opportunity to work out the parameters of Federation governance. In this perspective, the city was positioned as a key "base laboratory" for working out principles and practices of inter-group coexistence that then could be applied at higher levels (J. Berthoud, interview). This positioning of Mostar as a critical fulcrum around which larger progress in BiH was dependent created for the city both a unique opportunity and a heavy burden.

An unintended partner

The international community, despite its best intentions to counter the development of a "parallel" city, has been used as an unintended partner in the creation of just such a city. A report in 2003 concludes, "Virtually all of the provisions of the Interim Statutes of the City of Mostar remained mere declarations on paper, and none were implemented in accordance with the original intentions of the Statute" (Commission for Reforming the City of Mostar 2003: 13). It declares further, "The City of Mostar has never come to life nor exercised the basic predispositions of its competencies." There has been an enduring contradiction between the "diplomatic/political fiction of ethnic integration and the reality on the ground of ethnic partition imposed by war and reinforced by post-war political decisions and popular enmity" (Burg and Shoup 1999: 376). The articulation by EUAM of the central zone strategy in early 1996 is indicative of innovative and strategic thought applied to local political organization and urban space relatively early in the post-war process. Given the pervasive war territoriality of the urban system, a central zone that would belong to neither side, or rather to both, and which would be governed by an ethnically balanced city council, was a pioneering concept by the IC applied to the micro-scale.

The international community had a vision for a unified Mostar and the tools and powers to begin the process of normalization. In practice, however, the frequent ignoring and bypassing of the central zone's urban planning authority by ethnic municipalities was seldom met by sanctions or construction stoppages by the OHR. In other cases, such as the illegal allocation of land by ethnic municipalities to developers, the ability of OHR to review for adherence to international regulations was hampered by lack of staff capacity and urban expertise. For the larger urban area, the OHR made limited and ineffective efforts to compel the six ethnic municipalities to turn over to the city-wide administration the numerous competencies of local governance (including urbanism, finances, education, taxation, infrastructure, and housing), as spelled out in the Interim Statute.

The capture of financial and taxation powers by ethnic municipalities allowed for the funding of ethnic-promoting activities; concurrently, the ethnic fragmentation of urbanism powers provided an essential key to making money for these municipalities (N. Moore, interview). There existed in reality three urban planning departments – one in the west, one in the east, and one in the central zone having limited powers. The capture of education by the municipalities prevented Bosniak and Croat children coming together in the same school, and instead enabled the creation of ethnic-specific curricula that will propagate myths and stereotypes among today's young.

An inherent difficulty in the implementation of international goals was the relationship the IC developed with ethnic political leaders. The IC tended to act in ways that made implementation of its formal power ambiguous and dependent upon the cooperation of local leaders' actions (J. Mier, interview). In their consultations with the local level, the IC would commonly deal with political party representatives or their delegates and not with individuals outside organized political bodies (P. Palameta, interview). This established a disastrous relationship for the IC in a setting where the nationalist prerogatives of the political parties were so dominant, resulting often in the impeding of international intentions by local war-hardened objectives and tactics. Actions by EUAM and OHR at times ignited the fury of one or both ethnic groups who criticized third-party imposition of foreign will, such as when the international community unilaterally created the boundaries of the six city municipalities and the central zone. Yet, in actuality, these international arrangements regarding local governance were agreeable to Croats and Bosniaks because each group perceived that they could keep their social and political bases intact (P. Vilanova, interview). Despite Dayton requirements that political representation in the municipalities be based on 1991 Census ethnic population distributions, the municipality councils were largely still dominated in 2003 by mono-ethnic parties. For the central zone, there was tacit agreement among local political leaders that this district would be exploited for ethnic gain. Proactive planning and development sponsored by the EUAM/OHR to advance the mixed neutrality objectives of the zone came too late in the post-war period and in too limited a scope.

The six-municipality structure of local governance created the framework for the ethnic fragmentation of the city-wide interest; what was needed to bring the parallel city into reality, however, was a set of locally elected ethnic politicians who would use this local governance framework to sabotage international goals. The fact that these individuals gained post-war political power shows how the international community was an unintended partner to ethnic obstruction and fragmentation. Because the city was deemed essential to larger peace-building efforts, the IC thrust local democracy upon Mostar earlier than anywhere else in Bosnia Herzegovina. The Dayton Accord specified that local elections would take place in the city by mid-1996 and the voting in Mostar became the first post-war democratic election in all of Bosnia.[29] In sharp contrast, local elections in the rest of Bosnia were delayed until Fall 1997 due to logistical challenges about where displaced persons and refugees would vote and be counted. In Mostar, with about 60 percent of registered voters participating, elections produced clear majorities for a Bosniak coalition of political parties (led by the SDA) for eastside municipalities, and for the main Croat party (HDZ) for westside municipalities.[30] The intention of the IC was that the holding of municipal elections would be a concrete and positive first step toward the city's democratization and normalization. In effect, however, democracy's early emergence in the city locked in obstructive ethnic elements that would then act to retard the city's normalization. N. Moore (interview) believes that early elections, occurring as they did during instability and threat, helped legitimize those individuals who prosecuted and profited from the war and put into office nationalist die-hards strongly resistant to change. J. Saura (interview), who was an organizer for the elections, states that it clearly was too early for a mature democratic election to take place in Mostar. With the successful holding of elections on the heels of open warfare in Mostar came democracy, but along with it a cementing of power and influence for war leaders, nationalist entrepreneurs, and criminalized elites that would obstruct the potential benefits inherent in democracy.

The inability of the international community to implement actions that physically and socially integrated the urban area is explainable, in part, by EUAM's and OHR's desire to avoid reinflaming ethnic and nationalist tensions (N. Moore, interview). Although the IC had clear goals of unification and strong tools of implementation for post-war Mostar, it is understandable that the IC would want to pursue some modicum of urban stability after a period of inter-ethnic trauma and thus not support or take actions that disrupted war territoriality. It may also be attributable to a hesitancy of the IC to micro-manage what it probably perceived as the banal minutiae of local development, regulation, and planning. The IC is clearly more accustomed in post-war environments to addressing challenges at larger political geographies and at broader programmatic levels (Chesterman 2004). In contrast, its challenge in Mostar was focused on the micro-geography of a city and its districts, and this required oversight of project-specific details. OHR personnel have strong credentials in the political and legal fields and have special sensitivity to these issues; however, such expertise does not lend itself to dealing with ethnic issues at the micro-scale, project-specific level. Nonetheless, by allowing for an ethnically circumscribed and parallel city to develop, the IC's actions, or lack of them, helped create separate ethnic universes that will beget nasty legacies over the long haul, including inter-group intolerance, lack of understanding, and radicalization.

Reconstruction and return

Another avenue to the normalization of Mostar is through interventions by the international community to physically reconstruct the city and to encourage the return of displaced persons and refugees to their pre-war place of residence. Reconstruction and returns policies are difficult to implement after war. They become trapped in the same ethnic dynamics that distorted local government power and, in addition, can present disturbing moral questions pertaining to the meanings and limits of normalization in a post-war environment. The challenges faced by the international community in seeking to normalize Mostar lead to unsettling conclusions about the nature of ethnic space, displacement, and inter-group stability.

The damaged buildings that characterized the old front line ten years after the war obscure more positive indicators of reconstruction. During the EUAM early period, a large amount of money (approximately 150 million Euros) was allocated for rebuilding for the two-year period. M. McCullough (interviews) remembers how the EUAM staff rapidly identified projects, engineers would write up their assessments in one week, and the EUAM head would approve the expenditure the following week. Because of its key role in anchoring the Muslim-Croat Federation, the IC felt that reconstruction needed to be fast and dramatic in order to bring people back into the area (N. Moore, interview). With money and streamlined processes, the infrastructure foundation was rebuilt early in the process – McCullough estimates that reconstruction of 32 schools, 8 bridges, and the entire water, electrical and phone systems was completed during this period, along with about 10,000 dwelling units. Many projects had a political dimension to them, especially those joint projects intended for use by both Bosniaks and Croats (such as a landfill and hydroelectric plant). The overall rebuilding strategy and allocation of funds became embroiled in political debate and criticism by both sides (Garrod 1998). Bosniaks complained that too much was being spent on lesser damaged west Mostar, while Croats claimed funds were going disproportionately to east Mostar. The EU head attempted to allocate funds based on need. By the end of the EUAM period, Garrod (1998) asserts that 65 percent of funds went to the east and 35 percent to the west.

Some in the international community believed that reconstruction should be used as leverage to require nationalists to move forward on the political front. However, Garrod (1998: 11) takes issue with the view, stating "money would never buy the hardliners, to whom political objectives and national identity, culture, language and so on were infinitely more important than any amount of money. They would just say 'take your money and go.' " A few observers note that reconstruction may even have locked in and reinforced divisions. N. Moore (interview) recalled the working premise that services and facilities needed to be duplicated on either side of the river so that reconstruction would not stimulate renewed conflict; this pattern of expenditure "in hindsight likely further created and emphasized divisions." Reconstruction becomes particularly inflammatory when the international community links rebuilding of houses and apartments to the return of ethnic minorities to their pre-war residence (which means primarily Bosniaks to west Mostar areas and Croats and Serbs to east Mostar neighborhoods). Local Mostar political leaders (especially in the west) have been unwilling to support the returns process and worked successfully to have little or no conditionality placed on EU reconstruction money (N. Moore, interview). A further difficulty in linking reconstruction to return is that much "bilateral aid" from specific country donors was spent with limited requirements dealing with the location of reconstruction and its intended beneficiaries[31] (N. Moore, interview). In the end, internationally funded reconstruction of housing not linked to return of ethnic minorities has often reinforced rather than moderated division, entrenching the new ethnically segregated nature of the urban area.

The return of ethnic minorities to their former pre-war locations is a key plank in the international management of Bosnia, with Annex 7 of the Dayton Accord guaranteeing the right of return for displaced individuals and refugees. For the city of Mostar, return appears as an essential condition for reconstituting the pre-war city of mixed nationality. Despite the signing of numerous agreements and memoranda of understanding between the IC and municipal authorities, "action has failed to keep pace with rhetoric" (Commission for Reforming the City of Mostar 2003: 54). West Mostar political leaders were particularly obstructive to returns for most of the 1995–2000 period. This blockage began to come to a head in 1998, when Bosniak resident groups intentionally selected certain areas in west Mostar for return that were contentious politically, thus forcing the IC to open up these neighborhoods (N. Moore, interview). For the next three years, the IC removed local officials from office who were opposing the return process.

After many years of minimal returns, data on registered minority returns for the 2001–2003 period indicate some progress on returns. About 5500 Bosniaks returned to west Mostar, about 2200 Croats to east Mostar, and an estimated 5600 Serbs returned to east and west Mostar during these years.[32] Yet, these data also illuminate the limits of the returns program. The Bosniak returns constitute about 35 percent of the estimated 15,000 Bosniaks who were expelled from west Mostar; this percentage return is below the approximately 50 percent rate for Bosnia Herzegovina overall. The Croat return, meanwhile, is estimated at only about 10 percent of displacees and has been strongly discouraged by local Croat political leaders who are seeking to maintain ethnic spatial separation.[33] More telling about the partial nature of returns is that despite some progress, "the demographics of the city remain quite different than they were before the outbreak of war" (Commission for Reforming the City of Mostar 2003: 54). Reconstruction of Mostar has occurred frequently without minority return, and "this great gap between minority return and physical reconstruction remains ten years after the war" (Sanja Alíkalfic, UN High Commissioner for Refugees, Mostar, interview).

Unsettling predicaments

In seeking to "normalize" Mostar and resurrect its pre-war conditions, the international community faces a difficult quandary in pursuing what undoubtedly is a morally justified objective. Allowing the war-created ethnic separation to continue (by building residential units to accommodate the displaced in their resettled areas) would in essence accept the logic and goals of the war-makers. As poignantly brought out by Javier Mier (OHR, Mostar, interview), "if we accept what this terrible war has produced, then what is the sense of our being here?"

Yet, on the other hand, encouraging or forcing return of minority residents to their pre-war neighborhoods can destabilize micro-geographies and create shaky foundations for the larger peace-building endeavor (Nigel Moore, political advisor, EU Police Mission, interview).

In actuality, what is happening in Mostar (and elsewhere in Bosnia, as we saw in Sarajevo) is moderate minority return of households and significant return of properties to pre-war occupants. The Property Law Implementation Plan (PLIP) had legally returned by early 2004 residential properties to over 9,000 pre-war occupants in Mostar, constituting about 94 percent of all claims (OHR 2004b). This is a necessary step toward breaking the post-war segregated structure of the city, yet it is not a sufficient step in itself. In reality, many households who have repossessed their pre-war units do not go back and reoccupy but rather rent their units out, sell them, or trade them with those who have repossessed units on the other side of the ethnic divide (N. Moore, interview). Property is successfully repossessed but it initiates a process that, in the end, solidifies post-war segregation of the city. A household is certainly made better off by the repossession of its pre-war unit and by selling, renting, or trading it the household gains a stream of income that can initiate a process of economic self-sufficiency. However, what is good from an individual household perspective runs counter to the IC's objectives of desegregating the city. The PLIP experience points to the difficulties, when seeking to normalize a city, of balancing individual and private interests with a collective and public interest. It also likely represents a middle-ground outcome for the international community between the hope of bringing back pre-war residential ethnic distributions and the despair of accepting all the losses imposed upon ethnic households during the war. PLIP's return of properties runs the risk of inflaming local ethnic tensions significantly less than the return of ethnic minority households; yet, at the same time, its accommodation to local segregating pressures runs counter in spirit, if not in law, to Dayton and international goals.

The difficulties of normalizing Mostar present disquieting insights into the relationship between war-induced displacement and post-war inter-group stability. It may well be that war-induced ethnic migration produces a more sustainable and stable situation of post-war inter-group relations. At the city-wide level, majority cities may be more workable than mixed ones. Sarajevo is a strong majority Bosniak city while Mostar is composed of antagonistic sides that reside within the same small-scale urban system. This has produced a situation where Sarajevo can get on with the normal problems of a big city, while Mostar has been hemmed in by the constant barrage of nationality-based problems and is unable to address normal problems of public administration (G. Wochein, interview). For a city that hosts a mixed population, it may be that segregation rather than integration is more sustainable in post-war years. The highly segregated environment of Mostar has likely produced over the ten post-war years some inter-group stability by minimizing ethnic tensions.

In seeking to reconstitute a workable society, some within the international community have begun to accept, implicitly and reluctantly, that the war-created purification of urban space provides a foundation that promotes its larger security goals. The OHR maintains its explicit focus, for public consumption, on the high moral ground provided by the Dayton objectives of the right of minority return and the demographic re-creation of pre-war Bosnia. Yet, several interviewees spoke of the reality underneath this moral high ground, midst the fact that after ten years the return of displaced persons and refugees in Bosnia may have reached its practical end with about one-half of this population remaining in its displaced and resettled locations.[34]

Yet, an implicit acceptance of ethnic separation as constituting a workable foundation for short-term inter-group stability must come to terms with the costs of this accommodation. With ethnically sorted and sterilized environments come political and social dynamics that perpetuate and harden ethnic compartmentalization. Separation begets a sense of group identity that can lead to radicalization. In Mostar, the city-wide public interest has been sacrificed for the illusion of ethnic righteousness, and along with the death of the city has come the blocking of inter-group tolerance, interaction, and understanding that can only be achieved by living in shared spaces that nurture more inclusive notions of humankind than nationality.

Mostar ten years after

Mostar is a significant missed opportunity to work out at the micro-scale the key parameters of shared governance and territory needed for the effective functioning of Bosnia Herzegovina at the macro-scale. Management of the city has not provided a model for the Federation or the country. Mostar provides lessons for BiH, but they are largely negative ones, entailing how ethnic war-profiteers are able through relentless efforts to obstruct development of a cross-national public interest, and how the IC has unintentionally accommodated these assertive ethnic actions. The fragmentation and internal walls that divide Mostar mimic those that have haunted the Bosnia state. In both city and state cases, agreements to stop the fighting accepted ethnic partitioning that since has stood in the way of societal progress.

"From a ten-year point of view," states M. McCullough (interview), "one would likely view the international effort here as a failure." Yet, he also counsels the use of a longer time horizon to judge what has been done: "If one looks at it only at ten years it would be the wrong point of view. Rather, we should look at our impact from a 50-year perspective. From that view, we have put in an excellent foundation and we are one-half the way up the ground floor, although certainly not on the second floor or working on the roof." He puts the tragedy that happened here in human terms and pleads for patience:

> *I have 3 sons ... If I was here during the war and lost one of my sons, I am not going to look at my watch after 10 years and say, "well, he didn't count. Let's forgive and forget." It is unrealistic and unfair to think that deep and fresh scars will disappear after 10 years. This expectation irritates me greatly.*

This longer-term perspective tempers an overly critical view of the IC in Mostar by recognizing the limits that thwart it in the immediacy of war trauma, that there is a long learning curve involved (both for international officials and local leaders), and that even limited progress on stability and reconstruction today sets an important context for more visible progress and normalization further down the road.[35]

Impatient with the stunted progress of public authority in ethnically fragmented Mostar, the High Representative of Bosnia imposed through a January 28, 2004 unilateral decree the political unification of the city of Mostar. Beginning in March 2004, Mostar was to have a single city administration for the entire pre-war area of the city. The six ethnic municipalities were extinguished as separate units of local government and reconstituted as "city areas," which would function as electoral areas and sites for branch offices for the city administration. For a second time, similar to the early post-war years, Mostar is positioned as a potential model and building block for figuring out multinational governance in BiH. As described in ICG (2003: 1), "The rationalization of Mostar's governance could point the way towards overcoming the ethnonational barriers and redundant administrative structures that plague BiH." Viewing Mostar as part of a package of "local remedies to the national-administrative partition that has characterized the post-war period," effective governance in this difficult city "would offer both a template for other segregated towns and encouragement for BiH in general" (ICG 2003: 1). The unification decree itself considers the "resolution of the Mostar question as crucial to the sustainable and peaceful development of Bosnia and Herzegovina" (OHR 2004a: 1). It creates a city of Mostar that could potentially provide a "push from below" to political reform at the Bosnia state level. Two characteristics that could contribute to larger peace-building envision unified Mostar as a place of electoral compromise and of administrative reform.

In terms of electoral compromise, the 2004 political unification of Mostar limits the ability of the demographic majority to rule and imposes through its electoral rules a shared governance model. It also includes safeguards to protect vital national interests. The city council is to be composed of 35 members, 18 selected from the "city areas" (three per district) and 17 from a city-wide list of candidates. No more than 15 city councillors can come from a single nationality group. Two-thirds majority will be required for decisions deemed of greater importance (including approval of urban plans, budgets, amendments to the city statute, and issues of symbolic importance). Further, decisions that affect the "vital national interests" of a particular ethnic group will require majority approval within each of the ethnic groups represented in the council. Through the creation of a city-wide democracy that at the same time draws partially upon ethnic districts for electoral representation and has procedures for protecting group interests, the unification decree seeks to "build a foundation for a progressive future predicated upon, inter alia, protection of national vital interests" (OHR 2004a). This constitutes an effort to move forward a communal cross-ethnic public interest for Mostar while remaining cognizant of group interests.

The electoral system created by decree was not acceptable to either of the major political parties in Mostar; "the middle way chosen was perfect because the party proposals were each biased toward protecting and promoting their party aspirations" (J. Berthoud, interview).[36] This middle way builds in electoral attributes that seek to reduce the risks that might occur under straightforward majority rule and thus to safeguard multinational local governance. The IC hopes that such a model of shared governance encourages political parties and voters not to think of democracy as a way to take political power, but as an ongoing process of compromise across nationalistic groups (J. Berthoud, interview). While district-based voting for about one-half the seats would assure that each part of the city is represented, the city-wide representatives reflect more the principles of unification and proportionality. Because city-wide candidates would need to appeal to broader constituencies, there is the hope that they would be more supportive of issues and projects of city-wide benefit.

Limitations on the rule of the demographic majority that are now part of Mostar's charter[37] set the city apart from other cities and municipalities in Bosnia, where rule by the demographic majority is allowed[38] (ICG 2003). There is a legitimate point to be made that if this shared governance model is appropriate for Mostar, it should also be used in many cities and municipalities in Bosnia that have been traumatized and ethnically sorted during the war. In this model of peace-building from below, a multitude of local experiences with shared governance and political compromise would be a better foundation for creating a viable multi-ethnic state than an assemblage of war-sorted municipalities, each led by the demographically superior group.

In its eradication of municipalities and the lodging of full power in the unified "city" of Mostar, the decree may be foreshadowing another peace-building path in Bosnia – the employment and empowerment of "city" governments as foci for inter-group interaction and the development of transnational mores and values (ICG 2003). The "city" is rare, ill-defined, and has been kept outside the legal frame of Bosnia for decades (J. Berthoud, interview); only Mostar and Sarajevo are "cities" in the Federation. Instead, the typical local government is the municipality, which tends to be smaller in geographic area and, in post-war Bosnia, is likely to be ethnically homogeneous. By creating or empowering city governments as overarching public authorities, appeals to a set of broader city-wide and trans-ethnic issues can more easily occur than in smaller municipalities. The Mostar unification decree brings all the interests of the six formerly ethnic municipalities together into a single decision-making forum. The ICG (2003) sees the potential for "cities" to be key building blocks in Bosnia, suggesting that cities could be established in Bosnian towns that are over 50,000 population. Most of these towns (according to the 1991 census) were mixed, with one group not having much more than 50 percent of the population and others not having much less than 15 percent. The state could provide financial incentives to local leaders to encourage cityhood, but in order to become a "city" local leaders would need to adopt certain shared governance provisions (such as limits on rule of demographic majority and minimum requirements for seats for non-majority groups).

Administrative reform is the second plank of bottom-up peace-building. A necessary supplement to electoral restructuring in Mostar is reforming the system whereby public employees are hired and public services delivered.[39] A city administration in Mostar is now to be responsible for the entire city territory. Public employment is an important source of glue in rebuilding the multinational quality of a city, constituting an opportunity where Bosnian Croats and Bosniaks can work together. Before unification, the ethnic fragmentation of the urban area by the six municipalities meant that such joint work experiences were limited. Even the central city administration existed until 2000 as physically separated ethnic workforces. For ten years, public employment in the six municipalities and the city administration has been bloated and a favorite tool of ethnic patronage. The IC estimates that only about 30 percent of the approximately 700 public employees before unification had more than secondary education (P. Barrera, interview). The biggest need in urbanism, according to M. Raspudic (interview), is to establish an institutional system that uses experts, education, and accountability. Through the more professional management of urban processes, there can be a "reduction in the possibilities for corruption and the return of dignity to people in the city" (M. Raspudić, interview).

With unification, the IC started a civil service system to guide hiring decisions in Mostar; it is the hope that, with time, merit and qualifications will replace ethnic quotas as the main criterion for public employment.[40] The intent of administrative reform in Mostar is that the allocation of public land, planning and development decisions, appointments of administrative officials, public employment, and the administration of public companies, all before exposed to the fragmenting effects of ethnic partisanship, will now be guided by criteria of merit, reason, and appeals to a transnational city-wide interest. Mostar's administrative reform presents an important model and path breaker for the Bosnian state; what is to occur over a condensed three-month period in Mostar (due to its special circumstances) will take about one year elsewhere in Bosnia.[41]

Proving it on the ground

It is ironic that Mostar, a retrogressive example for ten years, is considered by the IC as a potential template for political reform elsewhere in Bosnia. Yet, there is some logic to the thinking that if Mostar can be solved, Bosnia can be too – much the way that after one solves the hardest knot among a tangle of yarn, the remainder of it easily comes free. In this view, the ornery and special case of Mostar becomes not an outlier to the central Bosnian problem because it is so different, but a core part of solving it because in its difficult nature it exposes more clearly than any other government unit the problems fundamental to creating a post-war multinational society. It may be a coincidence of historical timing, yet Mostar's political unification decree did turn out to be a precursor to progress on the national front, predating by two years an agreement among Bosnian state leaders to move toward a stronger, more unified form of national government.[42]

Yet, the future of Mostar is likely more pedestrian and less inspiring than the IC hopes. The years of gridlock, rampant parallelism, and wholesale ethnic obstruction may come to an end after ten years. In its place will be continuing issue-by-issue disagreements between the antagonistic sides over specific city problems (J. Berthoud, interview). The hope is that over time there may be the development of more complex patterns of political interest that moderate the hardened Croat-Bosniak divide. For example, on financial issues, Croat leaders likely would want some compartmentalization or division of the tax base because there is a greater revenue base per capita on the west side. In contrast, they may advocate for a more open, unified approach to urbanism so they can gain access to north Mostar land reserves that are east of the river. For their part, Bosniaks would want a more unified approach to tax base sharing in order to access west Mostar resources but a more compartmentalized approach to urbanism in order to protect their land. In this scenario of more nuanced political stands depending upon the issue at hand, there is potential for the breakdown of the political walls of intractability. Nonetheless, there likely will be no grand breakthroughs of inter-group cooperation, but rather incremental steps along a multi-decade path of normalization.

The seemingly mundane issues of city building are those that will determine the future of this city. Negotiated rules and procedures of a post-war society and the structuring of its public authority are absolutely necessary for the moving forward of a city and its society. Yet, in the end, only the urban policy component of public authority has the ability, and responsibility, of making genuine positive contributions to peoples' daily lives in their neighborhoods, stores, markets, and public settings. Planning and urbanism have indispensable roles in proving that there is a viable and shared future in Mostar. For unification to be successful, electoral rules and administrative policies must be matched by concrete positive outcomes felt by Croats and Bosniaks alike in economic opportunity, social assistance, educational quality, and police reform. Only through improved quality of life and opportunities will there be more open and expressed support for unification.

Economic development in the Mostar area is vital to prevent an ethnically based economic divide. In a traumatized city, there should not be a winning side economically because material inequality and relative deprivation can reignite ethnic group-based tensions. post-war poverty has hit Bosniaks more than Croats (N. Moore, interview). And, Bosnian Croats are increasingly connecting to the economic opportunities in Croatia, at one point only 15 miles away from Mostar. It is in the future, however, that there are more worrisome possibilities of an economic disparity that will harden ethnic cleavages. As early as 1993, Bosnian Croats could obtain Croatian as well as Bosnian citizenship. With Croatian passports, Bosnian Croats since the war have been able to travel into EU countries without a visa and with few constraints. In contrast, Bosniaks who hold only a BiH passport face greater obstacles and have needed an annual visa for travel to many countries since the war.[43] This greater ability of Bosnian Croats to travel for business, school, or recreation has linked them to a web of European opportunities less available to Bosniaks. Because the country of Croatia is closer to EU membership standards, Croatia may gain EU membership quicker than Bosnia. Should this happen, the economic divide will further widen in Mostar as Bosnian Croats will have an EU passport and Bosniaks will not.[44] In a poignantly stated prognosis of what this means for Mostar, W. Herdt (interview) worries that the old town will develop into a Bosniak tourist-type "Indian reservation" amidst a city otherwise economically managed by Bosnian Croats.

Urbanism and reconstruction must move forward in a situation where the anger and bitterness is still present. Urban interventions cannot directly lead to reconciliation of peoples damaged and abused by war. A process of reconciliation between the peoples was solely lacking ten years after the war and is desperately needed. Further, this city may not ever again have the special quality that it did before hostilities; that is probably gone forever. No one can likely put this city back together again. Yet, still, urbanism is a path by which concrete achievements can be produced on the ground that might contribute over time, in association with inter-group dialogue, to less trauma and feelings of loss and hopelessness. In this city captured and subordinated to war and institutionalized ethnic hatred, Jaroslav Vego (interview) helps us understand the simple yet potent role of urbanism,

> This is our home, this is our country. We need a chance to stay. My grandfather and great grandfather lived here and I would like for my children to have a chance to live here. The state of mind of the common person today is confusion and uncertainty. What urbanism can do is to give a chance to common citizens of Mostar to stay and live here, to show them that there is a future here.

Notes

1. The name *Mostar* is linked to the importance of this transportation link across the river. The word *most* means bridge in South Slavic languages and *Mostar* means bridge-keeper.
2. Source: Nigel Moore, formerly OHR Mostar, interview.
3. Counting population, especially its ethnicity, after war is problematic. Usually, no formal census is done by the state for many years due to the sensitivity of the issue and fear that it may be implicitly recognizing new demographic realities. The International Crisis Group used an admittedly inexact science by scanning voter registration records and categorizing voters into ethnic groups based on the likely national identity of each name.
4. In the wider region, OHR estimated that about 36,000 dwellings out of about 75,000 total had been damaged to some extent (OHR 2002b).
5. The agreement also anticipated the creation of cantons in the Federation as a way to protect ethnic self-rule and envisioned a future confederation between this Bosnian Federation and the country of Croatia.

6. The signatories included the Presidents of the Republic of BiH, the Representative of the Croat people in BiH, the mayors of Mostar West and Mostar East, and the EU.

7. The Mostar annex to the Dayton accord was signed, on the Croat side, by the then hard-line mayor of west Mostar and by Croatia's defense minister. On the Bosniak side, signatories were the mayor of east Mostar and the foreign minister of BiH.

8. "Others" would be primarily Serbs and minorities, although for most of the interim period more than one-half of these seats remained vacant.

9. The IC favored the creation of six rather than two municipalities to avoid the appearance of a straight two-way split of the city into Croat and Muslim halves, a situation which de facto existed at the end of hostilities in mid-1994 (Garrod 1998).

10. EUAM transferred responsibilities to the Office of High Representative (United Nations) and a set of other international agencies.

11. This Bosnian party is an extension of the HDZ party in Croatia, whose leader was Tudjman and has been the ruling party in Croatia between 1990 and 2000 and since 2003.

12. Specifically, west Mostar municipalities refused to adopt "vital interest" clauses of the Interim Statute that are required when voting on certain sensitive issues (affecting culture, education, religion, national monuments, and housing) that a majority of minority group (i.e. Bosniak) councilors must approve.

13. Organization for Security and Cooperation in Europe (OSCE) Mission to BiH Public Administration Reform Unit, reported in Commission for Reforming the City of Mostar (2003: 62).

14. The Washington Agreement (1994) gave cantons the right to determine education policy in post-war Bosnia Herzegovina. De facto split Ministries of Education in Herzegov-Neretva Canton have instituted split educational systems and curriculum (International Crisis Group 2000).

15. Attempting to increase ethnic proportions in an urban area is one of two main methods of urban territoriality, which equates land control with political control (Sack 1986). The other method, also evident in Mostar, is manipulation of jurisdictional boundaries to fit ethnic group aspirations. For another example of manipulation of subgroup populations in a city for political purposes, see Bollens (2000) [Israel's strategy in Jerusalem since 1967].

16. Decision Enacting the Statute of the City of Mostar, Office of the High Representative, Sarajevo, 28 January 2004.

17. Raspudic and Aiello (1996).

18. The zone also includes key common infrastructure not spatially contiguous with this common strip of land – three power plants on the Neretva, three fresh water sources, and the Mostar airport.

19. Koschnick's support of this larger central zone led to a violent public demonstration by local Croats against him at the EU's headquarters at Hotel Ero, surrounding and attacking Koschnick's armored car before he escaped.

20. EU's accommodation to Croat stipulations – a reduction in central zone size and removal of Croat population from it – ironically allowed Croat politicians to subsequently criticize the central zone as a de facto seventh, and Bosniak, municipality. The EU overturning of the Koschnick central zone plan also led to his resignation as EU administrator.

21. The largest such company was Inter-Invest, an enterprise connected with the HDZ.

22. Mostar City (2004).

23. Estimate by author based on data in Mostar City (2004).

24. Reconstruction does not guarantee return of previous tenants. Some may choose to sell their reconstructed units or rent them. Thus, although reconstruction creates the possibility of recreating past ethnic mixing, in the end the neighborhood can shift to one of mono-ethnicity.

25. Most of the residential blocks along Šantića Street and the Boulevar also had ethnic mixing within their structures; in only 3 of the 21 buildings along Šantića, and in none of the 12 buildings along the Boulevar, was one ethnic group occupying more than 90 percent of the units (Mostar City 2004).
26. The Aga Khan Trust for Culture, based in Geneva, seeks to improve built environments in Muslim societies. The World Monuments Fund is a New York-based non-profit that preserves and protects endangered works of historic art and architecture.
27. While east Mostar is historically Bosniak Muslim, the area on the west bank of the Neretva River near the Stari Most Bridge is a Bosniak Muslim neighborhood.
28. There was an unsuccessful effort to create in 1948 a "corpus separatum" for Jerusalem under direct international rule. Gdansk and Trieste are historic examples of free cities under international rule (Hepburn 2004); in both cases, such a status was used to deal with competing national claims to those cities. Elsewhere in Bosnia, for the city and district of Brcko, negotiators created a special status that includes an international supervisory regime and local power sharing. Located at a geographically sensitive area at the border of the Muslim-Croat Federation and Republika Srpska, Brcko's level of direct international intervention is greater than that created for Mostar after the war.
29. National elections in Bosnia occurred September 1996.
30. These 1996 elections in Mostar were followed up by September 1997 elections after the IC decided that Mostar should hold elections along with all other municipalities country-wide. Both Bosniak and Croat political parties threatened a boycott of this election, which in the end produced similar results for the municipalities and a 14–10 Bosniak majority for the City Council (Garrod 1998).
31. The Return and Reconstruction Task Force (RRTF) of the IC has had less influence on the parameters of bilateral aid than it has had on European Union money. However, the advantage of bilateral aid was that it could reach the ground more quickly than EU aid (N. Moore, interview).
32. Office of the High Commissioner for Refugees data, in Commission (2003: 54).
33. The relatively higher percentage of Serb returns may be due to two factors: (1) the onus of conflict in the area not being focused directly on them, and (2) the fact that Republika Srpska, the location of most war-displaced Serbs from Mostar, is only about 15 miles from Mostar city.
34. Interviewees' identities not listed due to the sensitivity of the subject matter.
35. McCullough points out that the realities of Bosnia have taught the IC lessons about the long-term nature of post-war peace-building. Compared to EUAM's mandate of two years in Mostar, provisional reconstruction teams (PRTs) staffed by US-led coalition forces in Afghanistan have ten-year mandates.
36. The Croat HDZ party wanted proportional representation with no maximum quotas on seats, which they anticipate would reward them with the most council seats, based on their estimates that they are now the demographic majority in the city. The Bosniak SDA desired equal representation of each nationality group (fixed quotas) in order to prevent them from minority status should Croats now have the demographic advantage (Commission for Reforming the City of Mostar 2003).
37. Before the unification decree, the city council worked under a shared governance, fixed quota system but its de facto power for the full urban area was minimal. The unification decree imposes a model of shared governance upon the entire "unified" city area.
38. With the exception of Brcko, which has been controlled through a special international arrangement.
39. The unification decree concentrated on political-electoral aspects and intentionally did not clarify how administrative reform was to proceed (J. Berthoud, interview).

40. After the war, the IC imposed nationality quotas on public employment (based on 1991 Census data). This clearly has been hard to implement in local governments where demographic percentages have changed drastically due to the war. In the new civil service system, when nationality quotas conflict with merit criteria, the civil service requirement will prevail (P. Barrera, interview).
41. The difficulty of seeking both ethnic balance and civil service/merit goals remained a fact of life for the IC one and one-half years after the unification decree's start date (OHR 2005).
42. Agreement, November 21, 2005. Signed by the eight leading political parties of Bosnia Herzegovina.
43. Many Bosnian Serbs, similar to Bosnian Croats, have passports with their mother country – in this case, Serbia.
44. This circumstance would be similar to residents in the country of Cyprus, which in 2005 was brought into the EU. Because of lingering political disagreement and walled separation on the island only the southern, Greek Cypriot, part of the island gained EU advantages, not the northern, Turkish Cypriot, part.

7 Urbanism, inter-group conflict, and political transition

This book has focused on the physical, social, and political-institutional attributes of urban peace building amidst democratic transitions and competing nationalist agendas. Each of the areas has been exposed to periods of significant political uncertainty, caused by violent nationalistic conflict (in the cases of Sarajevo and Mostar) or nonviolent regime change (in the cases of Barcelona and the Basque Country cities). These are "extreme" illustrations of how urbanism is influenced by, and itself influences, processes of democratization and inter-group accommodation. Yet, hundreds of cities across the globe will address and negotiate in incremental and evolutionary ways the great challenges of inter-group coexistence, tolerance, and multinational democracy that have been thrust upon these four case studies because of political and nationalistic pressures. War, conflict, and societal breakage in Bosnia and Spain provide lessons for how political leaders in cities throughout the world can cope constructively with societal uncertainty and multinational tension.

This concluding chapter synthesizes what I have found. First, I discuss the specific characteristics of the four case studies. Second, I induce from these findings a set of broader implications and lessons for the ability of cities to be local contributors to the building of peace in a society.

CITY AS LABORATORY: THE FOUR CASES

Barcelona

- *Transitional urbanism.* Urbanism performed a crucial formative role in the political transition, helping to construct the urban terrain for democracy. Urbanists did not wait for the formal beginning of democracy to create a new development vision for Barcelona. The timing of urbanist intervention early in the transition increased planning's effectiveness as a tool for building a more equitable and livable post-Franco city. Large-scale urban planning was used to change the prevailing logic of unregulated speculation in the city, institute a collective project that distributed urban benefits more fairly, and educate the populous about the potential of democratic action. There appears to be a window of opportunity during political transitions when planning can facilitate the discussion of new city-building options, further the trajectory of innovative ideas, and help consolidate these new options. Such urbanism anchors and foreshadows broader societal changes to come and illuminates a formative path toward progressive social goals and outcomes.

- *Political opposition.* Planning in the form of the long-range plan had the ability to outline an alternative urban future and thus played an important role in catalyzing and consolidating political opposition during transitional uncertainty. Planning provided a useful template and platform for consensus that was badly needed midst transitional uncertainty. The General Metropolitan Plan (GMP) of 1976 was a key planning intervention that allowed, amidst unsettled conditions, consensus to be built among groups with different ideas on how to reform society. The power of planning was constructed out of the many years of criticism of the Franco city by progressive urbanists who helped work out the rudiments of what democracy would look and feel like on the ground. The criticism of urban chaos and speculation during the latter Franco years – in essence an indirect advocacy for a more democratic city – had a catalyzing effect on political opposition by providing it with a vocabulary that related to people's everyday needs.

- *Urbanism amidst uncertainty.* Political transition and uncertainty created prime conditions in Barcelona for planning support and effectiveness, while times of political stability have tended to generate conditions deleterious to planning support. In the uncertain first years after Franco, there was a fundamental disruption of societal relationships while the need for a template to guide urban society was at its greatest. In stark contrast, there was a significant lessening in support for long-range planning after Catalan society normalized and democracy was consolidated in the 1980s. In Barelona, the capacity of large-scale planning was greater at points of major disruption in society than at times when there has been a smooth change in governing ideology (such as through elections). During times of fundamental regime change, societal relationships become sufficiently scrambled that those seeking political power look for avenues and vehicles for expression. One such avenue for the expression of power is urban planning and its legitimized face of rationality.

- *Connecting urban and political.* Urban professionals linked problems of urban deficiencies to lack of political voice during the latter Franco years. Political repression created the conditions that catalyzed neighborhood organizing and vitalized public planning linked to neighborhood issues. Urbanists played a significant role in connecting urban problems with larger political issues for neighborhood movements. They also helped link neighborhood group activities to democratic initiatives in workplaces and clandestine opposition parties. Urbanists helped neighborhood organizations analyze local problems and possible solutions, but also provided a political orientation with democratic objectives of local democracy and freedom. Criticism of the unequal and under-serviced metropolis was by extension a criticism of the disempowered political condition of residents under Franco; nevertheless, urban criticism advanced by conscientiously and wisely staying within the convenient occupational blinders supplied by its reputation as a technical and non-political profession. Despite this compartmentalization and dismissal by the regime, urbanist thought and practice continued, advanced, and strengthened its links to other forms and sectors of growing political opposition to the authoritarian regime in the mid-1970s. The Franco regime's suppression provided a catalyst for creative public sector interventions and for the societal consensus in the city that burst out during and after the political transition.

- *Scale of urban intervention.* To effectively express post-transition urban democracy in the short-term, targeted architecture and design interventions had greater influence than large-scale planning. In the early years of democracy beginning in 1980, context-sensitive and small-scale interventions of architects and designers were more valuable in imprinting democracy upon the Barcelona landscape than were more abstract and broader-scale urban plans. The importance of large-scale, comprehensive planning was at its height not in the early years of democracy but earlier during the transition. At that time, planning in the form of the General Metropolitan Plan (GMP) constituted an important starting point, or framework, for moving forward and helped shape the important design interventions to come in the 1980s. Once formal change occurred and democracy was institutionalized, however, architecture and design took over as key agents. Based on a belief that urban design changes could help translate cultural values, project-specific improvements in the 1980s educated the public about the potential of democratic and collective action.
- *Urbanism and nationalism.* Urbanism in Barcelona has been a fundamental component in the development of Catalan nationalism in two respects: (1) the city has been a cultural crucible where a place-based and inclusionary (rather than an ethnic-specific and exclusionary) nationalism has developed; and (2) the city has been in the post-Franco years a crossroads between two differing Catalan nationalist political projects. Barcelona and Catalonia are simultaneously nationalistic and porous, indicating that regional nationalism can survive, and even thrive, amidst significant and prolonged periods of in-migration. Spanish immigration into the region in the 1950s through the 1970s led to a social hybridization of Catalan society, and this has led to a relatively open and inclusive Catalan nationalism. The allowance for regional autonomy in Spain has likely facilitated a Catalan nationalism based on sense of place in competition with other regions. This place-based identity has been more welcoming of non-native residents of Catalonia than one based on ethnic origin and place of birth. Substantial internal migration, the autonomous

structure of the Spanish state, and shrewd regional political leadership have created a society that combines nationalism and cultural identity with a fluid dynamism. Today, amidst globalization and Europeanization, the political projects of center-right Catalan nationalists and leftist socialists may be converging on a re-conceptualized notion of Catalan nationalism that combines both regional identity and cosmopolitan fluidity. Such ideological movement would transcend the traditional political postures of these two groups, one predicated more on the rural heartland, the other on development of the metropolitan sphere.

Sarajevo

• *Partitioning of ethnic space.* The ability of urbanism and local policy to reinstate Sarajevo's multicultural environment has been constrained by diplomatic agreements. The 1995 Dayton peace accord "misplaced" Sarajevo, providing little space to integrate and assimilate peoples. In Sarajevo and Bosnia, ethnic differences have been accommodated and reinforced geographically through the drawing of political boundaries. Such ethnic partitioning retards the capacity of Sarajevo and other local governments to act as grassroots foundations upon which multi-ethnic democracy can advance. The Dayton redrawing of political space in Bosnia into two ethnic entities, and on the Federation side, into ten cantons that reinforce ethnic separation, leaves little room for cities like Sarajevo to act as societal transformation agents. Ironically, while the war damaged but did not eradicate the multicultural spirit of Sarajevo, the political boundaries drawn to stop the war may over time slowly deplete the city's ability to spawn cross-ethnic integration and tolerance. The City of Sarajevo is crippled institutionally. Its functional urban system is separated into two ethnic entities, the Federation side of the city is hemmed in by a powerful and ethnically homogeneous canton, and the city is further divided internally by the four more empowered municipality governments that lie within its borders. The effect of such ethnic demarcation and gerrymandering has been to tighten the screws on Sarajevo city's ability to act as an opportunity space for multiculturalism in the future. The political boundary that is within Sarajevo's urban sphere lacks a physical or intimidating presence. It is nonetheless a line of separation that has influenced – and will continue to do so – where people live and how and where they choose to interact.

• *Peacemaking and urban peace-building.* Peacemaking decisions made during transitional uncertainty and urgency can adversely narrow the grounds upon which a society is reconstituted. In this sense, peace-making can obstruct peace-building down the road. In Bosnia, diplomatic peace-making created conditions that will handicap the emergence of a genuine multicultural democracy. Diplomatic agreements to stop the fighting can ran at cross-purposes to goals of normalizing and re-constituting a society. International negotiators, in their desire to stop the brutal war, contained Sarajevo behind negotiated political boundaries built upon ethnic differences. An open, specially administered, and shared Sarajevo died at the peacemaking table. During the war and before Dayton, there were several diplomatic efforts to preserve the special quality of Sarajevo. Yet, when peace finally came to Bosnia, in the form of the Dayton Accord of December 1995, peacemaking set off processes that unraveled efforts to create Sarajevo as a multicultural space amidst a fracturing state. Distributing power in a post-war state in order to facilitate ethnic self-rule responds to a compelling logic of how to reconstruct a collapsed state, yet the costs of this strategy seem higher and longer term than the benefits. International community acceptance of ethnic self-rule has established an institutional and geographic framework that has compartmentalized and entrenched ethnic identities territorially and has thus obstructed compromises necessary for the success of a multi-ethnic state. The foundation upon which Dayton is based – the partitioning of political space to accommodate ethnic difference – is more likely to prolong, or at best suspend, ethnic conflict than to solve it. Sarajevo represents a significant missed opportunity in efforts to reconstitute Bosnia. In peace accords that necessarily emphasize conditions that stop a shooting war, peacemakers should be cognizant of the potential of urban areas to help, with time, to reconstitute and rediscover multi-ethnic tolerance.

- *Urban displacement.* Although morally repugnant in its acceptance of the outcomes of war, accommodation of relocation (versus return) heightens urban and group security in the short term. However, this security is purchased at tremendous likely long-term costs in terms of society building. Encouraging returns to pre-war locations has moral weight behind it because it may reanimate pre-war multi-ethnic integration. Yet, it also may stimulate inter-group tension and conflict and cause hardship on returnees if they are disconnected from social and economic support in their pre-war locations. On the other hand, accepting non-return of substantial persons who now are abroad or internally displaced within Bosnia is a pragmatic response amidst the difficulty of seeking to recreate pre-war demographic geographies and may produce over the medium term a more stable inter-ethnic situation. Yet, such ethnic partition based on relocation likely has long-term negative consequences on inter-group understanding and tolerance.

- *The challenge of multiculturalism.* The effort at facilitating returns of displaced persons and refugees to Sarajevo is not restoring the multi-cultural residential fabric of the city. Legal return of pre-war properties has occurred, but not the return of pre-war occupants. The ability of minority returnees (Serbs and Croats) who have come back to the city to remain in their pre-war neighborhoods is tenuous. In Sarajevo, well-intentioned on-the-ground policies by the international community regarding return of displaced populations and legal recovery of property face significant obstacles in their goals of recovering the city's multi-cultural heritage. Sarajevo continues to have a considerable part of its original population displaced elsewhere in Bosnia and Serbia. After ten years, it is likely that many Bosnian Serbs will not return to the Federation part of Sarajevo. The "success" of the property repossession law in Bosnia, meanwhile, hides a dispiriting reality – the selling of repossessed units that, in aggregate, is cementing the ethnic sorting of post-war Sarajevo. The sustainability of ethnic re-integration that has occurred is tenuous and vulnerable to relapse. To sustain and increase ethnic re-integration of the city, the international community and municipalities must manage micro-geographies and territoriality in ways that respect group identities for the city's minority communities. Yet, such sensitivity to fine-grained urban dynamics is not a common competency of the IC. Rather than ask whether Bosnia's pre-war demographic structure can be restored, a more realistic question is whether the notable but partial level of ethnic reintegration can be sustained. I conclude that Bosnia's ethnic reintegration is not of sufficient strength or momentum. Absent significant economic and political improvement in Bosnia in the future, the extent of ethnic reintegration that has occurred in the first ten post-war years, due in part to extraordinary efforts by the IC, may not be sustainable over the next generation.

- *Transcending ethnic boundaries.* To move forward and transcend the ethnic partitioning of Dayton, practical on-the-ground strategies must be used to complement and reinforce political reform at the diplomatic level. Bottom-up peace-building tactics that positively affect peoples' daily life provide a necessary foundation upon which larger diplomatic advances can proceed. This constitutes the most important challenge facing the IC and Bosnian authorities in the second post-war decade. There is an emerging acknowledgement and hope within the international community that functional and economic linkages may increasingly transcend and de facto erase the Dayton boundary lines in BiH and that trust-building within the Sarajevo region can be an important building block in the overall state building project. By the time of the 10th anniversary of Dayton, the international community, increasingly aware of the con-straining effects of Dayton's geography, had already given some attention to practical strategies that would transcend Dayton's partitioned political space. These programs by the IC, emphasizing economic reform and revitalization, do not seek to explicitly overturn or rewrite Dayton's political lines, but to establish new functional partnerships and alignments that will gradually minimize the importance of these lines in determining household and political party behavior. Because these techniques have the ability to positively affect people's daily lives, they hold promise for complementing any advancement achieved at the higher, diplomatic level to reform and strengthen Bosnian governance.

Bilbao, San Sebastian, Vitoria (Basque Country)

- *Pragmatism and political gridlock.* Planning and urbanism have provided a space of rationality and even consensus in a society where political debate has been constrained by militant nationalism and distorted by violence. Pais Vasco combines dynamism and capacity at the urban level with stasis and disability at the larger political level. In urban affairs, the major Basque cities have public sectors that are active, partnering, internationally connected, and financially able to affect change on the ground. At the same time, violence has constrained political debate and options. Dynamism at the urban level and disability at the larger political level have coexisted as parallel tracks, each with semi-autonomous trajectories. Urban programs and policies have spawned cooperation between public agencies that has transcended differences on larger nationalistic issues. Overlaying this functional level of urbanism in Pais Vasco is a regional politics that has been disabled and distorted by extremists' threat of violence. Such a socially traumatic environment has led to responses by the central state and Basque municipalities that have empowered Basque governments and stimulated urbanism and innovation. The difficulties of engaging in inter-group dialogue (between nationalists and non-nationalists, and between nationalists) midst the threat and reality of radical violence have been real and debilitating. Absent the possibility of inclusive political negotiations, strategic planning has provided a forum wherein political violence can be discussed in terms of its impact on the medium- and long-term development of Basque society and its cities. In a circumstance where political parties are hamstrung by nationalist dynamics, consensus-based projects provide places for constructive dialogue and actions.
- *Nationalism and urbanism.* Nationalism and urbanism have influenced each other in the Basque Country; at times they have been synergistic, other times antagonistic. Basque nationalism and urbanism affect each other in myriad ways, involving issues of state-regional politics, culture, intra-regional politics, economics, and image-making. There are cases in which the goals of nationalism and urbanism are aligned or when the pursuit of nationalism has helped urbanism. In post-Franco negotiations over the new Spanish state, Basque nationalistic aspirations helped to create a space of substantial regional autonomy and significant financial resources within which proactive urban interventions have occurred for over 20 years. However, urbanism and nationalism can also be at odds. Culturally, there is a tension between the traditionally rural-based Basque national ideology and the growing urban reality of Pais Vasco. Moderate nationalists may view increased urbanization as an effective counter to the rural- and small-town-based political power of extreme nationalist groups. In this spirit, urbanism is an important counter to Basque extremist nationalism. There are also cases in which nationalism and

urbanism are neither compatible nor hostile, but where urbanism is redefining the terms of nationalist debate, possibly leading to transcendence of traditional views and toward a region that is cosmopolitan, open, and future-oriented rather than closed in ideologically and economically.

- *Narrowing the ground for violence.* An effective urbanism, as the operational and most concrete form of regional autonomy and self-government, can help increase public condemnation of violence and radical nationalism. The functional basis of urbanism can lead to shared understandings of coexistence and to trust and belief in political means toward addressing inter-group conflict. In addition, planning provides a space of reflection pertaining to the challenge of Basque political violence. Urbanism may narrow the ground for violence but does so over a long period of time and indirectly by lessening public support for politically motivated violence. The Basque Country case shows how urbanism and city development, and the ability to change and improve quality of urban life and opportunities, can shuffle the decks in a region that otherwise would remain obstructed by political gridlock and societal violence. Urban policy can ameliorate nationalistic tensions in two respects: 1) it allows for opportunities for consensus-building and partnering; and 2) it can increase the public's allegiance and trust in local government and thus public buy-in to political, rather than violent, means toward resolving conflict.
- *Underneath functional urbanism.* The functional basis of planning can be used to pursue cultural goals aligned with regional nationalism; in this case, planning is used to support and foster one side's aspirations. The protocol of functional objectivity, and the public face of urbanism as neutral and technical, provide effective anchors and tools for those who are empathetic to the nationalist perspective. Astute nationalists who understand that urban models and processes can set agendas and psychological frames of reference can commandeer urbanism and its traditional functional foundations for service to their nationalist project. Urbanism can be used to support and rationalize a cultural bias, and functional arguments can be assets to nationalistic projects.
- *Recovering regional identity.* Basque planners during the Franco years constituted pockets of opposition to "official" planning. After the transition, planners have been instrumental in facilitating the reemergence of Basque regional and cultural identity. In opposition to Franco regime urbanism, nationalist planners provided alternative visions regarding growth that countered the expansionist and speculative orientation of regime urbanism. These efforts sought to preserve town centers and cultural attributes of the built landscape by minimizing building intensity and spatial sprawl. Through the professional association, Col-legio Architectura Pais Vasco, Basque urbanists were able to connect to nationalist urbanists in Catalonia and with European professionals.

These pockets of opposition-based dialogue provided small theaters within which innovative ideas and strategies could be discussed. With the establishment of democracy, planning's role shifted to helping recover urban and regional self-esteem in the face of industrial decline and political trauma. The decades of conflict and Francoist repression of the Basque Country was an assault on a people's collective memory, in the forms of their cultural identity, traditions, and language. Planning, in its treatment of land, territory and culturally historic places, is a central process toward helping to recover Basque collective memory. In this recovery process after political repression, planning's importance lies in the close alignment between nationalism, land, and territoriality.

- *Urbanism in the new Europe.* Increased urban connectivity due to internationalization and Europeanization are providing new footholds for Basque nationalism in its efforts to gain greater autonomy vis-à-vis the Spanish state. Regional nationalism and globalization can be compatible rather than contradictory. New international and European connections can support, not erode, the Basque nationalism project. The European platform creates new networks of interaction and interdependency, and new territorial scales, which compete with the region's traditional links with Madrid. With Pais Vasco increasingly interconnected with other regions throughout the continent, Basque Country enhances its ability to redefine what sovereignty means in an interdependent world, entailing a region not bound by state dictates but free to interact across European space.

Mostar

- *Urbanism manipulated.* Urbanism and urban governance in post-war Mostar have been primary means by which war profiteers have solidified their power and reinforced nationalist divisions. Urbanism aimed at the general city-wide interest has been relegated and subordinated to the goals of specific political interests. War by means other than overt fighting has been carried out in Mostar for ten years after the open hostilities of 1992–1993 and 1993–1994. These other means – parallel institutions, demographic manipulation, obstruction of city-wide integration, and corruption of public power for private and ethnicized gain – have brutalized the city and its collective sphere. The spoils of war have included the city itself, its inhabitants, and its institutions. The ten years of institutional and political division of Mostar has hardened antagonisms between the two sides, stimulating and cementing greater inter-group economic, religious, and psychological differences.
- *The parallel city.* Mostar is an uncommon example of direct international management of a city. The international community has sought to counter the development of a "parallel" city, but has ultimately been used as a partner in the creation of just such a city. Mostar represents a

significant missed opportunity to use an urban accommodative strategy as a local remedy and model for inter-group relations at broader political levels. Allowance by Dayton for the creation of six ethnic municipalities – three on each side of the ethnic divide – created a key and debilitating framework for the ethnic fragmentation of the city-wide interest. The system of integrated and city-wide governance intended by Dayton has operated for ten years mostly as a shell and artifice. The disparity between international intent and local reality is more bedeviling because the international community had the vision and tools to help normalize the urban fabric. The development of the "parallel city" was not due to lack of foresight on the part of the international community. Rather, the IC did not commonly use tools available to them to combat behavior by ethnic municipal leaders that reinforced within their municipality borders the dual and parallel nature of the urban area and, within the central zone, intruded upon that district's neutrality goals and instead carved it up for ethnic gain. The OHR made limited and ineffective efforts to compel the six ethnic municipalities to turn over to the city-wide administration the numerous competencies of local governance (including urbanism, finances, education, taxation, infrastructure, and housing). Further, the IC acted in ways that made implementation of its formal power dependent upon the cooperation of local leaders' actions. This established a disastrous relationship for the IC in a setting where the nationalist prerogatives of the political parties were dominant, resulting often in the impeding of international intentions by local war-hardened objectives and tactics.

- *The frozen seed of neutrality.* The "central zone" strategy in post-war Mostar represents both the promise and lost opportunity of using "neutral" planning and spatial buffers as a means of reconstituting a city of extreme division. The international community presented a well-developed conceptualization of how planning and urbanism would contribute to bridging the nationalist divide. Consisting of a common strip of land along the former confrontation line where joint Federation, Canton, and City institutions and administrations would be located, the central zone was to act immediately as a spatial buffer between the two sides and to indicate to both Croats and Bosniaks that no land would be allocated based on wartime positions. It would be a place of neutrality and ethnically balanced control and administration. In reality, both Croats and Bosniaks have acted repeatedly to freeze this seed and obstruct its ability to grow roots that would connect the two sides. The same forces that captured the six municipalities for ethnic gain also were able insidiously to warp and dismantle the integrative goals of the central zone. The central zone, intended to be neutral space immune from the ethnic compartmentalization occurring elsewhere in the city, became in the early post-war days a target of ethnic territorial ambitions and remained that way for ten years. Croats, more than Bosniaks, have strategically built ethnically exclusive institutions in the central zone, an outcome directly opposed to the intended use of the zone as an area of joint and mixed use. If the city administration, supported by the international community, had enforced the rules of governance laid out in the Interim Statute, then urbanism may have had an increased chance to change the ethnically distorted logic of Mostar's development.

- *Paradoxes of urban "normalization."* The difficulties of "normalizing" Mostar lead to unsettling conclusions about the nature of ethnic space, displacement, and inter-group stability. Reconstruction and returns policies present disturbing moral questions pertaining to the meanings and limits of normalization in a traumatic urban environment. Allowing the war-created ethnic separation to continue (by, for instance, building residential units to accommodate the displaced in their resettled areas) would in essence accept the logic and goals of the war-makers. On the other hand, encouraging or forcing return of minority residents to their pre-war neighborhoods can destabilize micro-geographies and create shaky foundations for the larger peace-building endeavor. Despite clear goals of unification and strong tools, it is understandable that the IC would want to pursue some modicum of urban stability after such a period of inter-ethnic trauma, and thus not support or take actions that would disrupt war territoriality. The IC must come to terms with the possibility and implications that war-induced ethnic relocation may produce a more sustainable and stable situation of post-war inter-group relations. Any countenancing by the IC of ethnic partitioning, however, creates separate ethnic universes that are likely to beget nasty legacies of inter-group intolerance and radicalization.

- *City as critical fulcrum.* Mostar has been, and continues to be, a potential model for figuring out multinational governance in Bosnia and Herzegovina. International negotiators were cognizant that Mostar was a special case that deserved exceptional attention. This positioning of Mostar as a critical fulcrum around which larger progress in BiH was dependent created both a unique opportunity and a heavy burden for the city. Ten years after the war, impatient with the stunted progress of public authority in ethnically fragmented Mostar, the High Representative of Bosnia imposed by unilateral decree the political unification of the city of Mostar to begin in March 2004. Again, as in the early post-war years, Mostar is positioned as a potential model and building block for figuring out multinational governance in BiH. The unification decree creates a city of Mostar that could potentially provide a "push from below" to political reform at the Bosnia state level. Two characteristics that could contribute to larger peace-building entail unified Mostar as a place of electoral compromise and of administrative reform.

- *Proving the future.* Planning and urbanism have key roles in proving "on-the-ground" that there is a future in Mostar for all nationality groups. The seemingly mundane issues of city-building will make or break the future of this city. Negotiated rules and procedures of a post-war society and the structuring of its public authority are absolutely necessary to move forward. Yet, in the end, it is the urbanism component of public authority that has the ability, and responsibility, to make genuine positive contributions to peoples' daily lives in their neighborhoods, stores, markets, and public settings. For unification to be successful, electoral rules and administrative policies must be matched by concrete positive outcomes felt by Croats and Bosniaks alike in economic opportunity, social assistance, educational quality, and police reform. Only through improved quality of life and opportunities will there be progress on societal and psychological reconciliation and a more open and expressed support for a shared Mostar.

CITY AS CRUCIBLE: CITIES, NATIONALISM, AND DEMOCRATIZATION

Urban interventions are able to play distinct roles in the creation of inter-group co-existence and societal peace building, and they can constitute a bottom-up approach able to complement top-down peacemaking negotiations. Urban policy and governance are unique and essential peace-building resources. However, whereas Spanish cities played this peace-constitutive role, the Bosnian cities of Sarajevo and Mostar have not. Explaining why some cities play a progressive role in shaping new societal paths while others do not will help us understand how this peace-constitutive city function comes about, and how this role of urbanism may be misplaced or neglected.

A city is a channel between national political goals and the on-the-ground psychological and material welfare of a society's citizens. In the national transitions encountered in Spain and Bosnia, goals pertaining to the construction of multinational democracies need to be proven in the streets and neighborhoods of cities. In these cases, urban interventions seek to reinforce and actualize abstract new governing ideologies that assert democracy, multinational tolerance, and openness. In Barcelona, public squares were constructed that opened up the city and invited public access after years of stultifying Francoism, and urban policy implemented a program that required greater public capture of the positive economic benefits of private development projects. In Bilbao, urban development partnerships between local, regional, and central state levels constituted mechanisms of cooperation that actualized transition from Franco authoritarianism and created opportunities for political advances amidst extremist violence. In Sarajevo and Mostar, urban policies by the international community seek returns of minority households in efforts to reconstitute and actualize multinationalism. Further, in Mostar, the international community endeavored to delineate a central zone spatial buffer that would engender cross-ethnic and ethnically neutral activities.

Yet, the power of the city lies not only in the potential of urbanism to implement and actualize new governing ideologies. The flow of influence carried by the city runs also in the other direction – from the city upward to extra-urban and state levels. A city has certain spatial and political dynamics that differentiate it from the state level, and these can provide opportunities for concrete and innovative interventions that affect peoples' lives more immediately and meaningfully than state actions. Barcelona was ahead of the Spanish state and the Catalonia region in its ability to actualize multinational democracy in its built and institutional landscapes. Actions by Basque Country cities have catalyzed a dynamic urban track and a reconsideration of political nationalism amidst an otherwise lagging and sclerotic regional politics. Despite missed opportunities and slow progress, the roles of Sarajevo and Mostar in not only anchoring the Bosnia state but also constituting multinational models able to stimulate further state-level inter-group integration, reveal the bottom-up potential of the city organism in a transitioning national setting. Cities not only anchor but also can catalyze the reconstitution of multinational societies.

Cities are critical agents and outcomes in the development of a multinational democracy. They constitute a necessary and stringent test of whether, and how, group identity conflicts can be effectively managed during a time of political uncertainty amidst a shift toward democracy. As a crucible of difference, the city is a test of whether different nationalistic groups can coexist within the proximity, interdependency, and shared geography of the urban sphere. As a crucible of democratization, the city is where the values and processes of democracy are pragmatically translated onto the urban spatial and political landscape and where shared governance and local negotiations can prefigure national peace-building advances. When cities are able effectively to engage in a peace-constitutive function, the larger peace can be anchored and catalyzed in the urban sphere. When cities are unwilling or unable to engage in peace-building, the advancement of societal peace is restricted.

This work advances our understanding of the role of urbanism vis-à-vis the state beyond the dichotomous state-versus-city debate to appreciate the interdependent nature of this relationship. I view urbanism as both shaped, and shaper, in such settings. Cities are constrained by their larger societal contexts and by the actions of their nation-states, but at the same time through their actions can influence change in this larger context and help a society progress from conflict to stability. I synthesize now the broader implications and lessons for the ability of cities to be local contributors to the building of national and regional democratic peace.

Urbanism, conflict, and stability

Urban interventions that improve inter-group coexistence in a city have distinct roles to play in societal peace building. These roles necessarily differ depending upon whether the city and society are experiencing active conflict, a fragile and tenuous peace, or a degree of urban normalcy after conflict.

The ability of cities to contribute to effective accommodation of group differences, and to help further a societal democratization process, depends on the

Figure 7.1   Case study cities on the conflict–stability continuum.
Note: Urban regions studied in this book are in bold.

larger context of societal stability or conflict. As the potential for a society to make fundamental progress on peace advancement increases, the scope and breath of urban peace-building contributions becomes greater.

I revisit Figure 2.4 and in Figure 7.1 locate each of the four case study areas (in bold) along the continuum from active, unresolved conflict to stability and normalcy. Mostar, Sarajevo, Basque Country, and Barcelona exist along different points of the urban conflict-instability continuum. Positioning them conceptually along this continuum allows me to induce from the specifics of the cases a set of broader implications and lessons for the ability of cities to be local contributors to the building of peace in a society. The more rightward placement of Barcelona and Basque Country suggests that there may be elements of the 25-year history of Spain's democratic evolution that are lessons for Bosnia and its cities. Further, there exists intra-state variation: Barcelona appears farther ahead than Basque Country, and Sarajevo more so than Mostar. This means that, independent of national context, Barcelona and Sarajevo have features that put them closer to stability and normalcy than their respective sister cities.

Mostar – suspension of violence

Mostar is a category [2] city, not psychologically far from open conflict and existing in a suspended state of nonviolent ethnic division. Genuine movements toward multinational peace have not yet taken place ten years after military hostilities. Overt violence has stopped, yet war by other means has continued a decade after the war's end. There is deep segregation and political partitioning, and Croats and Bosniaks have institutionally constructed and lived in parallel worlds that further cement and reinforce ethnic fragmentation. The city is in gridlock and urban policymakers (both international community and indigenous) have failed over the ten-year period to create urban conditions that enable tolerance, openness, and genuine democracy.

In cities like Mostar (and Nicosia) where the beginnings toward peace are fragile and uncertain, and in cities like Jerusalem that live midst active conflict, urban policymakers are strongly restricted by their political environment, yet their potential importance is high because they may provide one of the few avenues through which to break out of political gridlock. The formulation in Mostar of the central zone strategy illustrates how urbanism can conceptualize potential integrative influences amidst severe fragmenting forces. Urbanists can propose and debate conceptual urban models and principles of mutual co-existence for the city. Because the physical city may be too difficult an environment for such discussions to occur in a formal way, these urban deliberations may need to take place outside the region and in the form of unofficial, "second-track" negotiations not officially sponsored by the two sides.[1] In addition, the involvement of a third-party government or foundation as financier and leader of these urban talks is likely. The important goal of these talks is to envision an alternative urban scenario and set of guiding principles of co-existence for a time when the antagonistic sides can live in the same urban system in non-belligerent, more normalized ways. In cities experiencing active or unresolved conflict and instability, planning should seek to maintain future flexibility to the extent possible and to continue inter-group dialogues that consider alternative futures and on-the-ground peace-building models. The development of such conceptual models of urban peace-building are certainly not sufficient; the Mostar case shows clearly how a well-conceptualized model (the central zone) can be sabotaged during implementation. Nevertheless, it is essential that cities that experience political gridlock have access to peace-building principles and models that can be used when local politics allow for them or when international overseers unilaterally impose them.

Sarajevo and Basque cities – movement toward peace

Sarajevo and the Spanish Basque cities are in category [3] because they have experienced sufficient progress toward urban stability that there is less possibility of a relapse to active violence. The urban areas show greater signs of normalcy, and local policies are not hamstrung by political and ethnic gridlock. In cities experiencing progress amidst a fragile peace, urbanism can advance and implement peace-promoting strategies that move in parallel with the political implementation of peace agreements and technical side-agreements. In this way, peace accords are operationalized so that city residents feel tangible and positive peace dividends. Greater experimentation and innovation in urban policymaking is possible. Decisions regarding development, provision of economic opportunities, and delivery of public services can be made in ways that create and promote urban spaces that start to repair the ethnic-nationalist divide both physically and psychologically. Opportunities are more available for the building of flexibility into the urban landscape that breaks down old ethnic territorial markers. Chances increase for the creation of cross-nationalist cooperative ventures, establishment of economic enterprises in areas that provide equal opportunity across the ethnic-nationalist divide, provision of public spaces that bridge ethnic territories, and post-war reconstruction and relocation policies that seek to overcome wartime geographies.

In Sarajevo, policymakers deal more with the normal problems of a big city than explicitly ethnic and nationalistic ones; it doesn't experience ethnic gridlock like Mostar nor is it hampered by physical partitions as in Nicosia. Greater attention can be paid to managing the micro-geographies of ethnic minorities in order to sustain them in the urban sphere, although progress in Sarajevo on issues of mixing has been limited. Sarajevo also has greater ability to explore new ways to functionally transcend the ethnic boundaries that war created and the Dayton Accord validated. Recent attempts to create regionalized economic development forums and strategies that functionally transcend ethnic boundaries are signs of policy innovation and movement.

In Bilbao, the spawning of new state-regional-local intergovernmental cooperative ventures that transcend differences on nationalistic issues, and the prioritization of functional over political issues, is indicative of progress toward more workable governance amidst nationalism. The extent of innovation, dynamism, and engagement by urbanism in the Basque Country suggests that urban interventions may be outpacing the rate of political progress in the region overall. It is also the primary reason why I place Basque Country more advanced on the conflict-stability continuum than the other category [3] cities of Sarajevo and Belfast, where urban actions remain more constrained by ethnic-nationalistic political factors.

Although type [3] cities show movement toward normalcy, local peace-building efforts remain experimental in the sense that full urban stability has not yet been reached. Remembrances of trauma and conflict remain close to the surface, and they can be inflamed by local public policies that are not sufficiently sensitive to ethnic and nationalistic group needs. Peace-building advances haltingly as distrust between the antagonistic sides remains a reality. For instance, Sarajevo today is not an ethnically mixed city as before. Due to the placement of war-created ethnic entity boundaries through the urban region of Sarajevo, it is politically fragmented into Federation and Serb Sarajevo parts, with heavy psychological baggage associated with that delineation. Type [3] cities, because they exist in a status prior to the consolidation of a multinational democracy, are also susceptible to setbacks and threat. Up until the time when democracy is seen as the only game in town (category [4] cities), the movement of cities toward peace can be held hostage by the threat of political violence by paramilitary groups. This threat amidst an otherwise progressing and normalizing society is a distinguishing characteristic of the Basque Country case study.

The observation that Sarajevo is further along the continuum toward peace than is Mostar presents a moral dilemma – that Sarajevo's relative manageability is due to it being now a city having a strong ethnic majority, compared to Mostar's status as a city of approximate and competitive demographic parity between antagonistic groups. The implications of this judgment are troubling for those that wish to advance peace-building in an urban environment. I assert, contrary to such implications, that the appropriate goal of urban peace-building is to manage competing group rights within the same shared urban system. Thus, any increase in the manageability of urban governance that results due to the ethnic homogenization of a city's population is sidestepping the larger society's need to genuinely accommodate different groups in a space of shared governance. This is another reason I place the Basque Country to the right of Sarajevo. While Basque nationalists and non-nationalists share the governance of Basque cities and engage in their collective enterprises, Dayton political boundaries initiated processes that relegated Serbs and Croats to the sidelines of Sarajevo city governance.

Barcelona – stability/normalcy

Positioned farthest to the right on the continuum is the city of Barcelona. Compared to cities in the three categories to the left on the continuum, category [4] cities have an absence of political violence, and inter-group nationalistic conflict is focused almost exclusively through political channels. Cities in this category begin to resemble a normal, pluralist city more than one hosting competing and antagonistic nationalisms or ideologies. As cities enter into category [4], urban peace-building efforts can be amplified and consolidated in ways that bring the full fruits of societal stability to the city. After a negotiated resolution of political conflict and the beginnings of urban normalcy take hold, urban peace-building has a primary role in solidifying and extending that peace, addressing adverse physical and socioeconomic legacies of the active conflict stage, and legitimizing and expressing new societal goals in the urban landscape. Urban values of public access, equality, and democracy can be implemented in the city with lesser concern about possible inflammation of ethnic or nationalistic issues. The society, and city, has moved sufficiently beyond this ethnic sensitivity, and options for urban intervention are relatively unrestricted.

Barcelona, with its international reputation as an innovative, cosmopolitan city in the New Europe, does not appear at first glance to belong in this group of troubled urban areas. But Barcelona's success must not allow us to forget its history. Barcelona has been a site of enduring conflict between a regionalist Catalan nationalism and a centralist Spanish nationalism. Amidst the historic and contemporary potency of regional nationalism, Barcelona and Catalonia present a case that effectively combines strong group-based identity with democratic inclusiveness. Urban policymakers in Barcelona have helped to fundamentally change the assumptions and logic of the city's growth, and along with it, its attendant winners and losers. In addition, city planners and architects consciously oriented their works toward making a difference in the daily life of residents, first by seeking small changes, then larger ones that expressed what democracy is and could mean to daily life. This is an urban society that has absorbed a substantial number of non-Catalan Spanish immigrants since the 1950s and one that supports an open and increasingly cosmopolitan urban region. This inclusive nationalism – integrating international linkages and openness with its cultural heritage and history – appears an enviable goal for the many cities in the world that battle with competing, exclusionary forms of nationalism.

The promise of the city

> In their potential to build peace, cities are constrained by their larger context, but at the same time through their actions can influence change in this larger context and help an urban society progress from conflict to stability. Cities and urbanists are not in a dependent and derivative position vis-à-vis larger societal forces, but can be catalytic of progress along an evolutionary path of normalization. Urban planning and policy can play key formative roles during political and societal transitions through their abilities to articulate alternative urban futures. Because cities are microcosms of larger conflict in these societies, city policies have the capacity to devise urban models and strategies that can complement, help formulate, and actualize larger political accords. If cities are left unprotected and unplanned, however, ethnic antagonists can submerge and fragment the peace-constitutive potential of cities in pursuit of their group aspirations.

The capacity of urban intervention is dependent upon the larger conditions of conflict and stability that exist in a city and society, as described above. However, what is more revealing about the power of planning and urbanism is that, while the type of urban strategy is dependent upon context, urbanism can also help change that context. Urban actions can catalyze movement of a city along the conflict-stability continuum. Certainly, urban actions cannot be contrary or irrelevant to the necessities imposed by larger societal conditions. If urbanism acts outside its context (such as an effort to implement a westernized regulatory plan upon a city of fragile inter-group relations), such actions will be irrelevant at best and counterproductive to peace-building, at worse. However, if urbanism actions are property anchored to contextual realities, the ability of urban interventions to contribute positively to the advancement of inter-group tolerance increases.

Cities can operate as dynamic loci of change that constitute the cutting edge of social and economic policies. Urban interventions and policies can thus play formative and shaping roles amidst political and societal transitions. Most understandings of urbanism would view its influence as necessarily occurring only after political transition periods stabilize and clarify; city actions would then seek to solidify and consolidate new political goals and objectives. This was evident in the Barcelona case and by the significant revitalization efforts in Bilbao in the Spanish Basque Country. Yet, I find that urbanism's more significant value lies further upstream and takes place when a societal transition is at its most uncertain and unstable. Cities – and the policymakers and grassroots community groups in them – can play key formative roles in the reconstitution of a political and social order.

The city is a crucible of democratization, wherein the values and processes of democracy are pragmatically operationalized in the urban spatial and political landscape. Urban interventions amidst political transition and uncertainty can establish key parameters that shape new and more equitable relationships between different sectors in society and establish the importance of public interest considerations that were submerged under predemocratic regimes. Cities can be laboratories of democracy wherein new solutions to social and economic problems are tested on a limited geographic scale before they are possibly extended to broader societal application. The way cities are structured can either facilitate or jeopardize continued democratic progress. A city such as Mostar that has parallel and fragmented ethnic governance and urbanism constitutes a faulty foundation that will stunt and restrict societal peace. In contrast, a city such as Barcelona where the ground rules for a new urban democracy are formulated during a society's political transition not only brings democracy to the streets sooner and more robustly, but also constitutes an urban foundation that is a catalyst for progress at broader political geographies. The reconfiguration of a city – through the building of shared public squares and housing for all sectors, the assertion of public interest urbanism, flexibility and integration of urban form, citizen and neighborhood participation, and governance based on cross-ethnic compromise – is not solely a product or residual of larger societal progress on peacemaking, but can be an organic and formative part of the process of democratization.

Urban policies and directives negotiated during and in the early years of transition can set important precedents that shape long-term urban and political development, either to the benefit or detriment of subsequent democratic development.[2] Because urbanism changes people's lives more directly and visibly than national political agreements, it can be a leading-edge force in creating new peace-constitutive development parameters and processes before the procrustean bed of societal relationships dries and organized resistance to change develops. As a democracy progresses and matures, the urban system will likely move from relative simplicity in institutional relationships and clarity in urban transformation goals toward more complex institutional relationships and compromised urban goals as aspirations meet reality.[3] After the shock of political transformation, societal elements will likely move back to some equilibrium; societal actors and organizations will slip back to stability and actively resist further change. There thus exists a window of opportunity where urbanism can establish new parameters and pathways from which subsequent development will evolve. In the early more uncertain stages of transition, urbanism concretizes aspects of the cityscape affecting territoriality, openness, and equality, having long-term effects on inter-group relations. Urban interventions during the early years of transition start a momentum toward healthy and flexible urban functioning or distorted and rigidified city life. The interventions foreshadow and operationalize democracy and the public interest. Barcelona's is a story of policymakers being able to change the prevailing urban development logic as a way to structure further urban change processes. By the time societal elements and relationships congealed (and resisted planning changes) in the 1990s, the basic ground rules of urban democracy and public interest planning had been set.

In the Basque cities, urban interventions came later in the democratization process, but have played a role over the long term in eroding the legitimacy of those extremist groups resistant to the normalization and democratization of that region. In the Bosnian cities of Sarajevo and Mostar, the dangers of not protecting the urban sphere as a place of transformation and multiculturalism are revealed. Amidst this vacuum, Bosnian nationality groups that benefit from ethnically delineated state, cantonal, and city boundaries entrenched themselves in segregated spatial and institutional compartments and became formidable agents actively resistant to societal change.

Urbanism is a valuable contributor to broader societal progress. Before new relationships between societal and urban interest sectors are set in stone, urbanism can articulate critical and peace-constitutive paths forward for the city's development and articulate how relationships will be structured between numerous and competing interests. During the political transition in Barcelona, planners provided an avenue for the building of consensus and for formulating alternative urban futures. Urbanism did not wait for the transition to settle out politically, but helped pre-figure a new society in the city and region. Urbanists in Barcelona composed a platform upon which diverse interests opposed to the Franco regime could stand together. In addition, planners and architects played a primary role in "changing the prevailing logic" of speculative, private sector-led city growth by asserting public interest obligations in both long-range plans and site-specific project design. Further, urbanists helped to connect the concerns of neighborhood associations in the city to the larger political issues of disempowerment being articulated by labor unions and clandestine political groups. Barcelona urban professionals during the transition were capable of connecting the problems of daily urban life to the more far-reaching root political problems of the society. Neighborhood associations, labor unions, and trained professionals combined their efforts and linked protests over poor neighborhood conditions to issues of political and economic disempowerment under the Franco regime.[4] Between 1968 and 1973, there was an evolution of neighborhood association consciousness – from daily consumption issues to broader questions of local politics and political representation. Urban professionals provided technical and political assistance to neighborhood associations and helped them analyze their urban/daily problems and possible solutions. This shows the ability in transitional phases for urban policymakers to link grassroots peace-building principles and programs with larger political debates, and through these means to introduce innovative and new ways of thinking about how to advance societal peace.

In Barcelona, urbanists helped move urban society from the "gray" and static Franco city to, initially, the fragile and emerging democratic city, to, eventually, the stabilized and robust multinational city of today. Urban interventions have changed in tactics and focus of scale to respond to the changing needs of the city during this process of transformation, and have included emphases on large-scale planning frameworks, smaller site-specific architectural interventions, and larger-scale development projects that restructure the metropolis. The city has also influenced the progression of the Catalan nationalist project toward greater openness and assimilation of cultural difference, processes that have moderated inter-group cleavages and tensions.

In the Spanish Basque cities, there was not the magnitude of direct urban engagement during the transition from Franco to democracy that was evident in Barcelona. It was a more difficult transition due to the incidence and potential for political violence; urbanism, along with other prosaic aspects of Basque life, was slow in normalizing. Yet, the Basque case still portrays the capacity of urbanism to affect a broader societal context. The effectiveness of urban interventions during the 1990–2005 period may have moderated political violence as a societal option over the long term and helped to advance a regional politics more able to overcome its disabilities. Urbanism has been a significant contributor to the normalization of that region, opening up new shoots of growth and new institutional relationships that have shuffled the cards of an area that would otherwise be locked in stasis due to political violence. Planning and urbanism have been able to provide a space of rationality, agreement, even consensus in a society where political debate has been constrained by militant nationalism and distorted by violence. Urbanism is an important operational element of regional government that is leading to increased trust and belief in nonviolent means for addressing the nationalist issue. Urban dynamism and betterment have been an effective counter-argument to those who advocate more extreme forms of Basque nationalism.

Internationally sponsored Sarajevo urbanism has made significant strides in reconstructing and normalizing the physical fabric of the city. However, there is an important caveat here – the city has made this progress because its post-war majority Bosniak status likely allows greater space for urban action than if the city had retained its pre-war mix of Bosniaks, Croats, and Serbs. Sarajevo and its municipalities on the Federation side are not hampered by ethnic considerations when making public administration and policy questions, and this has provided a certain degree of freedom. Yet, the ethnic compartmentalization of Sarajevo (by entity and Cantonal lines) that has facilitated post-war urbanism is also a major constraint in reconstituting the multicultural essence of the pre-war city. Sarajevo's potential to be a catalytic agent in Bosnian normalization (to be a multicultural springboard for the country) is dependent upon a functional and political reconstitution of its urban sphere so that the city can operate as a magnet and engine for multinational connectivity and integration. Finally, the Mostar case shows an urbanism stopped dead in its tracks and captured by ethnic/nationalist dynamics metamorphosed from wartime. It is an urbanism unable to move the city and the Herzegovina region forward beyond a status of simple absence of overt conflict. A conscious design of a "central zone" urban strategy meant to catalyze urban normalization met its demise as international urban managers, and their indigenous counterparts, failed to use their statutory powers to safeguard the zone from ethnic predators. The political unification of the Mostar urban area in 2004 establishes on paper a new realigned and de-ethnicized political landscape, yet it remains to be seen whether integrative urban policies, buried for ten years in the

Mostar region, can possibly arise. If such were to emerge and make a difference, the city could then begin the long-dormant movement from non-war stasis to peace building.

The retarded pace of Bosnian urban peace building in the early post-war (1995–2000) years, compared to the performance of Spanish city cases in the first five post-Franco years (1975–1980), is surely attributable in part to the debilitating effects of active warfare in one transition and not the other. There is another related facet, however, that distinguishes Bosnian society from Spanish society in these years – the absence of a national Bosnian unity. Rustow (1970), in his democratization model, surmised that ethnic divisions that lead to basic questioning of national unity must be resolved *before* a transition to democracy becomes feasible. There must be some semblance of national unity in place where a "vast majority of citizens in a democracy-to-be ... have no doubt or mental reservations as to which political community they belong to" (Rustow 1970: 350), meaning in this case allegiance to a larger, cross-ethnic political community. Without this, Rustow asserts it is not possible to conceive of a transition toward democracy. Where no such national unity existed, such as Bosnia, democratization has proceeded in ways that are ethnically purified (Sarajevo) or are stagnant (Mostar).[5] In such circumstances, the role of urbanism as stabilizers along a path of democratization is certainly put to its greatest test. The years ahead in these two cities in Bosnia will indicate whether urbanism is capable of moving an urban system out of conditions of ethnic gridlock (Mostar) and ethnic partiality (Sarajevo). The effectiveness of urbanism in Basque Country amidst different, but also challenging, conditions of an ongoing threat of extremist violence suggests that we should not underestimate urban policy and governance as key agents amidst division.

There is mutuality between urban peace-building and societal peacemaking. On the one hand, urban peace-building is difficult without a larger societal evolutionary process. Barcelona urbanists' abilities in the 1970s and 1980s to affect physical change in urban public spaces and to recalibrate the relationship between private and public interests would not have occurred without the death of Franco and his regime. In the Basque Country, urbanism's innovativeness and ability to engage in fundamentally changing the physical landscape would not have been possible without regional autonomy and financial agreements that provided the institutional foundation and fuel for such engagement. Yet, on the other hand, advancement in national peacemaking that occurs without urban progress on the ground is not rooted in the practical and explosive issues of inter-group and territorial relations. For Bosnia, national peacemaking efforts are hampered by the fact that diplomatic agreements did not use urbanism and the city sphere as structuring devices that could anchor and catalyze broader societal progress. The Dayton and Washington Agreements focused on stopping the fighting, but put much less emphasis on building grassroots peace, setting up local institutions, and drawing new local political spaces that would foster and fortify multiculturalism. The peace accords allowed urban institutions and logic to be shaped by ethnic and nationalistic divisions, directly and obviously in Mostar, more subtly and indirectly in Sarajevo. Unfortunately, the urban political and institutional world created by Dayton and Washington was such that it has pushed the society toward an equilibrium characterized by ethnic-nationalistic partitioning and fragmentation, rather than one less rigidly nationalistic and more accommodative of change and flexibility.

Cities as strategic foundations

> Cities are necessary and strategic foundations on which to build a sustainable and integrated society. They are unique and essential peace-building resources in societies that are reconstituting themselves politically. Cities that are empowered will be able to engage in activities that increase the potential for integration and assimilation of peoples across group identity boundaries.

The city is a crucible of difference, constituting an omnipresent test of whether different nationalistic groups can coexist midst the proximity, interdependency, and shared geography of the urban sphere. In its structure of political representation, territorial development, delivery of public services, and regulation of ethnically salient land uses, the city is consistently faced with the challenge of balancing the accommodation of group rights and expression with the advancement of cross-group civic allegiance. It is in the city that the abstract goals of equality, tolerance, empathy, and justice are given meaning as people connect, or not, with the city (Merrifield and Swyngedouw 1997).

By the nature of what it is and what activities it enables, a city is an integrative influence for individuals and activities within its borders. After the trauma of a war, this integrative effect will be minimal or nonexistent as antagonistic groups stay far away from each other in terms of residential and work life. However, if properly configured so that its jurisdictional space includes multiple groups, a city will over time constitute a container within which economic and social interactions start to take place across the ethnic divides. The nature of city life is that it brings people together. Drawing on a common tax resource pool, a single city government that represents multiple ethnic groups in a fair way may at first divide up the city resources and allocate to their respective groups based on patronage and favoritism. However, over time and as younger and more accommodating political leaders take over from war-traumatized ones, negotiations about how to most effectively spend public tax money may aspire to collective city-wide goals instead of ethnic-specific objectives. Cities do not always produce social and economic integration. Indeed, and especially pertinent in my analysis of Mostar, cities must be institutionally and geographically configured in ways that create opportunities for these positive inter-group effects to occur. Recourse to a collective cross-ethnic interest in a multicultural setting is only possible if municipal political geographies reach across and encompass all ethnic group interests within a single urban government system set up to fairly represent each of these group interests. Such a local governance framework sets the necessary condition for war hatreds and antagonisms to be moderated at the local level, likely over considerable periods of time.

In societies with stringent territoriality, cities can be the only places where the necessities of economic need and interdependence bring peoples together in a dynamic and mixed way. In contrast, neighborhoods, cantonal regions, and even states can become demarcated ethnically and susceptible to the protective strategies of ethnic politics. In Barcelona and the cities of Pais Vasco, nationalists and non-nationalists are more mixed at the city scale than they are in small towns or rural places. Languages and cultures mix in an urban setting and open up a space of dialogue. The political empowerment of these cities and their regions in the new Spanish Constitution provided opportunities for public planning, in pursuit of a collective interest submerged under Franco, to illuminate and operationalize the democracy of the new regime. The collective, public interest in Barcelona has been robust, vital, and catalytic of inclusive nationalism. In the Basque cities, the collective spirit of city governance has provided an alternative and competing non-violent path for that society.

In contrast, the Bosnian cities were not empowered but submerged and marginalized and even exploited (in the case of Mostar). The collective spheres of these cities have been damaged. In Sarajevo's case, its collective identity is fragile and susceptible because it is constrained by new post-war ethnic geographies. In Mostar, its collective identity as a city has been destroyed along with much of its social capital. Without an active urban governance system in Bosnia, international community efforts to build a democratic Bosnia lack the local foundational level of democracy from which to build. Instead, Bosnia's political geography of ethnically demarcated cantons and autonomous regions will reinforce and advance fragmenting impulses in the new country. Only at the geographic level of the city, where the potential for ethnic mixing and economic interdependence exist, do the forces of ethnic integration have a chance to counter over time those of ethnic separation.

Cities, nationalism, and Europeanization

> Cities constitute key pivots in the developing nexus between regional nationalism and European and global integration. European and global integration provide opportunities for the forging in cities of transcendent group identities and revised nationalisms.

In looking at the city, one identifies footholds and crevices where nationalist projects are being negotiated and modified in a contemporary global world. It is in a city where nationalist political projects must take stands on concrete and complex urban processes and issues that can clarify and refine its view of the world. Since urban processes in places like Spain and Bosnia are increasingly connected in the contemporary world to European political integration and economic globalization, this means that nationalist projects must create a workable relationship with these integrative phenomena.

In the Barcelona and Basque city cases, urbanism has inspired elements of inclusiveness and innovativeness that are transforming nationalist projects into more modernized and sophisticated projects. Barcelona is a city where cosmopolitanism has fostered a non-essentialist Catalanism that is more conducive to inter-group tolerance and plurality than a more rigidly defined nationalism would be. Its active engagement in EU activities and in Mediterranean regional programs further connects this urban-based Catalan nationalism to global opportunities and partnerships. In Bilbao, San Sebastian, and Vitoria, urban aspirations to connect economically with each other within the region and to European opportunities is producing an urbanized and functionally connected Basque nationalism that presents a clear alternative path to the rural-based ideology of militant Basque nationalism. In cases such as Catalonia and Basque Country, European Union inducements and mechanisms may allow these regions to develop some independence in programmatic and financial relationships with the EU, relative to the central (Spanish) state. To the extent that these new channels satisfy some aspirations for greater autonomy, political moderation within these substate regions may occur.

The approach by Europe seeks to acknowledge regional cultural autonomy within a web of a more integrated economic and political system. The Spanish experience in balancing central and peripheral imperatives is informative. Antagonistic political rhetoric notwithstanding, we saw in both Barcelona and Basque cases how regional self-government, when property structured and supported by Spanish central government, can be a stimulus and catalyst toward economic development. Region-based nationalist and cultural vibrancy supported by a central framework can result in robust outcomes; witness Bilbao's impressive city revitalization (enabled by the transitional agreement on Basque financial autonomy and actualized through local-regional-central government partnerships) and Barcelona's robust post-Franco economic recovery and its inter-governmentally supported event-driven urbanism. Despite the fundamental political disagreements that remain between the state and the Catalan and Basque regions, these programmatic successes display how regional autonomy can thrive and be catalytic when operating with the support of the central Spanish state. The emergence of the EU as a political and economic unit means that should there be political gridlock between Spain and the two historic regions in the future, it may well be the EU that increasingly provides a central framework to support historic regional nationalism.

In the Bosnian cities of Sarajevo and Mostar, group identities for now are likely being rigidified rather than transcended. Derived from the hatred and antagonism of wartime, the hardening of nationalistic allegiance is further being influenced by the ethnic strangulation and division of political space through the drawing of entity and municipal boundaries. Accordingly, the influence of Europeanization in addressing inter-group conflict and nationalism will necessarily operate through different mechanisms than in Spain. Whereas the EU seeks to acknowledge Catalan and Basque substate nationalism within an overarching central framework, its mission in Bosnia is to transcend and counter Serb, Bosniak, and Croat substate regionalism. Encouragement of direct relations between each of the ethnic regions of BiH – the Muslim-majority parts of the Federation, Republika Srpska, and the Croat-majority Herzegovina region – and the EU would be counterproductive to the future viability of the Bosnian state. Instead, EU intervention is being structured to actively counter the emergent substate nationalism that was accommodated and unintentionally strengthened by Dayton.

Ascension to the European Union is being extended to the Balkans states as leverage to inspire change at multiple levels of governance. At the inter-state level, EU is using a multiple-state regionalism approach (inclusive of Bosnia, Croatia, Serbia and Montenegro, Macedonia, and Albania) that is structured to induce cooperation across ethnic geographies and to promote stability amidst nationalism in the western Balkans. Within Bosnia itself, the EU is structuring its financial and programmatic interactions using five economic development regions: Sarajevo, Herzegovina (Mostar), northeast (Tuzla), northwest (Banja Luka), and central (Zenica). These regions use functional and historic economic boundaries that transcend Dayton's ethnic boundaries and seek to reverse the continued hardening of an erosive substate ethnic nationalism. In the Sarajevo economic region, 18 municipalities from the Federation and 13 municipalities from Republika Srpska now participate in regional development planning for an area of over 700,000 residents. In the Herzegovina economic region, there is the effort by the EU to induce cooperation across Bosniak and Croat sectors of Mostar city by increasing connections to regional opportunities that exist in Croat, Bosniak, and Serb territories of both the Federation and Republika Srpska.

The EU is thus programmatically restructuring substate regionalism in Bosnia to overcome the deficits of Dayton's political boundaries – ethnic fragmentation, excessive layering of government, and limited empowerment of local governance. A new EU functional governance architecture is being grafted upon an obstructive ethnic architecture, with cities at the center of these cross-ethnic functional spaces. Each of these regions has an urban economic engine, and within each area regional development strategies will need to be coordinated (presumably across ethnic geographies) before EU funds are dispersed to an urban region. In this way, the EU is bypassing the top-down political structure created by Dayton and endeavoring to regenerate Bosnian society through the creation of city-centered functional relationships. If this functional institutional architecture takes hold, there will be greater opportunities for new forms of nonessentialist identities to be rediscovered amidst the complexity of economic and social life. At that time, Sarajevo and Mostar could emerge as key connectors to Europe, economic anchors that hold together Bosnia, and crucibles where war-torn relations between nationalities can be slowly repaired and reconstructed.

Urbanism and peace building

Planning and urban design professions have within their power the capacity to illuminate and articulate on the ground and in the streets what a multinational democracy means. Urbanists revitalize and redevelop public spaces, neighborhoods, historic areas, and other urban public assets in ways that either promote or discourage healthy inter-group and interpersonal life.

The power of planning comes from its pivotal position in society between political goals and concrete actions. Planning is in the position of operationalizing political strategies. In some cases, such as Franco's Spain, planning structured urban environments in ways that supported the economic and industrial programs of an authoritarian regime. In other cases, such as Mostar, the power of public planning is captured and subordinated by ethnic/nationalistic groups as a way to further carry out their nationalist agendas.[6] In unstable urban environments, evidence of the power of public planning is the importance that antagonistic nationalistic groups attach to controlling and shaping the planning apparatus (and thus development) to fit their own needs and ends.

More positively, this power of planning can be utilized to help support and substantiate democratization efforts. Planning and urbanism can concretize new democratic goals and policies through how they construct and revitalize public spaces, neighborhoods, historic areas, and other urban public assets. In Barcelona, the ability of planning to articulate and implement the post-Franco democratic city had a pedagogical quality to it, informing and teaching the city's residents about the physical and social-psychological characteristics of an open society. The close alignment of the interests of the new democratic administration and the citizenry facilitated a mutual social learning process about the relationship between political change and urban betterment. In the Basque case, physical revitalization and restructuring of Bilbao is promoting a new and transformed sense of city identity that is competing with negative industrial and political images. Even in the failed planning case of Mostar, a well-conceptualized strategy of how spatial planning could be linked with social inter-group objectives showed the creative potential of planning amidst ethnic hatred.

I believe, as described by Borja and Castells (1997), that cities are privileged places for democratic innovation. Urbanist and planning interventions can constitute the most visible and meaningful edge of such democratic innovation. They can close or open up a city physically, fragment or integrate a city socially, submerge and dominate cultural identities or support them in ways that nurture diversity within unity, and build cities that reinforce and harden group identities or seek to transcend them. Planning and development decisions in multinational and contested cities can establish bridges and links between competing ethnic communities or they can build boundaries and figurative walls. Such actions will send emotive symbols to future generations about what the city either aspires to in hope or accepts in resignation.

Planning actions and principles amidst group conflict and political uncertainty can create and support urban conditions that are necessary for sustained peace and the mediation and reconciliation of inter-group conflict over time. These grassroots actions will not turn around a society that is splintered or unraveling; they cannot create peace where it does not exist in people's hearts and souls.

What urbanism can do, however, is create physical and psychological spaces that can help actualize larger peacemaking and inter-group reconciliation. Economic development, humanitarian, and institution-building strategies operating at the ground level can be, and should be, full partners with diplomatic peacemaking and larger-order considerations of societal reconstitution.

Based on what I have seen in the four case study urban areas (and earlier research), I put forward for consideration by local government administrators and nongovernmental organizations a set of city-building and urban design principles that can mitigate socio-economic and political tensions in situations of inter-group conflict.

Flexibility and porosity of urban form

Urban planning and policy should maintain as much flexibility of urban form as possible, choosing whenever possible and practical spatial development that maximizes future options. Walls, urban buffers, and other urban forms that delineate physical segregation of groups or facilitate psychological separation should be discouraged. Fulfillment of this goal will allow for future mixing of populations (if and when members of the respective groups choose) and normalization of urban fabric after active conflict abates. It must be emphasized that this is not integration or coercive assimilation, but rather the creation of an urban porosity that allows normal, healthier urban processes to occur.

Engagement in equity planning

Urban strategies and interventions should be targeted in ways that address the local manifestations of the long-term structural causes of conflict and tension. Development and humanitarian interventions should counter individual and group-based feelings of marginality, disempowerment, discrimination, and unequal access to services and goods. Project design and interventions should empower those groups in the city working toward peaceful solutions and co-existence, and the process of project design should be structured in order to increase communication across different urban groups. In cities of robust group identities, public participation from the start is vital in urbanism processes. Independent of the project's benefits themselves, this participation in deliberations is of vital significance in reconstructing a traumatized or torn city because it demonstrates how the democratic process works.

Sensitivity to urban ethnic homelands and frontiers

Local authorities should, through their regulatory powers, locate sensitive land uses having cultural and historic salience (churches, mosques, private schools, cultural community centers) within urban neighborhoods identified with those specific cultural groups. They should encourage in interface, or boundary, areas between cultural neighborhoods those types of land uses that encourage mixing of different groups in a supportive environment. Planners should prepare systematic assessments of cultural effects for proposed land uses of certain types (those having cultural importance) and in certain spatial areas (areas of interface and mixing). "Ethnic impact reports" should explicitly account for the social-psychological impacts of proposed land uses on the respective cultural communities of the city, and should be used in the decision-making process regarding development proposals. For agencies and individuals who are involved in project interventions amidst socio-economic and political tensions, they should assure institutional openness so that there can be continuous learning and institutional adaptability to reflect increased understanding about how urbanism can best be sensitive to ethnicity and conflict.

Protection and promotion of collective public sphere

For the seed of urban peace building to grow, the public sphere in both physical and institutional forms should encompass and respond to all competing identity groups in the city. Physically, planners should revitalize and redevelop public spaces as places of democracy, inter-group interactions, and neutrality. These spaces permit and contribute to cohesion and social equality and encourage activities that are the grounds for remaking an urban citizenship that is cross-ethnic in nature. Institutionally, local governance and urbanism should be reconstituted, and geographically configured, in ways that span ethnic divides and promote inclusiveness, dialogue, and negotiations. This institutionalization of the collective public sphere at the local level should be created as part of larger societal peace agreements; if not, ethnic fragmentation of local institutions will likely occur and set in motion processes that obstruct opportunities for positive inter-group effects. Municipal political geographies should reach across and encompass all ethnic group interests within a single urban government system that is designed to represent fairly each of these group interests.

Diffusion of grassroots peace building

In order to extend the impact and enhance the sustainability of innovative and progressive urban strategies, institutional linkages should be developed that diffuse peace-building knowledge both horizontally (to other urban areas in the country) and vertically (to regional and state governments, and to international organizations). Associations of local governments should engage proactively with negotiators during transition periods and seek to incorporate local grassroots lessons and partnerships into state-level diplomatic peace negotiations. These associations can develop and approve principles of tolerance and peace that can guide all participating local governments in a country, and provide through practical handbooks how urbanism can productively address conflict. Local government organizations that operate at the international level can be repositories of information about how municipal governments can facilitate and promote peace-building. With such local government and NGO advocacy, the chances that peace accords will recognize, rather than restrict, urbanism and local governance as peace-building assets is increased.

It is in a city where urban practitioners and leaders must do the hard work of creating the practical elements of a multinational democracy, one that avoids the extremes of an engineered and subordinating assimilation, on the one hand, and an unbounded and fracture-prone multinationalism, on the other. It is in a city where our greatest challenges and opportunities lie. Dewey (1916, 73) stated long ago, "A democracy is more than a form of government; it is primarily a mode of associated living" where one's decisions and actions must be made with regard to their effect on others. Such a balancing act between the interests of oneself and one's group with those of other people and other groups takes place most fundamentally in decision-making forums and lived experiences grounded in the city. Through our shaping of the city, we construct the contours of multinational democracy.

Notes

1. Jerusalem and Nicosia provide examples of such discussions by urban professionals and leaders amidst non-resolution of the political question. In Jerusalem, Israeli–Palestinian interaction at the level of urban professionals continued during times of great political tension. In March 2001, amidst hostilities that began November 2000, the author participated in a joint workshop of Israeli and Palestinian urban professionals examining the challenges and future options of planning a Jerusalem of mutual acceptance. This meeting was an offshoot of a larger joint effort, begun in 1995, which contributed technical support to the 2000 Camp David peace negotiations. Each group in the 2001 workshop had unofficial connections with their respective governments rather than formal and explicit sponsorship. In Nicosia, for 13 years, the mayor of the Greek Cypriot city (Lellos Demetriades) and his Turkish Cypriot mayor counterpart (Mustafa Akinci) met regularly on a clandestine basis, often driving through two checkpoints in a UN escorted car to do so. Such friendship helped bring about in the 1980s the development of a Nicosia Master Plan that disregarded the dividing line and planned for the city as a unified entity. This cooperation has facilitated the European Union-funded development of pedestrian areas in the commercial and historic centers on both sides of the line in ways that would enable them to be connected in the future. An important precursor of the Master Plan was the agreement by the two men to maintain a joint city-wide sewer system, also EU funded, encompassing both sides of the divide. Joint technical meetings of Greek Cypriot and Turkish Cypriot town planners, architects and engineers have formed an important bicommunal mechanism, although nationalist opposition sometimes obstructs their meetings in the buffer zone.
2. Similarly, Sorensen (1998) describes how political elite deals that secure the early stages away from authoritarianism can lead to restrictions that obstruct the further development and strengthening of democracy.
3. I borrow the idea of system complexity from the subdiscipline of human ecology and apply it to the process of democratization. Human ecologist Amos Hawley (1984, 2) described adaptation as an "irreversible process of cumulative change in which a system is moved from simple to complex forms."
4. I found in earlier research that in Johannesburg (South Africa), a catalytic alliance of urban organizations and issues took place during that country's transition from

apartheid, effectively connecting the goals and tactics of urban coexistence to the need to restructure the country's basic political parameters (Bollens 1999).

5. The question of national unity in Spain is by no means settled, but consensus over national direction in post-Franco Spain was significantly stronger than in post-war Bosnia.

6. Such partisan planning is also evident in apartheid-era South Africa and post-1967 Israeli planning in Jerusalem (Bollens 1999, 2000).

Appendix:
Interviews conducted

BARCELONA (55)

April 2003–July 200

Carol Perez	Consul General, United States Consulate, Barcelona (4/28/03)
Albert Broggi	Consultant, Ajuntament de Barcelona. Coordinator, AULA Barcelona (4/29/03)
Jordi Borja	Head, Urban Technology Consultant. Deputy Mayor, City of Barcelona, 1983–1995 (4/30/03)
Roser Viciana	Elected Councillor, City of Barcelona. Head, Councillor's Office for Civil Rights (4/30/03)
Eugenia Sanchez	Technical Advisor, Office for Civil Rights, City of Barcelona (4/30/03)
Oriol Nel-lo	Member, Catalonia Parliament. Professor, urban geography, Universitat Autónoma of Barcelona (9/15/03)
Pere Vilanova	Professor, Political Science, Department of Constitutional Law and Political Science. University of Barcelona (9/17/03)
Paul Lutzker	Consultant, Ajuntament of Barcelona (9/19/03)
Meritxell Batet	Director, Carles Pi I Sunyer Foundation for Local and Autonomous Studies (9/29/03)
Joaquim Llimona	Secretary of External Relations, Department of the Presidency, Generalitat of Catalunya (10/6/03)
Joan Miquel Piqué	Project chief, Institut D'Estudis Regionals i Metropolitans de Barcelona (10/8/03)
Elena Sintes	Researcher, Institut D'Estudis Regionals i Metropolitans (10/8/03)
Joan Lopez	Project chief, Institut D'Estudis Regionals i Metropolitans (10/8/03)
Eva Serra	Architect, Barcelona Regional (Metropolitan Agency for Urban Development and Infrastructures) (10/10/03)

Josep Carreras	Planner and Chief, Territorial Information Services, Mancommunitat de Municipis (10/10/03)
Maria Badia	Secretary of European and International Policy, Socialist Party of Catalonia (10/14/03)
Francesc Muñoz	Professor of Geography, Universitat Autónoma de Barcelona (10/17/03)
Ian Goldring	PhD student, Universitat Polytechnic de Catalonia. Instructor, Universitat Internacional de Barcelona (10/22/03)
Jesús Maestro	Secretary of International Policy, Esquerra Republicana de Catalonia Party (10/23/03)
Doménec Orriols	Secretary of Communications, Department of the Presidency, Generalitat de Catalonia (10/28/03)
Manuel Herce	Civil Engineer and Co-owner, Infrastructure Engineering and Management (private company). Professor, Universitat Polytechnic de Catalonia (11/6/03)
Francisco-Javier Monclús	Architect and Professor of Planning, Universitat Polytechnic de Catalona (11/6/03)
Montserrat Pareja	Professor, Department of Economic Theory, Universitat de Barcelona (11/13/03)
Julio Ponce	Lecturer, Administrative Law, Universitat de Barcelona (11/13/03)
Maria Teresa Tapada	Lecturer, Social Anthropology, Universitat Autónoma de Barcelona (11/13/03)
Marina Subirats	City Councilperson, Barcelona. Professor, Sociology, Universitat Autónoma de Barcelona (12/4/03)
Jordi Borja	Head, Urban Technology Consultant. Deputy Mayor, City of Barcelona, 1983–1995 [2nd interview] (12/9/03)
Eugeni Madueño	Chief, "Vivir Barcelona" section, *La Vanguardia* newspaper, Barcelona (12/11/03)
Lluís Permanyer	Columnist on Barcelona, *La Vanguardia* newspaper (12/11/03)
Albert Serratosa	President, Institut D'Estudis Territorials, Generalitat de Catalunya and Universitat Pompeu Fabra (12/12/03)
Francesc Carbonell	Director of Research and Studies. Institut D'Estudis Territorials, Generalitat de Catalunya and Universitat Pompeu Fabra (12/12/03)
Andreu Ulied	Planner, MCRIT Planning Support Systems (planning firm), Barcelona (1/14/04)
Carles Navales	City councilor, Cornella de Llobregat, 1979–1991. Trade unionist and activist in the 1970s, Barcelona (1/21/04)

Joan Antoni Solans	Director of Planning, Generalitat de Catalonia (1980–2001). Co-author, General Metropolitan Plan of Barcelona, 1976 (1/22/04)
Joan Subirats	Professor, Political Science, Universitat Autónoma de Barcelona (1/27/04)
Montserrat Rubí	Technical Coordinator, Strategic Metropolitan Plan of Barcelona, 2004 (1/28/04)
Josep Montaner	Professor, Architecture, Escola Technica Superior d'Arquitectura de Barcelona (ETSAB), Universitat Polytechnic de Catalonia (1/29/04)
Rafael Suñol	Investment banker (industrial), Banco Sabadell. Formerly with Banco Industrial, Government of Spain, Madrid (1983–1995) (2/17/04)
Paul Lutzker	Consultant, Ajuntament de Barcelona [2nd interview] (2/17/04)
Ignacio Pérez	Principal Architect, A-Plus Architecture (private company), Barcelona. Assistant Director, Escola Técnica Superior d'Arquitectura, Universitat Internacional de Barcelona (2/18/04)
Alexandre Karmeinsky	Architect, Wortman Bañares Arquitectos, Barcelona (2/18/04)
Joan-Anton Sanchez	Foundacio Ramon Trias Fargas. Formerly policy analyst, spatial planning, Department of the Presidency, Generalitat de Catalonia (3/16/04)
Joan Trullén	Professor, Applied Economics, Universitat Autónoma de Barcelona. Academic Director, Consorcio Universitat Internacional Menéndez Pelayo de Barcelona (CUIMPB) (4/1/04)
Ferran Requejo	Professor, Political Science, Universitat Pompeu Fabra (4/5/04)
Juli Esteban	Director, Territorial Planning Program, Secretary of Territorial Planning, Department of Territorial Planning and Public Works, Generalitat de Catalonia (4/19/04)
Enric Fossas	Associate Director, Institut d'Estudis Autonomics, Generalitat de Catalonia. Professor of Law, Universitat de Barcelona (4/20/04)
Santiago Mercadé	Chief Executive Officer, Layetana Development Company, Barcelona (4/22/04)
Mario Rubert	Managing Director, Department of Economic Promotion, Ajuntament de Barcelona (4/26/04)
Salvador Rueda	Director, Agencia Local d'Ecología Urbana de Barcelona (4/28/04)
Oriol Nel-lo	Secretary of Territorial Policy, Generalitat de Catalonia [2nd interview] (5/18/04)

Francesc Morata	Professor and Director, Institut d'Estudis Europeus, Universitat Autónoma de Barcelona (6/9/04)
Oriol Clos	Director of Urban Plans and Programs, Urbanism Sector, Ajuntament de Barcelona (7/1/04)
Ricard Frigola	Director General, Instituto Municipal Urbanismo, Ajuntament de Barcelona (7/2/04)
Maria Buhigas	Architect and Urban Planner, Barcelona Regional (Metropolitan Agency for Urban Development and Infrastructures). Assistant to Josep Acebillo, Commissioner for Infrastructures and Urbanism, Ajuntament de Barcelona (7/2/04)
Manuel de Solá-Morales	Architect and Professor, Escola Technica Superior d'Arquitectura de Barcelona (ETSAB), Universitat Polytechnic de Catalonia (7/12/04)

SARAJEVO (17)

April, November 2003

Zdravko Grebo	Professor of Political Science, Faculty of Law, University of Sarajevo (4/26/03)
Pere Vilanova	Professor of Political Science, University of Barcelona. Head of Legal Office of EU Administration in Mostar, 1996 (9/17/03)
Bashkim Shehu	Albanian writer. Resident writer, Center of Contemporary Culture of Barcelona (CCCB), Barcelona (10/16/03)
Jesus Maestro	Secretary of International Policy, Esquerra Republicana Party of Catalonia. Formerly, Councilor for International Cooperation, Barcelona City Council, and participant in Barcelona-Sarajevo cooperative projects (10/23/03)
Jakob Finci	Head, Civil Service Agency, Bosnia and Herzegovina (11/18/03, telephone)
Said Jamaković	Director, Sarajevo Canton Institute of Development Planning (11/19/03)
Jayson Taylor	Deputy Head, Reconstruction + Return Task Force, Office of the High Representative (OHR), Sarajevo (11/19/03)
[confidential]	Writer and nongovernmental organization activist (11/19/03)
Muhidin Hamamdžić	Mayor, City of Sarajevo (11/20/03)
Gerd Wochein	Project Manager and Architect, OHR, Sarajevo (11/20/03)

Javier Mier	Criminal Institutions and Prosecutorial Reform Unit, OHR, Sarajevo (11/21/03)
Ozren Kebo	Editor-in-Chief, *Start Magazine*, Writer (11/21/03)
Richard Ots	Senior Business Development Advisor, OHR (11/21/03)
Morris Power	Sarajevo Economic Region Development Agency (SERDA); Formerly with Reconstruction and Return Task Force, OHR (11/22/03)
Ferida Durakovic	Writer P.E.N. International Center of Bosnia Herzegovina (11/24/03)
Dragan Ivanovic	Deputy Speaker, Sarajevo Canton Assembly; Member, Federation Parliament (Chamber of Peoples); Director, Center for Policy Research and Development (11/24/03)
Vesna Karadzic	Assistant Minister, Federation of Bosnia and Herzegovina Ministry of Physical Planning and the Environment (11/24/03)
[confidential]	Officer, international organization (11/24/03)

PAIS VASCO (15)

February 2004

Pedro Arias	Dialogue for Peace. Professor, Department of Chemical Engineering and Physical Environment, University of Pais Vasco, Bilbao (2/23/04)
Ibon Areso	Deputy Mayor and Councilor of Urban Planning, City of Bilbao (2/23/04)
Victor Urrutia	Professor, Sociology, University of Pais Vasco, Bilbao (2/23/04)
Francisco Llera	Professor, Political Science, University of Pais Vasco, Bilbao (2/24/04)
Pedro Ibarra	Professor, Political Science, University of Pais Vasco, Bilbao (2/24/04)
Martín Arregi	Director, Territorial Organization, Department of Territorial Management and Physical Environment. Pais Vasco Regional Government, Vitoria (2/24/04)
Sabin Intxaurraga	Minister of Planning and the Built Environment, Pais Vasco Regional Government, Vitoria (2/24/04)
Karmelo Sainz	Director, Basque Association of Municipalities [EUDEL], Bilbao (2/25/04)
Jose Manuel Mata	Professor, Political Science, University of Pais Vasco, Bilbao (2/25/04)

Jose Ramon Beloki	Councilor, Department of Territorial Management and Promotion, Diputación of Gipuzkoa, San Sebastián (2/26/04)
Jose Aranburu	Analyst, Department of Territorial Management and Promotion, Diputación of Gipuzkoa, San Sebastián (2/26/04)
Agustin Arostegi	Co-director, Eurocity project, Diputación of Gipuzkoa, San Sebastián (2/26/04)
Xabier Unzurrunzaga	Professor, Architecture, University of Pais Vasco, San Sebastián (2/26/04)
Kepa Korta	Director, Strategic Plan of San Sebastián (2/27/04)
Ana Rosa Gonzalez	Professor, Law, University of Pais Vasco, San Sebastián (2/27/04)

MOSTAR (22)

April 2002, November 2003, and May 2004

Neven Tomić	Deputy Mayor, City of Mostar (4/6/02)
Nenad Bago	Lawyer, Regional Office South, Office of the High Representative (4/7/02)
Pere Vilanova	Head of Legal Office, European Union Administration for Mostar, April/July 1996 (9/17/03)
Gerd Wochein	Architect, Regional Office, Office of the High Representative, Mostar, 1998-2001 (11/20/03)
Javier Mier	Regional Office, Office of the High Representative, Mostar, 1994-2001 (11/21/03)
Nigel Moore	Political Advisor, European Union Policy Mission, Mostar; Formerly, head of Return and Reconstruction Task Force, Mostar (5/5/04)
Wolfgang Herdt	Regional Director of Balkan Programmes, Malteser Hilfsdienst [NGO] (5/5/04)
Amir Pašić	Director, Aga Khan Foundation, Mostar (5/5/04)
Sanja Alíkalfic	Director, United Nations High Commissioner for Refugees (UNHCR), Mostar (5/6/04)
Murray McCullough	Head, Delegation of the European Commission to Bosnia and Herzegovina, Mostar Office (5/6/04)
Jaroslav Vego	Herzegov/Neretva Canton Ministry of Urban Planning; Professor of Architectural Engineering, Faculty of Civil Engineering, University of Mostar (5/6/04)
Marica Raspudić	Urban Planning Department, City of Mostar (5/7/04)
Haris Kovačić	Head, Urban Planning Department, City of Mostar (5/10/04)

Muhamed Hamica Nametak	Director, Puppet Theater of Mostar; Member, Pedagogic Faculty, University of Mostar (5/10/04)
Julien Berthoud	Head of Political Section, Office of the High Representative, Mostar (5/10/04)
Pablo Barrera	Political Officer, Office of the High Representative, Mostar (5/10/04)
Zoran Bosnjak	Architect, Urban Planning Department, City of Mostar (5/11/04)
Palma Palameta	Civil Engineer, Urban Planning Department, City of Mostar (5/11/04)
Semin Borić	Minister of Finance, Herzegov/Neretva Canton (5/11/04)
Zlatan Buljko	Head, Mostar Field Office, United Methodist Committee on Relief (5/11/04)
Marisa Kolobarić	Director, Abrasevic Youth Cultural Center, Mostar (5/11/04)
Jaume Saura	Member, Elections Monitoring Team, Mostar June 1996; Professor of International Law, University of Barcelona (5/20/04)

Bibliography

Ajuntament de Barcelona. (1953). *Urbanization Plan for Barcelona and its Area of Influence (The County [Comarcal] Plan).*

Ajuntament de Barcelona. (1999). *Breaking Walls: Conference on Divided Cities.* Proceedings from October conference. Sarajevo, Bosnia-Herzegovina: Adjuntament, Dictricte 11.

Ajuntament de Barcelona. (2003). *Barcelona, The Place to B.* Barcelona: Author.

Aga Khan Trust. (1999). *Reclaiming Historic Mostar: Opportunities for Revitalization.* 15 Donor Dossiers for Conservation of High Priority Sites in the Historic Core. New York: World Monuments Fund.

Aga Khan Trust / World Monuments Fund. (2000). *Rehabilitation of the Historic Neighborhoods: Action Plan.* March. 52 pages. Istanbul: AKTC/WMF Technical Support Office.

Agranoff, Robert and Juan Antonio Ramos Gallarin. (1997). Toward Federal Democracy in Spain: An Examination of Intergovernmental Relations. *Publius: The Journal of Federalism*, vol. **27**, no. 4: pp. 1–38.

Ajangiz, Rafael. (2001). *On the Value Added of Citizen Participation in a Context of High Political and Social Confrontation: The Case of the Basque Country.* Conference Paper–ECPR Joint Sessions. Grenoble. April 6–11.

Allen, John. (1999). Worlds within Cities. In Massey, Doreen, John Allen and Steve Pile (eds.) *City Worlds.* London, New York: Routledge.

Alonso, Andoni, Inaki Arzoz, and Nicanor Ursua. (1996). Critical Remarks on Rural Architecture and Town Planning in the Basque Country: The Case of Navarre, 1964–1994. *Philosophy and Technology*, vol. **2**, no. 1: pp. 16–26.

Alterman, Rachelle. (1992). A Transatlantic View of Planning Education and Professional Practice. *Journal of Planning Education and Research*, vol. **12**, no. 1: pp. 39–54.

Alterman, Rachelle. (2002). *Planning in the Face of Crisis: Land Use, Housing, and Mass Immigration in Israel.* London: Routledge.

Amin, Ash. (2002). *Ethnicity and the Multicultural City: Living with Diversity.* Liverpool: European Institute for Urban Affairs.

Anderson Consulting. (1990). *Plan Estrategico Para la Revitalización del Bilbao Metropolitano.* Bilbao.

Appadurai, Arjun. (1996). *Modernity at Large: Cultural Dimensions of Globalization.* Minneapolis: University of Minnesota Press.

Areso, Ibon. (2002). From the Industrial City to the Post-Industrial City. Pp. 170–172 in Gobierno Vasco 2002. *Euskal Hiria.* 2002. Basque Department of Territorial Planning and the Environment. Vitoria: Central Publishing Services for the Basque Government.

Ashkenasi, Abraham. (1988). *Israeli Policies and Palestinian Fragmentation: Political and Social Impacts in Israel and Jerusalem*. Leonard Davis.

Association of the Citizens-Returnees to the Sarajevo Canton. (2001). *Information on Status of Returnees to the Sarajevo Canton*. August. 11 pages. Sarajevo: Author.

Ball-Rokeach, S. J. (1980). Normative and Deviant Violence from a Conflict Perspective. *Social Problems*, vol. **28**: pp. 45–62.

Barberia, José Luis. (2002). Es Viable una Euskadi Independiente? *El Pais*. March 31.

Barcelona Regional. Undated. Agencia Metropolitana de Desenvolupament Urbanistic i D'Infrastructures, SA. Agency Portfolio. 69 pages. Barcelona: Author.

Basque Study Society. (2001). Deliberation sessions on the Basque Bayonne-San Sebastian Eurocity. Pp. 1–556 in *Azkoaga, Cuadernos de Ciencias Sociales y Económicas*, vol. 11.

Baum, Howell S. (2000). Culture Matters—But It Shouldn't Matter Too Much. Pp. 115–136 in Burayidi, Michael A. (ed.) *Urban Planning in a Multicultural Society*. Westport, CT: Praeger.

Beck, Jan Mansvelt. (2000). The Continuity of Basque Political Violence: A Geographical Perspective on the Legitimisation of Violence. *GeoJournal*, vol. **48**: pp. 109–121.

Benjoechea, Soledad. (2002). The Barcelona Bourgeoisie, the Labour Movement and the Origins of Francoist Corporatism. Pp. 167–184 in Smith, Angel (ed.) *Red Barcelona: Social Protest and Labour Mobilization in the Twentieth Century*. London: Routledge.

Benvenisti, Meron S. (1986). *Conflicts and Contradictions*. New York: Villard Books.

Berry, Brian and John Kasarda. (1977). *Contemporary Urban Ecology*. New York: MacMillan.

Bilbao. (1992/1994). *The General Urban Plan of Bilbao*. City Council.

Bilbao Ria 2000. (1998). *Memoria*. Bilbao.

Bohigas, Oriol. (1963). Barcelona: Between Cerda and Informal Shelter. *Edicion* 62.

Bohigas, Oriol. (1983). *Plans i Projects per a Barcelona 1981–1982*. Barcelona: Ajuntament de Barcelona, Area d'Urbanisme.

Bohigas, Oriol. (1985). Reconstruccio de Barcelona. *Edición* 62.

Bohigas, Oriol. (1996). The Facilities of the Eighties. Pp. 211–214 in Centre de Cultura Contemporania de Barcelona. *Contemporary Barcelona 1856–1999*. Barcelona: CCCB.

Bollens, Scott A. (1996). On Narrow Ground: Planning in Ethnically Polarized Cities. *Journal of Architectural and Planning Research*, vol. **13**, no. 2: pp. 120–139.

Bollens, Scott A. (1998). Ethnic Stability and Urban Reconstruction: Policy Dilemmas in Polarized Cities. *Comparative Political Studies*, vol. **31**, no. 6: pp. 683–713.

Bollens, Scott A. (1999). *Urban Peace-Building in Divided Societies: Belfast and Johannesburg*. Boulder, CO. and Oxford, UK: Westview Press.

Bollens, Scott A. (2000). *On Narrow Ground: Urban Policy and Ethnic Conflict in Jerusalem and Belfast*. Albany: State University of New York Press.

Bollens, Scott A. (2001). City and Soul: Sarajevo, Johannesburg, Jerusalem, Nicosia. *CITY: Analysis of Urban Trends, Culture, Theory, Policy, Action*, vol. **5**, no. 2: pp. 169–87.

Bollens, Scott A. (2002). Urban Planning and Inter-Group Conflict: Confronting a Fractured Public Interest. *Journal of the American Planning Association*, vol. **68**, no. 1: pp. 22–42.

Borja, Jordi. (1971). La Gran Barcelona. *Cuadernos de Arquitectura y Urbanismo*. Barcelona.

Borja, Jordi. (2001). The Pace and Extent of Barcelona's Urban Transformation. Unpublished manuscript.

Borja, Jordi. (2003). *El Espacio Publico: Ciudad y Ciudadanía*. Barcelona: Electa.

Borja, Jordi and Manuel Castells. (1997). *Local and Global: The Management of Cities in the Information Age*. London: Earthscan.

Brenner, Neil. (2004). *New State Spaces: Urban Governance and the Rescaling of Statehood*. Oxford: Oxford University Press.

Brown, Michael E. (ed.) (1996). *The International Dimensions of Internal Conflict*. Cambridge, MA: Massachusetts Institute of Technology Press.

Brugue, Quim, Ricard Goma, and Joan Subirats. (2004). Multilevel Governance and Europeanization: The Case of Catalonia. Unpublished manuscript.

Bryson, John. (1981). A Perspective on Planning and Crises in the Public Sector. *Strategic Management Journal*, vol. **2**: pp. 181–96.

Bublin, Mehmed. (1999). *The Cities of Bosnia and Herzegovina: A Millennium of Development and the Years of Urbicide*. Sarajevo: Sarajevo Publishing Co.

Buesa, Mikel. (ed.) (2004). *Secession Economy: The Nationalist Project and the Basque Country*. Madrid: Complutense University.

Burayidi, Michael A. (2000). Urban Planning as a Multicultural Canon. Pp. 1–14 in Burayidi, Michael A. (ed.) *Urban Planning in a Multicultural Society*. Westport, CT: Praeger.

Burg, Steven L. and Paul S. Shoup. (1999). *The War in Bosnia-Herzegovina: Ethnic Conflict and International Intervention*. Armonk, NY: M.E. Sharpe.

Burton, John W. (ed.) (1990). *Conflict: Human Needs Theory*. New York: St Martins.

Busquets, Joan. (1987). Centralitat i Implantacio Urbana. In various authors, *Arees de Nova Centralitat*. Barcelona: Ajuntament.

Busquets, Joan. (2004). *Barcelona: La Construcción Urbanística de Una Ciudad Compacta*. Barcelona: Ediciones del Serbal.

Cabre, A. and I. Pujades. (1988). La Poblacio: Immigracio i la Explosio Demográfica, in *Historia Económica de la Catalunya Contemporánea*, volume 4. Gran Enciclopedia Catalana. Barcelona.

Caldeira, Teresa. (2000). *City of Walls: Crime, Segregation, and Citizenship in Sao Paulo*. Berkeley: University of California Press.

Camos, Joan and Clara C. Parramon. (2002). The Associational Movement and Popular Mobilizations in L'Hospitalet: From the Anti-Francoist Struggle to Democracy, 1960–1980. Pp. 206–222 in Smith, Angel (ed.) *Red Barcelona: Social Protest and Labour Mobilization in the Twentieth Century*. London: Routledge.

Carrillo, Ernesto. (1997). Local Government and Strategies for Decentralization in the 'State of the Autonomies'. *Publius: The Journal of Federalism*, vol. **27**, no. 4: pp. 39–63.

Celik, Zeynep. (1997). *Urban Forms and Colonial Confrontations: Algiers Under French Rule*. Berkeley: University of California Press.

Centar Municipality. (2003). Summary Information Sheet. Opcina Center Sarajevo.

Central Intelligence Agency. (2005). *The World Factbook*. Online. Available HTTP: < http://www.cia.gov/cia/publications/factbook > (accessed 27 July 2005).

Centro de Investigaciones Sociológicas (CIS). *Post-Electoral Study 1992*. Number 1998. Madrid: CIS.

Chandler, David. (2000). *Bosnia: Faking Democracy after Dayton*. London: Pluto.

Chesterman, Simon. (2004). *You, The People: The United Nations, Transitional Administration, and State-Building*. Oxford, U.K.: Oxford University Press.

Clark, Robert P. (1979). *The Basques: The Franco Years and Beyond*. Reno: University of Nevada Press.

Coakley, John. (1993). Introduction: The Territorial Management of Ethnic Conflict. Pp. 1–22 in Coakley, John (ed.) *The Territorial Management of Ethnic Conflict.* London: Frank Cass.

Cohen, Michael A., Blair A. Ruble, Joseph S. Tulchin, and Allison M. Garland. (1996). *Preparing for the Urban Future: Global Pressures and Local Forces.* Published in cooperation with the United Nations Centre on Human Settlements (Habitat II). Washington D.C.: Woodrow Wilson Center Press.

Commission for Reforming the City of Mostar. (2003). *Recommendations of the Commission; Report of the Chairman.* Mostar. December 15.

Commission of the European Communities. (2003a). Stabilization and Association Report, Bosnia and Herzegovina. Commission Staff Working Paper COM (2003) 139 final. Brussels: Commission.

Commission of the European Communities. (2003b). The Western Balkans and European Integration. Communication from the Commission to the Council and the European Parliament. COM (2003) 285 final. Brussels: Commission.

Council of Europe. (2004). 2004 Municipal Elections – Bosnia and Herzegovina, Statement of Preliminary Findings and Conclusions. Sarajevo: Organization for Security and Co-operation in Europe, Office for Democratic Institutions and Human Rights Election Observation Mission. October.

Council on Foreign Relations. Terrorism Q&A – Basque Fatherland and Liberty. Online. Available HTTP: <http://www.terrorismanswers.com> (accessed July 31, 2002).

Covic, Boze (ed.) (1993). *Roots of Serbian Aggression.* Zagreb: Centar za strane jezike I Omladinski kulturni centar.

Cross-Border Agency for the Development of the Basque Eurocity of Bayonne-San Sebastian. Undated. A New City for Living Without Borders. San Sebastian: Author.

Dahl, Robert A. (1998). *On Democracy.* New Haven: Yale University Press.

Dewey, John. (1916). *Democracy and Education: An Introduction to the Philosophy of Education.* New York: MacMillan.

Diefendorf, Jeffrey M. (1993). *In the Wake of War: The Reconstruction of German Cities After World War II.* New York: Oxford University Press.

Diputacion Foral de Gipuzkoa. (2002). *Proceso de Reflexión Estratégica Guipúzcoa 2020: 4 Escenarios Para la Reflexion.* San Sebastián: Diputación, Departamento de Economía y Turismo.

Drazenovic, Ivana. (2000). Local Elections in B&H Completed. AIM Sarajevo. April 14. Online. Available HTTP: http://www.aimpress.ch/dyn/trae/archive (accessed December 1, 2005).

Dunkerly, David et al. (2002). *Changing Europe: Identities, Nations and Citizens.* London: Routledge.

Edles, Laura Desfor. (1998). *Symbol and Ritual in the New Spain: The Transition to Democracy after Franco.* Cambridge, MA: Cambridge University Press.

Elazar, Daniel J. (1994). *Federalism and the Way to Peace.* Reflections Paper No. 13. Kingston, Ontario: Institute of Intergovernmental Relations, Queen's University.

Eriksen, Thomas H. (1993). *Ethnicity and Nationalism.* London: Pluto Press.

Esteban, Juli. (1999). *El Projecte Urbanistic: Valorar la Periferia i Recuperar el Centre.* Model Barcelona. Quaderns de Gestio. Barcelona: Aula Barcelona.

Etherington, John. (2003). *Nationalism, National Identity and Territory: The Case of Catalonia.* Doctoral Thesis. Universitat Autónoma de Barcelona. June.

Etzioni, Amitai. (1968). *The Active Society.* New York: Free Press.

EUDEL (Association of Basque Municipalities). (2002). Civic Declaration in Defense of Democracy and Liberty, and with Respect for Plurality in the Basque Country. Bilbao: Association of Basque Municipalities. May 3rd.

European Stability Initiative. (2001). Reshaping International Priorities in Bosnia and Herzegovina: The End of the Nationalist Regimes and the Future of the Bosnian State. Sarajevo, March 22. Online. Available HTTP: < http://www.esiweb.org/docs > (accessed 21 October 2004).

European Union. (2003). The Thessaloniki Agenda for the Western Balkans: Moving Toward European Integration. June 16. 11 pages. General Affairs & External Relations Council.

European Union (and Sarajevo Economic Region Development Agency). (2004). *Regional Economic Strategy for Sarajevo Economic Macro Region.*

Euskobarometro. (2005) (May). Series Temporales. Bilbao: University of Pais Vasco.

Eustat (Basque Institute of Statistics). (2001). Population and Housing Census.

Federation Ministry of Displaced Persons and Refugees. (2003). *Plan for Arrival and Repatriation into Bosnia and Herzegovina Federation Territory.* April. 39 pages. Sarajevo: Author.

Flyvbjerg, Bent. (1998). *Rationality and Power: Democracy in Practice.* Chicago: University of Chicago.

Friedmann, John. (1987). *Planning and the Public Domain: From Knowledge to Action.* Princeton: Princeton University Press.

Friedmann, John and Clyde Weaver. (1979). *Territory and Function: The Evolution of Regional Planning.* Berkeley: University of California Press.

Friend, John and Allen Hickling. (1997) (2nd ed.). *Planning Under Pressure: The Strategic Choice Approach.* Oxford: Butterworth-Heinemann.

G-Gagnon, Alain and James Tully (eds.). (2001). *Multinational Democracies.* Cambridge: Cambridge University Press.

Garcia, Angel Maria Nieva. (2002). The Characteristics of Bilbao Ria 2000 and its Main Operations. Pp.159–166 in Gobierno Vasco 2002. *Euskal Hiria.* 2002. Basque Department of Territorial Planning and the Environment. Vitoria: Central Publishing Services for the Basque Government.

Garcia, Marisol. (2003). The Case of Barcelona. Pp. 337–358 in Salet, Willem, Andy Thornley, and Anton Kreukels (eds.) *Metropolitan Governance and Spatial Planning.* London: Spon.

Garrod, Martin. (1998). Report on European Administration of Mostar and Office of EU Special Envoy in Mostar, July 23 1994– December 31, 1996. Unpublished report.

Gausa, Manuel. (1996). A Leap in Scale: From Urban to Metropolitan Barcelona, pp. 225–240 in Centre de Cultura Contemporánia de Barcelona. *Contemporary Barcelona 1856–1999.* Barcelona: CCCB.

Ghai, Yash. (1998). The Structure of the State: Federalism and Autonomy. Pp. 155–68 in Harris, Peter and Ben Reilly (eds.). *Democracy and Deep-Rooted Conflict: Options for Negotiators.* Stockholm: International Institute for Democracy and Electoral Assistance.

Gjelten, Tom. (1995). *Sarajevo Daily: A City and Its Newspaper Under Siege.* New York: HarperCollins.

Gobierno Vasco (2002). *Euskal Hiria.* 2002. Basque Department of Territorial Planning and the Environment. Vitoria: Central Publishing Services for the Basque Government.

Godschalk, David R. (ed.) (1974). *Planning in America: Learning from Turbulence.* Washington D.C.: American Institute of Planners.

Gomez, M. (1998). Reflective Images: The Case of Urban Regeneration in Glasgow and Bilbao. *International Journal of Urban and Regional Research*, vol. **22**: pp. 106–121.

Guibernau, Montserrat. (2004). *Catalan Nationalism: Francoism, Transition, and Democracy*. London: Routledge.

Gurr, Ted R. (1993). Why Minorities Rebel: A Global Analysis of Communal Mobilization and Conflict Since 1945. *International Political Science Review*, vol. **14**, no. 1: pp. 161–201.

Gurr, Ted R. and Barbara Harff. (1994). *Ethnic Conflict in World Politics*. Boulder: Westview.

Gutmann, Amy. (2003). *Identity in Democracy*. Princeton, NJ: Princeton University Press.

Hall, Peter. (1998). *Cities in Civilization*. New York: Fromm International.

Harloe, Michael. (1996). Cities in the Transition. Pp. 1–29 in Andrusz, Gregory, Michael Harloe, and Ivan Szelenyi (eds.) *Cities and Socialism: Urban and Regional Change and Conflict in Post-Socialist Societies*. Oxford, UK: Blackwell.

Hawley, Amos. (1984). Sociological Human Ecology: Past, Present, and Future. Pp. 1–15 in Micklin, Michael and Harvey Choldin (eds.) *Sociological Human Ecology: Contemporary Issues and Applications*. Boulder and London: Westview Press.

Hepburn, A.C. (2004). *Contested Cities in the Modern West*. New York: Palgrave MacMillan.

Hommels, Anique. (2005). *Unbuilding Cities: Obduracy in Urban Sociotechnical Change*. Cambridge, MA: MIT Press.

Horowitz, Donald L. (1985). *Ethnic Groups in Conflict*. Berkeley: University of California Press.

Hughes, Robert. (1992). *Barcelona*. New York: Knopf.

Human Rights House Foundation. (2004). Bosnia—Local Elections 2004: Fair and Democratic, with Low Voter Turnout. Online. Available HTTP: < http://www.humanrightshouse.org > (accessed December 1, 2005).

Huntington, Samuel. P. (1997). *The Clash of Civilizations and the Remaking of World Order*. London: Simon and Schuster.

Infrastructures del Levant de Barcelona, SA. (2004). *Besos Seafront: A New Impetus for Barcelona*. Barcelona: Author.

Institut d'Estadistica de Catalunya. (2004). *Estadistica Basica Territorial*. Online. Available HTTP: < http://www.cbuc.es/5digital/idescat > (accessed June 18, 2006).

Institut d'Estudis Regionals i Metropolitans. (2002a). *Enquesta de la Regio de Barcelona 2000 Informe General (Survey of the Region of Barcelona 2000 General Report)*. Barcelona: Institut.

Institut d'Estudis Regionals i Metropolitans. (2002b). *Dades Estadistiques Basiques 2000 Districtes de Barcelona Volum 3*. Barcelona: Institut.

Internal Displacement Monitoring Centre. (1996). More Population Displacement in 1996. Oslo: Norwegian Refugee Council.

Internal Displacement Monitoring Centre. (2002). Minority Returnees Emerge as a Political Force after October 2002 Elections. *Global IDP Database*. Oslo: Norwegian Refugee Council.

International Crisis Group. (1998a). Rebuilding a Multi-Ethnic Sarajevo: The Need for Minority Returns. ICG Bosnia Project, Report No. 30, February 3rd, Brussels: ICG.

International Crisis Group. (1998b). Minority Return or Mass Relocation. ICG Bosnia Project, Report No. 33, May 14th. Brussels: ICG.

International Crisis Group. (2000). Reunifying Mostar: Opportunities for Progress. Europe Report no. 90, April. Brussels: Author.

International Crisis Group. (2003). Building Bridges in Mostar. Europe Report No. 150. November 20. Sarajevo/Brussels: Author.

Isin, Engin F. (ed.) (2000). *Democracy, Citizenship and the Global City*. London: Routledge.

i3 Consultants, IBK, and CODE. (2000). *Bayonne-San Sebastian Eurocity's White Paper* (Working Paper). June.

Jenkins, Brian and Spyros Sofos. (1996). *Nation and Identity in Contemporary Europe*. London: Routledge.

Jimenez, Jose Luis. (2004). PNV y PSE Anuncian Una Nueva Etapa en Euskadi. *La Vanguardia* newspaper. March 16, 44–45.

Joint Declaration. (2003). To Ensure Peace in BiH by Further Annexing of the Dayton Agreement. Issued by a group of 24 European political leaders. December 16. Sarajevo.

Jokay, Charles. (2001). Local Government in Bosnia and Herzegovina. Pp. 91–140 in Kandeva, Emilia (ed.) *Stabilization of Local Governments*. Budapest, Hungary: Local Government and Public Service Reform Initiative.

Karahasan, Dzevad. (1994). *Sarajevo, Exodus of a City*. New York: Kodansha.

Katz, Mishal. (1996). Utopias of the Sixties. Pp. 173–179 in Centre de Cultura Contemporania de Barcelona. *Contemporary Barcelona 1856–1999*. Barcelona: CCCB.

Keating, Michael. (1998). *The New Regionalism in Western Europe*. Cheltenham, UK: E. Elgar.

Keith, Michael. (2005). *After the Cosmopolitan? Multicultural Cities and the Future of Racism*. London: Routledge.

Knowlton, Brian. (2005). Bosnians Vow to Create a Stronger Government. *New York Times* November 22. Online. Available HTTP: http://nytimes.com (accessed November 22).

Kumar, Radha. (1997). *Divide and Fall? Bosnia in the Annals of Partition*. London: Verso.

Kurlansky, Mark. (1999). *The Basque History of the World*. New York: Penguin.

Kymlicka, Will. (1995). *Multicultural Citizenship: A Liberal Theory of Minority Rights*. Oxford: Clarendon.

Lake, David and Donald Rothchild. (1996). *Ethnic Fears and Global Engagement: The International Spread and Management of Ethnic Conflict*. Policy Paper No. 20. University of California, San Diego: Institute of Global Conflict and Cooperation.

Lamarca, Inigo and Eduardo Virgala. (1983). *El Estatuto de Autonomia del Pais Vasco*. Working Paper. Facultad de Derecho, Universidad de Pais Vasco, San Sebastián.

Lapidoth, Ruth. (1996). *Autonomy: Flexible Solutions to Ethnic Conflicts*. Washington D.C.: United States Institute of Peace.

Lefebvre, Henri. (1979). Space: Social Product and Use Value, pp. 285–96 in J.W. Frieberg (ed.) *Critical Sociology: European Perspectives*. New York: Irvington Publishers.

Le Gales, Patrick. (2002). *European Cities: Social Conflicts and Governance*. Oxford, U.K.: Oxford University Press.

Lijphart, Arend. (1968). *The Politics of Accommodation: Pluralism and Democracy in the Netherlands*. Berkeley, CA: Univ. of California Press.

Lijphart, Arend. (1977). *Democracy in Plural Societies: A Comparative Exploration*. New Haven: Yale University Press.

Lindblom, Charles E. (1977). *Politics and Markets: The World's Political-Economic Systems*. New York: Basic Books.

Lippman, Peter. (2000). Case Study: Democratic Initiative of Sarajevo Serbs. Washington D.C.: The Advocacy Project.

Llera, Francisco. (1999a). Basque Polarization: Between Autonomy and Independence. *Nationalism and Ethnic Politics*, vol. **5**, no. 3/4: pp. 101–120.

Llera, Francisco. (1999b). Franazo al Tren de Estella. *Claves de Razón Practica*. No. 95.

Loughlin, John. (2001). *Subnational Democracy in the European Union: Challenges and Opportunities*. Oxford, U.K.: Oxford University Press.

Lovrenovic, Ivan. (2001). *Bosnia: A Cultural History*. New York: New York University Press.

Lustick, Ian. (1979). Stability in Deeply Divided Societies: Consociationalisation vs. Control. *World Politics*, vol. **31**: pp. 325–44.

Lynch, Kevin. (1981). *Good City Form*. Cambridge, MA: MIT Press.

McGarry, John and Brendan O'Leary (eds.). (1993). *The Politics of Ethnic Conflict Regulation: Case Studies of Protracted Ethnic Conflicts*. London: Routledge.

McNeill, Donald. (1999). *Urban Change and the European Left: Tales from the New Barcelona*. London: Routledge.

Madariaga, Ines Sanchez de. (2004). Spatial Planning, in Fact Town Planning: From Regulation to Shared Visions. *QPE-Revista Electrónica*, vol. **6**: pp. 62–79.

Makas, Emily G. (2005). Interpreting Multivalent Sites: New Meanings for Mostar's Old Bridge. *Centropa*, vol. **5**, no. 1: pp. 59–69.

Marcet i Morera, Joan. (2000). Convergencia Democratica de Catalunya: The Only Spanish Liberal Reference at a European Level. Pp. 183–198 in De Winter, Lieven et al. (eds.) *Liberalism and Liberal Parties in the European Union*. Barcelona: Institut de Ciencies Politiques i Socials.

Marcuse, Peter and Ronald van Kempen (eds.). (2002). *Of States and Cities: The Partitioning of Urban Space*. Oxford: Oxford University Press.

Masser, Ian. (1986). Some Methodological Considerations. In Masser, Ian and Richard Williams (eds.) *Learning From Other Countries: The Cross-National Dimension in Urban Policy-Making*. Norwich, U.K.: Geo.

Mata, Jose Manuel. Terrorismo y Conflicto Nacionalista: La Debilidad de la Democracia in el Pais Vasco. Unpublished manuscript. Universidad de Pais Vasco, Bilbao.

Mees, Ludger. (2003). *Nationalism, Violence, and Democracy: The Basque Clash of Identities*. Houndmills, UK: Palgrave MacMillan.

Merrifield, Andy and Erik Swyngedouw (eds.) (1997). *The Urbanization of Injustice*. New York: New York University Press.

Metropolitan Corporation of Barcelona. (1976). *The General Metropolitan Plan for Barcelona*.

Ministry of Human Rights and Refugees, Bosnia and Herzegovina. (2003). Bilten 2003: Uporedni Pokazatelji. Sarajevo: Author.

Molinero, Carme and Pere Ysas. (2002). Workers and Dictatorship: Industrial Growth, Social Control and Labour Protest under the Franco Regime. Pp. 185–205 in Smith, Angel (ed.) *Red Barcelona: Social Protest and Labour Mobilization in the Twentieth Century*. London: Routledge.

Monclus, Francisco-Javier. (2003). The Barcelona Model: An Original Formula? From ''Reconstruction'' to Strategic Urban Projects. *Planning Perspectives*, vol. **18**: pp. 399–421.

Montalbán, Manuel Vazquez. (1992). *Barcelonas*. London: Verso.

Morata, Francesc. (1997). The Euro-Region and the C-6 Network: The New Politics of Sub-national Cooperation in the West-Mediterranean Area. Pp. 292–305 in Keating, Michael and John Loughlin (eds.) *The Political Economy of Regionalism*. London: Frank Cass.

Moreno, Luis. (1997). Federalization and Ethnoterritorial Concurrence in Spain. *Publius – The Journal of Federalism*, vol. **27**, no. 4: pp. 65–84.

Moreno, Luis. (2001). Divided Societies, Electoral Polarisation and the Basque Country. Unidad de Politicas Comparadas Working Paper, pp. 01–07.

Moreno, Luis, Ana Arriba and Araceli Serrano (1998). Multiple Identities in Decentralized Spain: The Case of Catalonia. *Regional and Federal Studies*, vol. **8**, no. 3: pp. 65–88.

Morley, David and Arie Shachar. (1986). Epilogue: Reflections by Planners on Planning. Pp. 142–151 in Morley, David and Arie Shachar (eds.) *Planning in Turbulence*. Jerusalem: Magnes Press, Hebrew University.

Mostar, City (Urban Planning Department). (2004). Reconstruction program for residential building in Šantića/T. Miloša Street and the Boulevar.

Mota, Fabiola and Joan Subirats. (2000). El Quinto Elemento: El Capital Social de las Comunidades Autónomas: Su Impacto Sobre el Funcionamiento del Sistema Político Autonómico. *Revista Española de Ciencia Política*, vol. **1**, no. 2: pp. 123–158.

Murphy, A.B. (1989). Territorial Policies in Multiethnic States. *Geographical Review*, **79**: pp. 410–421.

Murtagh, Brendan. (2002). *The Politics of Territory: Policy and Segregation in Northern Ireland*. New York: Palgrave.

National Statistics Institute. (2001). *Population and Housing Census*. Madrid: NSI.

Nel-lo, Oriol. (2001). *Ciutat de Ciutats: Reflexions Sobre el Proces d'Urbanitzacio a Catalunya*. Barcelona: Editorial Empories.

Nel-lo, Oriol. (2002). Equilibrium. In Subirats, Jaume (ed.) *Barcelona Acrostic*. Barcelona: Ajuntament.

Neill, William. (2004). *Urban Planning and Cultural Identity*. London: Routledge.

Neill, William, Dianna Fitzsimons, and Brendan Murtagh (1995). *Reimaging the Pariah City: Urban Development in Belfast and Detroit*. Aldershot: Avebury Press.

Newman, Saul. (1996). *Ethnoregional Conflict in Democracies – Mostly Ballots, Rarely Bullets*. Westport, CT: Greenwood Press.

Nordlinger, Eric A. (1972). *Conflict Regulation in Divided Societies*. Boston: Center for International Affairs, Harvard University.

Nordstrom, Carolyn. (1997). *A Different Kind of War Story*. Philadelphia: University of Pennsylvania Press.

Office of the High Commissioner. (2005). Interview with Martin Ney, Senior Deputy High Representative. September 12. Online. Available HTTP: < http://www.ohr.int/ohr-dept/pressi > (accessed November 8, 2005).

Office of the High Representative. (1996). *Protocol on the Organization of Sarajevo*. Federation Forum Meeting. October 25th. Sarajevo: Author.

Office of the High Representative. (2002a). *OHR Sarajevo Region – Plans 2003*. Sarajevo: Author.

Office of the High Representative. (2002b). *Reconstruction and Return Task Force Work Plan 2002*. Sarajevo: OHR.

Office of the High Representative. (2002c). RRTF Activities July 2002. OHR Briefing Memorandum from Head of RRTF.

Office of the High Representative. (2003). *Mission Implementation Plan*. Sarajevo: Author.

Office of the High Representative. (2004a). *Decision Enacting the Statute of the City of Mostar*. Mostar: Author.

Office of the High Representative. (2004b). *Statistics – Implementation of the Property Laws in Bosnia and Herzegovina*. Sarajevo: Author.

O'Leary, Brendan and John McGarry. (1995). Regulating Nations and Ethnic Communities. Pp. 245–289 in Breton, A., G. Galeotti, P. Salmon and R. Wintrobe (eds.) *Nationalism and Rationality.* Cambridge: Cambridge University Press.

Ombudsman Institution of the Federation of BiH. (2004). Implementation of Property Laws. Report on Human Rights Situation in the Federation. Sarajevo: Author.

Owen, David. (1995). *Balkan Odyssey.* London: V. Gollancz.

Palleres, Francesc, Joes Montero, and Francisco Llera. (1997). Non State-wide Parties in Spain: An Attitudinal Study of Nationalism and Regionalism. *Publius: The Journal of Federalism,* vol. **27**, no. 4: pp. 135–169.

ParWeb Solutions. (2005). Yugoslavia: Local Government and the Communes. Online. Available HTTP: http://www.reference.allrefer.com/country-guide-study/yugoslavia/ yugoslavia139.html> (accessed June 27, 2005).

Pejanovic, Mirko. (2002). Legislative and Executive Powers in the System of Local Self-Government in Bosnia-Herzegovina. Pp. 72–79 in Pintar, Rudiger (ed.) *Executive and Legislature at Local Level Structure and Interrelation in Countries of South-East Europe.* Zagreb, Croatia: Friedrich Ebert Stiftung.

Pizza, Antonio and Josep M. Rovira. (2002). *Desde Barcelona: Arquitecturas y Ciudad 1958–1975.* Barcelona: Col-legio d'Arquitectes de Catalunya.

Plunz, Richard, Mojdeh Baratloo, and Michael Conard (ed.) (1998). *New Urbanisms: Mostar, Bosnia and Herzegovina.* New York: Columbia University Press.

Preston, Paul. (2004). *Juan Carlos: Steering Spain from Dictatorship to Democracy.* New York: Norton.

Provincial Commission for Urban Development. (1968). *Master Plan for the Barcelona Metropolitan Area ("Provincial Plan").*

Przeworski, Adam. (1991). *Democracy and the Market: Political and Economic Reforms in Eastern Europe and Latin America.* Cambridge: Cambridge University Press.

Putnam, R. D. (1993). *Making Democracy Work: Civic Traditions in Modern Italy.* Princeton, NJ.: Princeton University Press.

Raspudic, Marica and Francisco Aiello. (1996). Data and Analytic Study, Structure Plan. Mostar: EUAM All-Mostar Strategic Planning Team.

Repatriation Information Centre. (1998). Municipality Information Fact Sheet—The City of Mostar (Central Zone). Sarajevo: International Centre for Migration Policy Development.

Rex, John. (1996). *Ethnic Minorities in the Modern Nation State: Working Papers in the Theory of Multiculturalism and Political Integration.* New York: St Martin's Press.

Rioja, Isabel Ramos. (2004). Trescientos Mil Problemas Por Resolver. *La Vanguardia* newspaper, June 6, pp. 32–33.

Rodriguez, Arantxa, Elana Martinez, and Galder Guenaga. (1999). Abandoibarra, Bilbao-Spain. Online. Available HTTP: < http://www.ifresi.univ-lill1.fr > (accessed January 16, 2003).

Rodriguez, Arantxa, Elana Martinez, and Galder Guenaga. (2001). Uneven Redevelopment: New Urban Policies and Socio-Spatial Fragmentation in Metropolitan Bilbao. *European Urban and Regional Studies,* vol. **8**, no. 2: pp. 161–178.

Roeder, Philip G. and Donald Rothchild (eds.) (2005). *Sustainable Peace: Power and Democracy After Civil Wars.* Ithaca: Cornell University Press.

Rogers, Richard. (1999). *Towards an Urban Renaissance.* Final Report of the Urban Task Force. London: Spon.

Romann, Michael and Alex Weingrod. (1991). *Living Together Separately: Arabs and Jews in Contemporary Jerusalem.* Princeton, NJ: Princeton University Press.

Rosenau, James N. (1990). *Turbulence in World Politics: A Theory of Change and Continuity*. Princeton, N.J.: Princeton University Press.

Rotberg, Richard (ed.) (2004). *When States Fail: Causes and Consequences*. Princeton, NJ: Princeton University Press.

Rothman, Jay. (1992). *From Confrontation to Cooperation: Resolving Ethnic and Regional Conflict*. Newbury Park, CA: Sage.

Rowe, Peter. (1997). *Civic Realism*. Cambridge, MA: MIT Press.

Rustow, Dankwart A. (1970). Transitions to Democracy. *Comparative Politics*, vol. **2**, no. 3: pp. 337–365.

Sack, Robert. (1981). Territorial Bases for Power. In Burnett, A. and P. Taylor (eds.) *Political Studies from Spatial Perspectives*. New York: John Wiley and Sons.

Sack, Robert. (1986). *Human Territoriality: Its Theory and History*. Cambridge: Cambridge University Press.

Sack, Robert. (1997). *Homo Geographicus: A Framework for Action, Awareness, and Moral Concern*. Baltimore: Johns Hopkins University Press.

Salaberria, Virginia Tamayo. (1991). *Génesis del Estatato de Gernika*. Onati: Instituto Vasco de Administración Publica.

Salvado, Ton and Josep Miro. (1996). The Appendages of the City of Kidneys. Pp. 135–149 in Centre de Cultura Contemporánia de Barcelona. *Contemporary Barcelona 1856–1999*. Barcelona: CCCB.

Sandercock, Leonie. (1998). *Towards Cosmopolis: Planning for Multicultural Cities*. Chichester: John Wiley and Sons.

Sarajevo Canton Government. (2000). *Sarajevo Canton 2000*. October. 112 pages. Sarajevo: Canton Planning Institute.

Sarajevo Canton Government. (2004). Canton webpage. Online. Available HTTP: < http://www.ks.gov.ba > (accessed February 26, 2005).

Sassen, Saskia. (1991). *The Global City: New York, London, Tokyo*. Princeton, N.J.: Princeton University Press.

Sassen, Saskia. (2000). *Cities in the World Economy*. 2nd ed. Thousand Oaks, CA: Pine Forge Press.

Savitch, Hank with Grigoriy Ardashev. (2001). Does Terror Have an Urban Future? *Urban Studies*, vol. **38**, no. 13: pp. 2515–2533.

Schon, Donald A. (1971). *Beyond the Stable State*. New York: Random House.

Schon, Donald A. and Thomas E. Nutt. (1974). Endemic Turbulence: The Future for Planning Education. Pp. 181–205 in Godschalk, David R. (ed.). *Planning in America: Learning from Turbulence*. Washington D.C.: American Institute of Planners.

Sennett, Richard. (1970). *The Uses of Disorder: Personal Identity and City Life*. New York: Norton.

Sennett, Richard. (1999). The Challenge of Urban Diversity. Pp. 128–134 in Nystrom, L. (ed.) *City and Culture: Cultural Processes and Urban Sustainability*. Karlstrona: Swedish Urban Environmental Council.

Serratosa, Albert. (2002). *Las Escalas Comarcal y Metropolitana en la Ordenación del Territorio*. Unpublished manuscript.

Sibley, David. (1995). *Geographies of Exclusion: Society and Difference in the West*. London: Routledge.

Simeon, Richard and Christina Murray. (2001). Multi-Sphere Governance in South Africa: An Interim Assessment. *Publius: The Journal of Federalism*, vol. **31**, no. 4: pp. 65–92.

Simmel, Georg. (1908). The Metropolis and Mental Life. Pp. 409–424 in Weinstein, D. from Kurt Wolff (Trans.) 1950. *The Sociology of Georg Simmel*. New York: Free Press.

Sisk, Timothy D. and Christoph Stefes. (2005). Power Sharing as an Interim Step in Peace Building: Lessons from South Africa. Pp. 293–317 in Roeder, Philip G. and Donald Rothchild (eds.) *Sustainable Peace: Power and Democracy After Civil Wars*. Ithaca: Cornell University Press.

Smith, Anthony D. (1993). The Ethnic Sources of Nationalism. Pp. 27–42 in Brown, Michael E. (ed.) *Ethnic Conflict and International Security*. Princeton: Princeton University Press.

Smith, David M. (2000). *Moral Geographies: Ethics in a World of Difference*. Edinburgh: Edinburgh University Press.

Smith, M. (1969). Some Developments in the Analytic Framework of Pluralism. In Kuper, Leo and M. Smith (eds.) *Pluralism in Africa*. Berkeley: University of California Press.

Snyder, Jack. (1993). Nationalism and the Crisis of the Post-Soviet State. Pp. 79–102 in Brown, Michael (ed.) *Ethnic Conflict and International Security*. Princeton: Princeton University Press.

Snyder, Jack. (2000). *From Voting to Violence: Democratization and Nationalist Conflict*. New York: Norton.

Solans, Joan Antoni. (1996). The General Metropolitan Plan of Barcelona. Pp. 203–206 in Centre de Cultura Contemporánia de Barcelona. *Contemporary Barcelona 1856–1999*. Barcelona: CCCB.

Sorensen, Georg. (1998) (2nd ed.). *Democracy and Democratization: Processes and Prospects in a Changing World*. Boulder, CO: Westview.

Stanovcic, Vojislav. (1992). Problems and Options in Institutionalizing Ethnic Relations. *International Political Science Review*, vol. **13**, no. 4: pp. 359–79.

Stark, D. (1990). Privatization in Hungary: From Plan to Market or from Plan to Clan? *East European Politics and Societies*, vol. **4**, no. 3: pp. 351–92.

Stark, D. (1992). Path Dependence and Privatization Strategies in East Central Europe. *East European Politics and Societies*, vol. **6**, no. 1: pp. 17–54.

Stokols, Daniel. (1996). Translating social ecological theory into guidelines for community health promotion. *American Journal of Health Promotion*, vol. **10**: pp. 282–298.

Strategic Metropolitan Plan of Barcelona. (2003). Strategic Plan of Barcelona Association and Ajuntament of Barcelona.

Subiros, P. (1993). El Vol de La Fletxa. In P. Subiros (ed.) *El Vol de la Fletxa: Barcelona '9: Crónica de la Reinvencio de la Ciutat*. Barcelona: CCCB/Electra.

Subirats, Joan, and Raquel Gallego. (2002). La Opinión Publica: La Diversidad de Una Nación Plural. Pp. 321–375 in Subirats, Joan, and Raquel Gallego (eds.) *Veinte Años de Autonomías en España: Leyes, Políticas Públicas, Instituciones y Opinión*. Madrid: Centro de Investigaciones Sociológicas.

Szelenyi, Ivan. (1996). Cities Under Socialism – and After. Pp. 286–317 in Andrusz, Gregory, Michael Harloe, and Ivan Szelenyi (eds.) *Cities and Socialism: Urban and Regional Change and Conflict in Post-Socialist Societies*. Oxford, UK: Blackwell.

Tajbakhsh, Kian. (2001). *The Promise of the City: Space, Identity, and Politics in Contemporary Social Thought*. Berkeley, CA: University of California Press.

Tiryakian, Edward A. and Ronald Rogowski (eds.) (1985). *New Nationalisms of the Developed West: Toward Explanation*. Boston: Allen & Unwin.

Toal, Gerard. (2005). Testimony before U.S. House of Representatives, Committee on International Relations, Subcommittee on Europe and Emerging Threats, 109th Congress. April 6.

Tobaruela, P. and Tort, J. (2002). *Darrere o'horitzo. Quinze converses per descobrir Catalunya*. Barcelona: La Magrana.

Toft, Monica Duffy. (2003). *The Geography of Ethnic Violence: Identity, Interests, and the Indivisibility of Territory*. Princeton, NJ: Princeton University Press.

Tronto, Joan. (1993). *Moral Boundaries: A Political Argument for an Ethic of Care*. London: Routledge.

Umemoto, Karen. (2001). Walking in Another's Shoes: Epistemological Challenges in Participatory Planning. *Journal of Planning Education and Research*, vol. **21**: pp. 17–31.

United Nations. (1999). *Towards a New International Financial Architecture*. Report of the Task Force of the Executive Committee on Economic and Social Affairs of the United Nations. New York: UN.

United Nations Development Programme. (2002a). *Human Development Report 2002, Deepening Democracy in a Fragmented World*. New York: Oxford University Press.

United Nations Development Programme. (2002b). *Human Development Report 2002: Bosnia and Herzegovina*. New York: Author.

United Nations Development Programme. (2004). From Emergency to Development: Assessing UNCP's role in Bosnia and Herzegovina. FMR 21, September. Sarajevo: Author.

United Nations High Commissioner for Refugees (UNHCR). (2001). UNHCR's Position on Categories of Persons from Bosnia and Herzegovina in Continued Need of International Protection. Sarajevo: Author.

United Nations High Commissioner for Refugees. (2004). Handbook – Canton 9, Sarajevo.

United Nations High Commissioner for Refugees. (2005a). *Statistical Summary*. May 31. Sarajevo: Author.

United Nations High Commissioner for Refugees. (2005b). *Update on Conditions for Return to Bosnia and Herzegovina*. January. Sarajevo: Author.

Urrutia, Victor. (2004). Bilbao, El Peso de un Contexto. Pp. 51–61 in Borja, Jordi and Zaida Muxi (eds.) *Urbanismo en el Siglo XXI—Bilbao, Madrid, Valencia, Barcelona*. Barcelona: Ediciones UPC.

Varshney, Ashutosh. (2002). *Ethnic Conflict and Civic Life: Hindus and Muslims in India*. New Haven: Yale.

Vivienda. (1969). *Evolución y Datos Fundamentales del Barraquismo en la Ciudad de Barcelona*.

Vucina, Srecko, and Borislav Puljic. (2001). *Mostar'92 Urbicid*. Mostar: Croatian Defense Council.

Ward, S.V. (2002). *Planning the Twentieth-Century City: The Advanced Capitalist World*. London: Wiley.

Weine, Stevan M. (1999). *When History is a Nightmare: Lives and Memories of Ethnic Cleansing in Bosnia-Herzegovina*. New Brunswick: Rutgers University Press.

Weisbrod, Carol. (2002). *Emblems of Pluralism: Cultural Differences and the State*. Princeton, NJ: Princeton University Press.

Williams, Robin M., Jr. (1994). The Sociology of Ethnic Conflicts: Comparative International Perspectives. *Annual Review of Sociology*, vol. **20**: pp. 49–79.

Wirth, Louis. (1938). *Urbanism as a Way of Life*. Chicago: University of Chicago Press.

World Bank. (2002). Bosnia and Herzegovina: Local Level Institutions and Social Capital Study. World Bank, Environmentally and Social Sustainable Development. June.

World Bank Group. (2004). Bosnia and Herzegovina Country Brief 2004. Online. Available HTTP: < http://web.worldbank.org > (accessed 5/31/05).

Yahya, Maha. (1993). Reconstituting Space: The Aberration of the Urban in Beirut. Pp. 128–166 in Khalaf, Samir and Philip Khoury (eds.) *Recovering Beirut: Urban Design and Post-War Reconstruction*. Leiden, The Netherlands: Brill.

Yarwood, John. (1999). *Rebuilding Mostar: Urban Reconstruction in a War Zone*. Liverpool: Liverpool University Press.

Yiftachel, Oren. (1992). *Planning a Mixed Region in Israel: The Political Geography of Arab-Jewish Relations in the Galilee*. Aldershot: Avebury.

Young, Iris Marion. (1990). *Justice and the Politics of Difference*. Princeton, NJ: Princeton University Press.

Young, Iris Marion. (2000). *Inclusion and Democracy*. Oxford: Oxford University Press.

Index

Printed and bound in Great Britain by
TJ International Ltd, Padstow, Cornwall